The Color of the Atmosphere with the Ocean Below:

A History of NASA's Difficult Journey to Successful Biological Remote Sensing of the Global Ocean

JAMES G. ACKER

ISBN: 1511963115
ISBN-13: 978-1511963114

DEDICATION

To Greg Leptoukh, who convinced me that I was the right person to write this history;

to Steve Varlese, who provided critical contact names for the CZCS effort at Ball Brothers;

and to Charlie Yentsch, André Morel, and Dennis Clark, without whom the science of ocean color measurements from space woud have been nearly impossible to accomplish.

I wish they were all still with us, so they could read it.

CONTENTS

ACKNOWLEDGMENTS

I imagine that every author who writes a history like this has an extraordinarily long list of people to acknowledge who helped him write it. That is certainly true in this case. First of all I want to acknowledge Greg Leptoukh, who was a mentor and friend at the Goddard Earth Sciences Data and Information Services Center (GES DISC). Greg's leadership made it possible to get Sea-viewing Wide Field-of-view Sensor (SeaWiFS) data to the users on the day it was supposed to get to the users, and he also valued my particular set of skills. When the Call for Proposals for the history came out, Greg encouraged me to submit a proposal. He convinced me there was no one better situated to be able to gather the information and write about it in a readable way. So with Greg's encouragement I submitted the proposal, and it was chosen for funding by NASA. Sadly, Greg passed away suddenly in 2012, so he is one of the persons to whom I dedicate this book.

The course of writing the proposal involved a lot of people. I had proposed (naively) that I wouldn't have to spend a lot of money on travel to interview people in person; instead, I did most of my interviews over the phone. I also proposed, uniquely, to have a "group interview" – an in-person meeting in which many of the important players in the ocean color missions would gather together and talk about it, while it was recorded both aurally and visually. This meeting took place at the University of South Florida College of Marine Science (my graduate school alma mater), and my former colleagues, Dr. Peter Betzer and Bob Jolley, acted as ideal hosts. Even though the meeting room was a little small and a little loud (due to construction outside), the meeting provided a great deal of information, and it also accomplished what I had hoped would happen – the discussions generated memories that might not have been remembered if it were not for the groupthink in the room. As one person remarked at the end of the meeting, and this was likely true, the meeting was the last time so many people involved with NASA's ocean color satellite missions were all gathered in the same place. I want to thank all of them for being there, and making it work.

I have to acknowledge that I wrangled some monetary support for the workshop from Ball Aerospace, Orbital Sciences Corporation, and Hughes Aerospace, the corporations that built or operated the CZCS, SeaWiFS, and the Moderate Resolution Imaging Spectroradiometers (MODIS-Terra and MODIS-Aqua). THANKS!

I also want to acknowledge Steve Varlese. I can't recall now how I got his name, but I was able to contact him, and he provided a lot of names of people that I could try to track down for more information on the fabrication of the Coastal Zone Color Scanner (CZCS). Steve gave me those names and valuable initial information, and then passed away suddenly. Were it not for him, I likely would have had much less information about the CZCS in this history. And he also sent me pictures, which are in this history, so he is too.

I did conduct several oral history interviews, all of which are available verbatim from the NASA History Division. So I want to thank everyone I interviewed: Dick Barber, Bob Kirk, Bill Barnes, Vince Salomonson, Charlie Yentsch, Jim Yoder, Mark Abbott, Janet Campbell, Bill Wallschlaeger, Randy McConaughey, Robert Woodruff, Gene Feldman, Mary Cleave, Alan Holmes, Dale Kiefer, Dennis Clark, Stan Wilson, Al Fleig, Bryan Franz, and Paula Bontempi.

At NASA Headquarters, Stephen Garber managed the compilation of the history according to NASA History guidelines, which was a new thing for me in a lot of ways. Steve needed a lot of patience to get my unguided efforts into the form of a history. I had thought that I would just write what happened. I was unfamiliar with footnoting to the extent that a history requires, compilation and preservation of sources, how to reference statements made in interviews, how to cite Web sites (and which Web sites not to cite), and how to deal with reviewers in the way that a historian does (not in the way that a scientist does). I also want to thank my reviewers – they were pretty tough on me, but they made the manuscript quite a bit better. Much of the context that you find in this history – the environmental movement, the oil spills, the impact on the public, etc., was added to this history because of their reviews. After the proposal funding ended, I had to rewrite and revise on my own time, and then wait for a spot in the publishing queue. Ultimately, due to the fact that this is a fairly narrow subject and getting it published by NASA would take even more time, we agreed to try self-publishing, which is now fairly easy to do in this information era. I'm sorry it took so long, but at least now it's in print.

At the end of the history you'll find Appendix 1, which I had initially thought would be the first chapter, explaining the basics of ocean optics and ocean color measurements from space. This Appendix is in large part authored by the prolific Dr. Howard Gordon, and I think that it does a great job of making this science understandable – even if it does have some long equations in it! I also want to thank Howard for being the main architect of the atmospheric correction process for ocean color data; as you will read, and as the title of the history implies, without atmospheric correction there would be no ocean color data.

Finally, I want to acknowledge the Internet, specifically the World Wide Web. Because of my familiarity with the Web, I had felt that a lot of the information that I needed could be found online. In that I was correct. What I didn't know was that the first decade of the 21st Century marked a real boundary between the traditional ways of compiling and writing histories, and a new way of doing them. The traditional time-honored way of gathering the information for a history is to identify sources, do interviews, go to libraries, make copies, find pictures in archives, and do quite a bit of interpolation and interpretation between the gaps that will certainly occur between the disparate sources of historical information. The Web makes it unduly easy to find information. It has been analogized as a firehose of information. It wasn't difficult for me to find a lot of information – in fact, I likely found too much. That's why this history is filled with names and incidents and events and anecdotes and technical details and cruise reports and meeting minutes and many other things – because I could find them. (I even found financial reports from Orbimage Inc. online, but I don't think they are online anymore.) Filtering all this information took a lot of time, and I wanted to put a lot of this information in this history. So the history might be long and technical and tedious in spots. I apologize for my zeal in packing in as much detail as possible. I hope that both the more entertaining anecdotes, and my penchant for occasional, hopefully humorous, commentary will make reading it interesting and entertaining, for the most part. This manuscript may not be a perfect or traditional history, but I think it tells the story of what happened and doesn't leave much out. And for that reason I hope it is valuable to NASA, and to science, and to the oceans.

Jim Acker, February 27, 2015

PREFACE

All in a hot and copper sky,
The bloody Sun, at noon,
Right up above the mast did stand,
No bigger than the Moon.

Day after day, day after day,
We stuck, nor breath nor motion;
As idle as a painted ship
Upon a painted ocean.

- Samuel Taylor Coleridge, *The Rime of the Ancient Mariner*, 1798

FOREWORD

Mankind seemingly has always wanted to know more about the mysteries of the ocean realm. In ancient Greco-Roman mythology, the oceans were ruled by a capricious god, Poseidon (Neptune if you prefer the Roman version), and the seas were the abode of dangerous monsters. The voyage of Odysseus acquaints us with the hideous multi-headed Scylla, who was deemed less dangerous than the whirlpool Charybdis – because while Scylla would only devour a few crew members, Charybdis would swallow an entire ship. The legends of mermaids or sirens, with beautiful voices beckoning mariners to their doom, are now believed to be based on encounters with lumbering dugongs or manatees – which bear scant resemblance to fair aquatic maidens. Because the oceans were impenetrable and unknowable beyond a few meters depth, essentially what could be seen through clear water from the deck of a ship or a breath-holding dive, they were realms of mystery and power.

Yet even as the gods were still believed to have dominion over the ocean realm, mankind was determining ways to investigate and understand them. The voyages of Odysseus in mythic poetry underscore the fact that the Mediterranean Sea was used for trade, and not so dangerous that it could not be navigated – though the wrecks of ancient trade ships found in the Mediterranean and Black Seas indicate that the seas were still capable of swallowing ships whole.[1] The Mediterranean Sea seems small in comparison to the Pacific Ocean, yet the Polynesians, a seafaring people comfortable on the scattered atolls and islands of the world's largest oceans, were able to safely navigate hundreds of kilometers by following "guide" or "pointer" stars to the various maritime nations of the Pacific.[2] To investigate the sea floor, diving bells were employed; supposedly Alexander the Great tried one, but more reliable history indicates that they were used in the 16th century.[3]

Thus, some of the earliest records of "oceanography" are ancient maps of the oceans, providing a method to navigate the trackless waters and to understand, at least primitively, their configuration on the globe. It is likely that ancient people familiar with the sea also knew something about how it functioned physically and biologically – when and where the trade winds would be most favorable, where various types of fish could be caught, the signs

---

[1] Jennifer McAndrew, "Phantoms of the Deep", http://www.utexas.edu/features/2008/10/27/shipwrecks/, accessed February 17, 2010; "Graveyard of Ancient Ships Found Deep in Mediterranean", National Geographic press release, http://www.nationalgeographic.com/society/ngo/events/97/ballardngt/release.html, accessed February 17, 2010.

[2] Sam Low, "Star Navigation", http://www.samlow.com/sail-nav/starnavigation.htm, accessed February 17, 2010.

[3] John Bevan, "Diving Bells Through the Centuries", South Pacific Underwater Medicine Society Journal, 29(1), 42-50, 1999.

of approaching storms, the boundaries of salt water and fresher water flowing into the sea from rivers, and where dangerous currents might lurk – for the myth of Scylla and Charybdis is associated with the Straits of Messina between Sicily and the toe of Italy, where fast and powerful currents are definitely found.[4]

So the next step in mankind's understanding of the oceans was to make better maps – and one of the best examples of the early scientific investigation of the oceans is Benjamin Franklin's famous map of the Gulf Stream. Franklin, responding to complaints that the mail took much longer coming from Europe to America than going from America to Europe, mapped the Gulf Stream with the aid of whaler Timothy Folger. Franklin realized that the Gulf Stream was defined by such indicators as temperature and color – the waters were warmer and clearer than the Atlantic coastal waters off the Colonies or the early United States of America.

Modern oceanography is generally considered to begin with the global voyage of the H.M.S. *Challenger*, from 1872-1876. The Challenger voyage covered the northern and southern Atlantic, parts of the Southern Ocean south of the Indian Ocean, the southeastern, central, and northwestern Pacific, and some of the waters around Indonesia. Routine sampling including bottom dredges, net trawls, depth soundings, measurements of temperature at various depths, samples of seawater, and observations of the atmospheric and meteorological conditions. The Challenger voyage allowed oceanographers the first systematic view of the geology, biology, and physical characteristics of the entire ocean, dramatically broadening scientific knowledge.[5]

There are other notable landmarks in the history of oceanography and ocean exploration: the descent to the abyss by William Beebe in his bathyscaphe; the invention of the "Aqua-Lung" by Jacques Cousteau and Emil Gagnon, substantially increasing the freedom of divers to explore underwater; the Gulf Stream drift voyage of the Grumman/Piccard PX-15 submersible *Benjamin Franklin* (occurring at the same time as the Apollo 11 moon landing mission); the ocean drilling projects that discovered evidence of the drying of the Mediterranean ocean; and magnetic maps of the ocean bottom that provided the proof that continents actually moved, the foundational discovery of plate tectonics.[6] These are only a few of many, many achievements in mankind's scientific exploration of the oceans.

---

[4] Rose Santoro, "The damned charm of Scylla and Charybdis", http://home.um.edu.mt/medinst/mmhn/rosa_santoro.pdf, accessed February 17, 2010.

[5] Steven M. Perry and Daphne G. Fautin, "Challenger Expedition (1872-1876)", http://hercules.kgs.ku.edu/hexacoral/expeditions/challenger_1872-1876/challenger.html, 2003, accessed February 17, 2010.

[6] William Beebe, "A Dark and Luminous Blue", http://seawifs.gsfc.nasa.gov/OCEAN_PLANET/HTML/ocean_planet_book_beebe1.html, accessed February 17, 2010; "The Historyof Scuba Diving", http://aquanaut.students.mtu.edu/Scuba/Scuba.htm, accessed February 17, 2010; "Deep Sea Sub Story Resurfaces", http://www.spacedaily.com/news/spacetravel-04zl.html, accessed February 17, 2010; Jeff Weissel, "MG&G History: A brief history of marine geology and geophysics at Lamont", http://www.ldeo.columbia.edu/research/marine-geology-geophysics/mg-g-history, accessed February 17, 2010; United States Geological Survey, "Magnetic Stripes and Isotopic Clocks", http://pubs.usgs.gov/gip/dynamic/stripes.html, accessed February 17, 2010.

Yet in all cases such as these, there is something missing. No matter how high the deck or bridge or lookout tower on a ship, the view of the ocean surface is limited to what can be seen to the horizon. Beyond that limit, there is no knowledge of change or connectivity – there is no *synoptic* understanding – a complete view of an entire ocean basin, even a hemisphere, all at the same time. There have been advances in the ways that knowledge is gained at point sources; moored or floating buoys can be deployed that provide much more information at the same time than a single ship voyage could ever accomplish. Yet still, these instrumented buoys only provide information at a single point in time and space.

The only way to achieve a broader, more comprehensive view of the oceans is to view them from greater and greater heights. The advent of air travel and such daring exploits as Lindbergh's crossing of the Atlantic provided a new way to view the oceans, from altitude – and this was a view that intrigued Gifford Ewing, who was both an oceanographer and a pilot.[7] Ewing advocated, pioneered, carried out, and in some ways underwrote the first preliminary observations of the oceans from altitude, and Ewing was perceptive enough to realize that even greater altitudes – particularly the altitude of a satellite orbiting the Earth – would truly provide the unifying global view of the oceans that "traditional" ship-based oceanography lacked.

It is the need to acquire a truly global view of the oceans that has engendered the field of satellite oceanography. The question which may then arise is: *why* is there such a need for this particular view? Satellites are expensive, and as history shows, when they fail it is rarely possible to send up a crew of repairmen to fix them. Why, then, is the somewhat less expensive alternative of more ships at sea or more buoys stretching to the horizon (and beyond) not better than the rigorous requirements of placing technically sophisticated instruments on a satellite in orbit around the Earth?

The answer to this question lies in the fact that the highly variable nature of the oceans was little suspected until satellite instruments were able to demonstrate it. Prior to the availability of images of the oceans from space, oceanographers had to "connect the dots" between discrete data points collected at individual station locations. So they assumed that between the points conditions remained basically the same, along a smooth continuum.

Images from satellites in space, particularly images of sea surface temperature and the concentration of the photosynthetic pigments in phytoplankton (the latter derived from accurate measurements of the optical characteristics of surface ocean waters), shattered the concept of a smooth continuum. The oceans were confirmed to be churning, spinning, surging, and swirling – filled with eddies and rings, jets and plumes and squirts, and water mass boundaries embellished with the rococo aquatic embroiderings of von Karman vortices[8]. An instrumented buoy or scientist in a ship could not sense or observe the

---

[7] Deborah Day, "Gifford Cochran Ewing Biography", http://repositories.cdlib.org/cgi/viewcontent.cgi?article=1197&context=sio/arch (PDF document, acquired 31 January 2009)

[8] NASA GES DISC, "Various Views of von Karman Vortices", http://disc.sci.gsfc.nasa.gov/oceancolor/additional/science-focus/ocean-color/vonKarman_vortices.shtml, accessed December 20, 2011.

changes that could be occurring just a few kilometers away, over the observational horizon. Furthermore, just as the oceans were variable in space, they were variable in time – changes in the same place could occur rapidly as a front or ring traveled over that particular location. A ship occupying a single station in the oceans could only observe the static conditions at a single time, not the changes that might occur at that station just hours or days after the ship departed.

However, just knowing that the oceans are highly variable is perhaps insufficient justification for the expense and suspense of putting expensive, technically sophisticated instruments into orbit. Indeed, the justification goes well beyond that. The only way in which the actual mechanisms of the Earth's geophysical systems can be ascertained, globally, is to map the oceans as they vary over time and space. The capability that satellites provide is this repetitive view from high altitude – allowing the creation of ever-changing maps of the ocean surface. This is the true view from space – not only looking down from a very high place, but looking down repeatedly, to allow categorization and recording of the variability of the oceans.

Satellite-borne instruments can now measure a wide variety of oceanic variables: currently, it is possible to measure sea surface temperature, sea surface height (with an astonishing resolution of mere centimeters), wind speeds, and optical characteristics, which allow derivation of biologically-related parameters. New satellite instruments will be capable of measuring changes in sea surface salinity – one of the basic parameters measured by the scientists on the HMS *Challenger*.

The history of satellite oceanography has demonstrated that one particular measurement is the most daunting – and that is the optical characteristics of the ocean's surface waters. Because these characteristics are very much related to the microscopic plants that are the foundation of the oceanic food web, this field has been termed "bio-optics", and more generally, "ocean color" (because not everything that changes the color of the oceans is biological). One of the most important variables that this field seeks to measure is the concentration of chlorophyll (from the flower goddess *Chloris*[9]), which indicates the presence and activity of phytoplankton – the primary producers that create organic carbon from sunlight and carbon dioxide, and in the process produce oxygen. Phytoplankton are thus a key, basic, and fundamentally important factor in the Earth's carbon system – a system to which all life on Earth belongs, and which also is an important determinant of the state of the climate in which all life on Earth exists.

The field summarized by the general term "ocean color" is a difficult scientific undertaking – and it is vital for a comprehensive understanding of the biological mechanisms of the world's oceans, how they are influenced by the physical components of the oceans (the winds and currents), and how the biological system of the oceans is related to all of the other flora and fauna on Earth, including humans – who by their own actions are causing remarkable changes to Earth's carbon system, including the biology of the oceans and chemistry of ocean waters. It is because of these rapid changes that ocean color is even

[9] "chloro-, chlor- +", http://www.wordinfo.info/words/index/info/view_unit/451, accessed February 17, 2010.

more important now than when it was first proposed and carried out, providing perplexed oceanographers with images that utterly inverted their comfortably naive understanding of how the oceans worked. Now that they are sure that the oceans are variable and changeable, they must understand the impact of the changes they observe, to provide humanity with a window on what may be changing irreversibly, and what could still be done to address those changes, devise strategies to reverse some of them, and to respond more knowledgeably and effectively to those changes which will continue inexorably. This scientific undertaking has quietly accomplished many things that were thought only a few decades ago to be unachievable – and in relating the historical accomplishments that made it possible, we also gain an understanding of why this particular scientific endeavour was necessary.

The quiet success of ocean color science, led by NASA's three ocean color missions, has not just supplied data that is being used by scientists investigating Earth's carbon cycle and climate system. The Coastal Zone Color Scanner (CZCS) demonstrated that this observational technique was possible, and very useful. The Sea-viewing Wide Field-of-view Sensor (SeaWiFS) showed that full-time observations of the oceans were the next necessary step in fully characterizing them. And the Moderate Resolution Imaging Spectroradiometer (MODIS) expanded the observational capabilities from space, providing increased information on their interconnected physical and biological systems. The success of these missions has initiated a variety of spin-off applications that provide tangible benefits to other branches of science, as well as to human society. These applications indicate why ocean color has become such an important branch of Earth remote sensing, despite the fact that the applications may not have been anticipated by the original mission designers, engineers, and scientists.

For science, ocean color lends insight into a variety of physical oceanographic processes. The patterns created by variations in phytoplankton concentration are adjunct data for other data types that indicate ocean circulation, particularly sea surface temperature and sea surface height. Ocean color is very effective in allowing the detection of eddies, due to the sharp delineation of high and low productivity areas associated with these features. Ocean color also provides insight into how storms and high winds mix and stir surface ocean waters, and how seasonal cycles of temperature and sunlight affect the timing of phytoplankton blooms. Ocean color and bio-optics also provide data on how ocean waters interact with sunlight, through the processes of absorption, scattering, and reflection.

Ocean color data also helps characterize the state of the many different subsystems in the ecology of the oceans. Ocean basins and sub-basins, marginal seas, enclosed seas, estuaries like bays and fjords – all of these have their own peculiar physical characteristics, and their own set of adapted organisms. Ocean color data indicates when blooms occur, and whether these blooms are regularly occurring events, or perhaps events which are suddenly anomalous, indicating important shifts in the ecosystem under study. Used in conjunction with other data, like collections of zooplankton or meteorological data, ocean color data can help "diagnose" the health and functioning of regional oceanic ecosystems.

One of the intersections of science and societal interest is the state of fisheries. (For those unacquainted with the term, a "fishery" refers to the stock of a particular fish; so there are tuna fisheries, lobster fisheries, anchovy fisheries, squid fisheries, shark fisheries, etc.) Ocean color data show clearly where fisheries are likely to be found – and the most popular commercial use of ocean color data has been to help fishermen find fish, a mixed blessing as the world's fish stocks are increasingly under stress. Still, helping fishermen find fish can sometimes lead to efficiency and increased profit, decreasing the need to catch more fish. Ocean color data is a vital component of fisheries studies, helping to find and characterize the oceanic state that supports the world's population of wild fish. Ocean color data has been particularly helpful to examine cases where fisheries have suddenly collapsed, leading to markedly reduced fish populations.[10]

Another area where science and society interact is when phytoplankton blooms turn "bad" – that is, when a particular form of phytoplankton becomes annoying, noxious, or even worse, toxic. These events are called collectively "Harmful Algal Blooms" or HABs – and they can be dangerous to humans and disastrous for fish. Toxic algae collected into the meat of fish and shellfish can poison the meat, leading to paralysis and death if the fish are consumed by humans. Toxic algae can cause massive fish kills, annoying to humans when the fish wash up and decay on shore. The same algae can even become airborne, causing pulmonary discomfort akin to asthma. Even if the algae are not toxic, massive blooms (which in some cases have been induced by human activities, such as increased fertilizer runoff from agricultural fields) can be very annoying – huge amounts of algae may wash up on shore and decay with a horrific stench, or cause incredible masses of foam to invade beaches used for recreation and sport. Algae dying and sinking to the bottom near the coast causes bottom waters to lose all their oxygen content, the process of eutrophication, creating bottom "dead zones" where nothing can live, including shellfish that are an important commercial harvest.

The indicators of phytoplankton blooms are one way that ocean color data can be used to monitor the health of ocean waters. The data can be used more directly to indicate water quality, by providing information on the clarity of waters in the ocean and along the coast. Coastal water quality is influenced by a variety of factors: phytoplankton, sediments, colored organic matter – and in many places, all three of these components are found together. Water clarity is a basic datum that informs scientists and decision-makers about the water quality in their particular region or jurisdiction of interest.

Related to the water quality use of ocean color data is the broader field of hazards and disasters. Ocean color data have been used to examine the effect of flood waters released from rivers; the effect of increased turbidity on coral reefs, both from storms and ongoing nutrient enhancement; sediments mobilized by high winds from typhoons and hurricanes; the aforementioned HABs, which can create economic disasters if they affect aquaculture activities like fish and shellfish farming; and even terrestrial disasters such as forest fires and dust storms caused by droughts. In these cases, ocean color data becomes part of larger investigations into the influence of natural and anthropogenic factors on weather and

[10] Laurie J. Schmidt, "Bloom or Bust: The Bond between Fish and Phytoplankton", http://news.eoportal.org/didyouknow/050425_fishbloom.html, April 26, 2005, accessed February 17, 2010.

climate, both regional and global. One of the most interesting applications of ocean color data has been to investigate the role of iron in oceanic phytoplankton biology, and ocean color instruments can be used to both watch dust storms and then the response of phytoplankton to the iron delivered from the airborne particles of dust as they fall to the ocean surface.

So now that some of the reasons for doing this particular branch of science have been described, it is time to explore the history of how NASA fostered and facilitated this remarkable endeavour in oceanography.

1

Oceanography and Early Ocean Optics Observations

Here is a brief description of a simple experiment anyone can do to investigate the fundamental optical properties of seawater: First, get some kosher salt, or laboratory grade sodium chloride (NaCl) – essentially salt free of impurities or additives. Then take a clear glass, and fill it with clear tap water. Take a few teaspoonfuls of salt and add them to the water in the glass. Stir the water and the salt with a spoon until the salt completely dissolves. Now hold the glass up, preferably with a lighted window or bright light behind it, and take a close look at it.

The salt water solution in the glass should be crystal clear, entirely transparent.

That's the color of the most of the water in the world's oceans.

Provided, of course, that there's nothing else dissolved or floating in the water.

However, even when ocean waters are at their limit of extreme clarity (to find such waters, you must travel to a region in the southern oligotrophic gyre of the Pacific Ocean, west of Easter Island, or gaze into an Antarctic polynya) there is usually something else present in ocean waters. The nature of the "something else" is why there can be a scientific endeavor investigating and understanding the changing colors of oceanic waters.

If you now add a few drops of milk to the glass, or a drop of food coloring, or a pinch of dirt from the garden – the appearance of the water in the glass will change rather markedly[11]. In actual ocean waters, there are substances that scatter light, in the same way that the milk does. And there are substances that absorb part of the visible light spectrum, the same way that the food coloring does. And there are substances that might absorb, scatter, and reflect light – all at the same time, depending on the physical state of these substances – like that pinch of dirt.

There is also another vital aspect to ocean waters that isn't simulated by adding milk or food coloring or dirt to a glass of water – and that aspect is the minuscule organisms that scatter, absorb, and reflect light, with various propensities to do that depending on what they are. Mixing everything together – small particles, big particles, dissolved substances, substances not quite small enough to be dissolved but barely big enough to be suspended, and especially those important small organisms – creates the amalgamation of "things" that affect the optical environment of seawater.

Given that the appearance of the ocean is a fundamental observation, it is very likely that the earliest systematic oceanographers took note of the apparent color of the oceans. We can only speculate that prior to the advent of systematic scientific oceanography, observers of the ocean – likely concerned about either their fishing success or whether an

[11] "Lesson Plans: Simple Light Scattering", http://education.arm.gov/teacherslounge/lessons/simplelight.stm, (accessed 31 January 2009).

angry oceanic god was about to wreak havoc on the coast – also noticed the color of the local ocean or sea and any noteworthy changes in it. Thinking in Old Testament terms for a moment, the plague on the Egyptians which turned the Nile waters red (no matter what the cause – volcanoes have been suggested) likely influenced the deltaic outflow into the Mediterranean, but there are no reports on file of that observation.[12] However, several centuries later (approximately), Aristotle supposedly noted the difference in the clarity of river water and ocean water.[13]

According to some narratives of the voyages of Christopher Columbus, on his first voyage to the West, impatient crew members spotted branches in the ocean. This "turbidity" was a sign that land might be nearby, which was confirmed by the sighting of land the next day.[14]

Defining the actual starting point of a nascent field of scientific endeavor is not always easy. Frequently, however, there are early milestones that can be identified to generally indicate the formalization of scientific inquiry. One of the earliest examples of systematic ocean observations is Benjamin Franklin's characterization of the Gulf Stream, which was compiled from the reports of whaling vessel captains and Franklin's own temperature measurements while on Atlantic-crossing voyages.[15]

The era of systematic oceanography may have begun with Matthew Fontaine Maury, who created a uniform method of recording oceanographic data, allowing charts of winds and currents to be made, primarily for the purpose of efficient ship navigation.[16] Maury was contemporaneous with Father Pietro Angelo Secchi (1818-1878) of the Vatican, a science advisor to the pope, who invented the measuring device named after him: the *Secchi disk*. At the behest of Commander Cialdi of the Papal Navy, who requested that Fr. Secchi come up with a way to measure the clarity of Mediterranean ocean waters, Fr. Secchi created a disk that was lowered until it was no longer visible, and this depth was recorded by means of markings on the rope to which the disk was attached. The depth of disappearance is now called the "Secchi depth."[17] The idea of lowering something into the water to characterize ocean clarity actually dates back to 1815, by Captain Kotzebue, and on a subsequent expedition onboard a ship with the lovely name of *Coquille*.[18]

As the lives of Maury and Fr. Secchi were nearing their end, the globe-circling

[12] "The BBC's Theory on the Biblical Plagues", http://www.christiancourier.com/articles/592-the-bbcs-theory-on-the-biblical-plagues (accessed 31 January 2009).
[13] Sophia Johannesen, "Two Thousand Years of Ocean Optics", *Optics and Photonics News*, April 2001, pages 30-36.
[14] "The Log of Christopher Columbus", http://www.columbusnavigation.com/diario.shtml, accessed 11 February 2011.
[15] Jerry Wilkinson, "History of the Gulf Stream", http://www.keyshistory.org/gulfstream.html, accessed 18 February 2011.
[16] U.S. Navy Museum, "Matthew Fontaine Maury (1806-1873)", http://www.history.navy.mil/branches/teach/ends/maury.htm, accessed 18February 2011.
[17] Minnesota Pollution Control Agency, "History of the Secchi Disk", http://www.pca.state.mn.us/index.php/water/water-types-and-programs/surface-water/lakes/citizen-lake-monitoring-program/history-of-the-secchi-disk.html, accessed 18 February 2011.
[18] Sophia Johannesen, "Two Thousand Years of Ocean Optics."

expedition which is generally taken to be the beginning of modern oceanography took place.[19] Now referred to commonly as the Challenger expedition, the voyage of the *H.M.S. Challenger* from 1872 to 1876 took numerous oceanographic measurements, collected biological samples, and also characterized the bathymetry (depth) of the world's major ocean basins. There is, however, no evidence that Secchi disk observations were part of the expedition's measurement protocols.[20]

Even if the *H.M.S. Challenger* scientists didn't employ it, the use of the Secchi disk indicated an increased awareness that water clarity could be quantified, and Secchi disk measurements form the first "database" of data relevant to ocean optics.[21] Early research cruises (the Secchi depth measurements in the Baltic Sea date back to 1902) did routinely measure the Secchi depth (Figure 1.1) as part of a suite of basic ocean parameters acquired at a research station, along with salinity, temperature, and depth (if the depth did not exceed the length of the sounding line).[22] As oceanic investigations became more sophisticated and quantitative techniques were developed and refined, the basic measurement suite expanded to include seston, dissolved oxygen, nutrients, conductivity, and increasing to a dizzying number of parameters amenable to measurement.

Figure 1.1 Secchi disk measurement. The disk is lowered incrementally to increasing depth until it is no longer visible from the surface. The disk where it disappears from sight is the Secchi depth; the clearer the water, the greater the Secchi depth will be.

In the field of ocean optics, the next noteworthy advance was provided by Swiss

---

[19] Some basic references on the history of oceanography include: Margaret Deacon, *Scientists and the Sea 1650-1900: A Study of Marine Science* (London, Burlington, VT, Ashgate, 1997); Anita McConnell, *No Sea Too Deep: The History of Oceanographic Instruments* (Bristol, CT, Adam Hilger, 1982); Eric L. Mills, *Biological Oceanography: An Early History, 1870-1960.* (Ithaca, NY, Cornell University Press, 1989).

[20] Dive and Discover, Woods Hole Oceanographic Institution, "The Challenger Expedition", http://www.divediscover.whoi.edu/history-ocean/challenger.html, accessed 18 February 2011.

[21] R.W. Preisendorfer, "Secchi disk science: Visual optics of natural waters," *Limnology and Oceanography* 31, 909-926, (September 1986); Thorkild Aarup, "Transparency of the North Sea and Baltic Sea - a Secchi depth data mining study" *Oceanologia* 44 (3),323-337 (July 2002).

[22] Thorkild Aarup, "Transparency of the North Sea and Baltic Sea - a Secchi depth data mining study"; Monica Bruckner, "Measuring Lake Turbidity Using a Secchi Disk", http://serc.carleton.edu/microbelife/research_methods/environ_sampling/turbidity.html, accessed 8 January 2010.

limnologist François-Alphonse Forel.[23] In 1887, Forel developed a color scale, and a way to assess the color of waters, by looking at the white background of a submerged Sechhi disk and trying to match the apparent color of the disk to a rack of vials containing water samples of different colors. (Forel wasn't just *any* limnologist; his studies of Lac Leman in Switzerland likely established the scientific field of limnology). Three years later, German limnologist Willie Ule refined the Forel color scale.[24] (Figure 1.2)

Certainly the archives of early oceanographic research institutions, compiled as more and more ships were sent to sea in search of unknown and unexplored data, contained Secchi depths and Forel colors measured at each research station. As noted earlier, Aarup systematically searched for Secchi depth measurements taken in the North Sea and Baltic Sea, in the process discovering measurements dating back to 1902 – but not many of them.[25] Even though there are not a lot of these early measurements, another study indicated that the Secchi depth in the Baltic Sea has been decreasing for 100 years.[26]

Figure 1.2 Forel-Ule color scale.

During the decades of the 20th century, science and technology in many fields leapt forward, and the field of ocean optics participated in this scientific advancement. In the 1930s, George Clarke made several important measurements for ocean opticians, including how sunlight penetrates ocean waters, and how seawater absorbs light.[27] Kurt Kalle also contributed to this field in the 1930s.[28] Another significant event which would prove to be critical to the advancement of ocean optics was the founding of the Visibility Laboratory at

[23] Roswell Austin, "Remote Sensing of the Oceans: BC (Before CZCS) and AC (After CZCS)." in Vittorio Barale and P.M. Schlittenhardt, editors, *Ocean Colour: Theory and Applications in a Decade of CZCS Experience* (Dordrecht, The Netherlands, Kluwer, 1993), pp. 1-16.

[24] Robert Arnone, Michelle Wood, and Richard Gould, "The Evolution of Optical Water Mass Classification", *Oceanography*, 17(2), 14-15 (June 2004); Genny Anderson, "Tools of the Oceanographer: Measuring Equipment", Excerpted chapter from an online course in Marine Science, http://www.marinebio.net/marinescience/01intro/tomeas.htm (accessed 8 January 2010).

[25] Thorkild Aarup, "Transparency of the North Sea and Baltic Sea."

[26] Laamanen, M, V. Fleming & R. Olsonen: "Water Transparency in the Baltic Sea Between 1903 and 2005", HELCOM Indicator Fact Sheet *Water Transparency, 2005* http://www.helcom.fi/environment/indicators2004/secchi/en_GB/transparency/?u4.highlight=Laamanen (accessed 31 January 2009)

[27] Clarke, George L., "Observations on the Penetration of Daylight into Mid-Atlantic and Coastal Waters," (1933); Clarke, George L., "Light Penetration in the Western North Atlantic and its Application to Biological Problems", 1936. Both references from Austin, "Remote Sensing of the Oceans: BC (Before CZCS) and AC (After CZCS)."

[28] K. Kalle, "Zum Problem der Meereswasserfarbe" (1938), referenced in Austin, "Remote Sensing of the Oceans: BC (Before CZCS) and AC (After CZCS)."

Massachusetts Institute of Technology (MIT) by Seibert Q. Duntley and MIT physics chairman Arthur Hardy in 1939.[29]

World War II resulted in a lot more ships at sea, and the deployment of warships and supply convoys also resulted in an increased number of ocean observations. Another result of the war was the development of underwater radiometers, instruments which could quantify the light intensity of the oceans, and measurements with these instruments were made around the world and compiled.[30] During the war, the National Defense Research Committee (NDRC) provided funding to the MIT Visibility Laboratory to improve technology for detecting underwater targets and obstacles and to aid search-and-rescue operations.[31]

What was lacking during the war era, however, was any sense of unification. Thus, the post-World War II era awaited a milestone in ocean optics. Danish oceanographer Nils Gunnar Jerlov provided just such a milestone in 1951. His landmark paper, which utilized several data sets, provided the first quantitative method to evaluate the optical characteristics of the world's oceans – and it also paved the way for Jerlov to eventually have an award in the field of ocean optics named after him.[32] (After all, Swedish dynamite magnates shouldn't corner the market on awards named for Scandinavians.) Jerlov's study used physics, specifically spectral diffuse attenuation coefficients, to classify the optical characteristics of different water masses. Jerlov continued his pioneering ocean optical work through the 1960s, culminating in the book "Marine Optics", published in 1976.[33] Along the way he found it necessary to expand his original set of three optical water masses to five, and then added nine different coastal water types. This expansion was an early indication of how vexing the study of ocean optics in coastal waters was going to be.[34]

The 1950s era also included the East Coast-to-West Coast migration of the "Visibility Laboratory" at Massachusetts Institute of Technology in 1952.[35] The Visibility Laboratory (VisLab) wasn't primarily focused on detecting chlorophyll in the oceans in the 1950s – as might be expected, the main role of the VisLab was to examine how to detect various "things" visually in the oceans and atmosphere, which one might surmise included submarines and high-altitude surveillance aircraft. Nevertheless, the VisLab became an

---

[29] Kenneth Voss and Roswell Austin, "An instrumental history of the Scripps Visibility Laboratory", *Proceedings of Ocean Optics XVI,* 2002, available from http://www.physics.miami.edu/optics/ken/OtherPapers/A25_VA_OOXVI_2002.pdf (accessed 5 September 2008)

[30] Robert Arnone, Michelle Wood, and Richard Gould, "The Evolution of Optical Water Mass Classification."

[31] Kenneth Voss and Roswell Austin, "An instrumental history of the Scripps Visibility Laboratory."

[32]Jerlov, N.G., "Optical Studies of Ocean Water." *Reports of the Swedish Deep-Sea Expedition.*, 3, 1-59 (1951).

[33] Nils Jerlov, "Optical Oceanography", *Oceanography and Marine Biology Annual Review* 1: 89-114, 1963; Nils Jerlov, *Optical Oceanography* (Amsterdam, The Netherlands, Elsevier, 1968); Nils Jerlov and E. Steeman Nielsen, editors, *Optical aspects of oceanography* (New York, Academic Press, 1974); Nils Jerlov, *Marine Optics*, (Amsterdam, The Netherlands, Elsevier, 1976).

[34] Robert Arnone, Michelle Wood, and Richard Gould, "The Evolution of Optical Water Mass Classification."

[35] Kenneth Voss and Roswell Austin, "An instrumental history of the Scripps Visibility Laboratory."

important research and development laboratory for the nascent scientific discipline of ocean optics.

Jerlov's papers in the 1960s heralded a marked increase in activity in the field of ocean optics, augmented by considerable technological advances for in-water instrumentation. The 1960s also ushered in the first satellites which provided images of the Earth from space. While Jerlov was working on ocean optics, Charles Yentsch, at the time located at Woods Hole Oceanographic Institute (WHOI), published a milestone paper that described conceptual ideas of what the oceans would look like from space. The paper, "Distribution of chlorophyll and phaeophytin in the open ocean", published in 1965, was the first discussion of how the actual perceived color of the ocean might vary due to differing concentrations of phytoplankton and their associated light-absorbing pigments.[36]

Yentsch, who was to become the bridge figure between ocean optics and observations of ocean biology from space, entered the field of optical oceanography and remote-sensing somewhat orthogonally. His first professional interest in the oceans was in the employ of the U.S. Navy, as a diver. Subsequent to that (and perhaps due to occasional closehand experience with ship hulls under the water) he researched barnacles at Florida State University and received a Masters degree in 1953. He went to the University of Washington to pursue a Ph.D., but the prospect of more active research lured him to WHOI in 1956.[37] In 1959, Yentsch and John Ryther published one of the first papers on using fluorescence to measure chlorophyll and phaeophytin in phytoplankton.[38]

Approximately a year before Yentsch published his important 1965 paper (and likely while he was working on it) WHOI hosted a conference sponsored by NASA that – even with the Space Age in its infancy and with NASA's manned spaceflight program poised between Mercury and Gemini – anticipated and predicted how satellites could perform oceanography from space. The conference was entitled simply "Oceanography From Space" and subtitled "The Feasibility of Conducting Oceanographic Explorations from Aircraft, Manned Orbital and Lunar Laboratories."[39] The convener of this conference was oceanographic entrepreneur Gifford Ewing.

Ewing began working on the oceans as the captain of a minesweeper, patrolling the Panama Canal Zone during World War II. After his military service, Ewing decided to pursue a Ph.D. in oceanography, studying under Roger Revelle at the Scripps Institute of Oceanography. One of his first research investigations concerned the currents in the lagoon of Bikini Atoll, related to atomic bomb testing. Ewing was both a motivator and an innovator, and he inspired many different individuals in the field of oceanography. As a

---

[36] Charles Yentsch, "Distribution of chlorophyll and phaeophytin in the open ocean", *Deep-Sea Research* 12, 653-666, (1965).

[37] Charles Yentsch, interview transcript, July 2008.

[38] John Ryther and Charles Yentsch, "The estimation of phytoplankton production in the ocean from chlorophyll and light data", *Limnology and Oceanography*, 2, 281-286, (1957).

[39] Oceanography from Space; Proceedings of the Conference on the Feasibility of Conducting Oceanographic Explorations from Aircraft, Manned Orbital and Lunar Laboratories. Gifford C. Ewing, editor. GC2.C748 1964 c.4. Reference Number 65-10, Woods Hole Oceanographic Institution, Woods Hole, MA USA, 1965. (Conference held August 24-28, 1965 at Woods Hole, Massachusetts, USA.)

sidelight, he also bought the La Valencia Hotel in San Diego in 1947.[40] Ewing counted airplane pilot as one of his skills, and he conducted whale surveys from his own private plane, and took aerial views of the SIO campus as it appeared in 1947. Ewing worked at Scripps from 1946 through 1964 before moving to WHOI.

While Yentsch's insights were circulating through the oceanographic community, Yentsch and Ewing teamed up. In the summer of 1967 (or possibly the summer of 1968) Ewing and Yentsch flew a spectroradiometer on WHOI's C-47 aircraft (the military version of the DC-3). A young oceanographer, Richard Barber, who had also come to WHOI to study under John Ryther, happened to be in the laboratory when Yentsch and Ewing came back from the first observational flight of the airborne spectroradiometer. With considerable excitement, they showed that the radiometer's signal had registered a signal due to chlorophyll. Examination of their flight path and timing indicated that the signal had occurred when they had flown over Cuttyhunk Island, the terminus of the chain of islands that has Woods Hole as its mainland anchor. The radiometer had detected a signal of chlorophyll absorption from patches of *Fucus*, a brown seaweed that grows in the intertidal zone of rocky shores. (Figure 1.3) Biological oceanographer David Menzel, also in the laboratory, commented wryly that the spectroradiometer hadn't detected chlorophyll in phytoplankton, it had only detected it in macroalgae (the biologist's way of calling it *seaweed*!) Nevertheless, this flight could be considered the birth of biological remote sensing in oceanography.[41]

Figure 1.3. *Fucus gardneri.* Photo by Daniel Mosquin.

Several landmark papers resulted from the research of Ewing, Yentsch, and collaborators. Ewing speculated on what exactly was needed for oceanographic remote sensing in 1967.[42] The early aerial observations led to the first paper to suggest that ocean color variability could be analyzed remotely, which was published in 1970 by Clarke, Ewing,

[40] Charles Yentsch, interview transcript, July 2008; Deborah Day, "Gifford Cochran Ewing Biography", http://repositories.cdlib.org/cgi/viewcontent.cgi?article=1197&context=sio/arch (PDF document, acquired 31 January 2009).

[41] Richard Barber, interview transcript, May 2008.

[42] Gifford Ewing, "Current and Future Needs for Remotely Sensed Oceanographic Data : A Speculation," *Woods Hole Oceanographic Institution Technical Report*, Reference Number 67-41.

and Lorenzen.[43] The third author of this paper was Carl Lorenzen, an oceanographer at the University of Washington with a variety of interests. One of his primary interests was improving the actual measurement of chlorophyll concentrations in seawater using fluorometry, (measurement of the fluorescence of chlorophyll that is stimulated by ultraviolet light). Lorenzen also wrote about how measurements of chlorophyll could be related to other oceanographic biological processes. Among his other interests were the rates of fecal pellet sinking from zooplankton – while the rate at which the excretions of zooplankton sink to the seafloor might not be a topic for formal dinner conversations, this is an important way that carbon is removed from the surface of the oceans to its depths.[44]

What Clarke, Ewing, and Lorenzen described in the paper was the flight of the spectroradiometer (provided by TRW – one wonders if Gifford Ewing's considerable powers of persuasion procured the device) – over different ocean areas. The oceanographic barnstormers flew at an altitude of 1000 feet (350 meters), an altitude that would likely scare the seagulls off the *Fucus,* and one at which Charlie Yentsch probably could have studied barnacles with binoculars. More importantly, the radiometer did indeed show that the data discriminated the optical characteristics of different water types. The research also determined something else very important – the ability to distinguish the water types, and in fact the ability to measure any light coming from the ocean surface at all – became markedly more difficult with increasing altitude.[45]

SCOR Working Group 15

One other event which occurred rather quietly in 1963 was the formation of a working group by the Scientific Committee for Ocean Research (SCOR).[46] The working group was number 15 (noteworthy because by September 2005, SCOR had convened working group 128, which examined incidences of hypoxia in the world's coastal zones, a topic in which ocean color data has been useful), and was jointly sponsored by the United Nations Educational, Scientific, and Cultural Organization (UNESCO) and the Inter-Agency Procurement Services Organization (IAPSO). The actual title of the group was "Photosynthetic Radiant Energy in the Sea." The committee was small but very international, consisting of eight experts on marine productivity and marine optics from

---

[43] George Clarke, Gifford Ewing, and Carl Lorenzen, "Spectra of backscattered light from the sea obtained from aircraft as a measure of chlorophyll concentration", *Science* 167 (3921), 1119 – 1121, (1970).

[44] O. Holm-Hansen, C. J. Lorenzen, R. W. Holmes, and J. D. H. Strickland. 1965. "Fluorometric determination of chlorophyll," *J. Cons. perm. int. Explor. Mer* 30: 3-15, (1965); Carl Lorenzen, "The in situ sinking rates of herbivore fecal pellets," *Journal of Plankton Research* **5**, 929–933, (1983).

[45] John Walsh and Dwight Dieterle, "Use of Satellite Ocean Colour Observations to Refine Understanding of Global Geochemical Cycles," in Thomas Roswall, Robert G. Woodmansee, and Paul G. Risser, editors, *Scales and Global Change, Spatial and Temporal Variability in Biospheric and Geospheric Processes,* http://www.icsu-scope.org/downloadpubs/scope35/chapter14.html (accessed 31 January 2009)

[46] Scientific Committee on Ocean Research, http://www.scor-int.org/ (accessed 31 January 2009)

seven different countries. The chairman of the working group was John Tyler of Scripps. [Members: Prof. Alexander A. Ivanoff (France), Prof. Nils Jerlov (Denmark), Dr. Harry R. Jitts (Australia), Dr. Yulen Ochakovsky (USSR), Dr. Yatsuka Saijo (Japan), Dr. John Steele (then in Scotland), and Dr. E. Steemann Nielsen (Denmark).][47] The working group was quite influential, creating some of the first conventions for performing optical and radiometric observations at sea. The working group's activities culminated in a research cruise in 1970 aboard the National Oceanic and Atmospheric Administration (NOAA) research vessel *Discoverer*, targeting the highly productive Peru Upwelling Zone.[48] However, according to André Morel:

"... due to diplomatic complications (fisheries conflict within the 200 nautical miles, between the South American nations and the U.S. fishing fleet), we were not authorized to approach the coast, nor the Galapagos islands… So we stayed midway between the Ecuador mainland and Galapagos (Stations 8 and 9, then 13 and 14); unfortunately, the highest Chl concentrations we met in the surface layers were hardly above 0.5 mg m^{-3}…"

17 scientists and five technicians took part in this research cruise.[49] The Working Group 15 cruise originated in Miami, Florida.

This picture in Figure 1.4 is from the Web site of Niels Kristian Højerslev, and apparently shows the *R/V Discoverer* around 1970.[50] Højerslev's site indicates he took part in the "UNESCO" cruise this year; it is stated that two Danish scientists participated in the cruise; they were Højerslev and Kjell Nygård. The picture shows the research vessel before it was repainted with the NOAA acronym and symbol; the number on the front is "OSS-02", and the words underneath are "U.S. Coast and Geodetic Survey."[51]

Howard Gordon recalled: "Interestingly, someone told me that there were a lot of well-known optics people on the *Discoverer* docked in Miami. I went over there and I met Tyler, Ochakovsky, and Ray Smith." [52]

---

47 Thomas Dickey and Raymond Smith, "The Jerlov Award of the Oceanography Society as presented to Raymond C. Smith, November 2002", *Oceanography*, 16 (1), pp. 31-32, (2003).
48 Author's note: The author was also fortunate to conduct oceanographic research on the *R/V Discoverer.*
49 John Tyler, "Report of SCOR Working Group 15", *Limnology and Oceanography*, 20(4), 680, (1975).
50 "Niels Kristian Højerslev: CV", http://www.gfy.ku.dk/~nkh/cv.html, (accessed 18 June 2009).
51 Roswell Austin, "Remote Sensing of the Oceans: BC (Before CZCS) and AC (After CZCS)"; André Morel, manuscript comments on Chapter 2, received December 2009.; NOAA, "Profiles in Time – C&GS Biographies: Dr. Harris B. Stewart, Jr.", http://www.history.noaa.gov/cgsbios/h_stewart.html, (accessed 18 June 2009). The "Profiles in Time Series" biographies were found in various official and unofficial publications including the Annual Report of the Superintendent of the Coast and Geodetic Survey, the Bulletin of the Coast and Geodetic Survey, The Buzzard, the NOAA Corps Bulletin, etc.
52 Howard Gordon, manuscript comments, received 17 December 2009.

Figure 1.4. *R/V Discoverer.* Photo from Web site of Nils Højerslev .

Ray Smith noted that the working group had conducted an earlier cruise in 1968 in the Gulf of California on a research vessel from Scripps. He said that he brought the U.S. instrument, André Morel brought the French instrument, and Ochakovsky brought the Russian instrument. As a side issue, Smith said that the Russians found it "scandalous that the Scripps ship had *Playboy* magazines on it." This concern was partly due to the fact that Ochakovsky's wife was the Education and Morals editor for *Isvestia.* At the end of the cruise, however, "all of the *Playboy* magazines had disappeared from the ship. Smith reported that "it turned out all of the magazines were packed around Ochakovsky's instrument." [53]

SPOC Program Office

While researchers were working in this area in the 1960s, national agencies were also taking an interest. Perhaps inspired by the 1964 conference at Woods Hole, NASA funded the U.S. Navy to start the Spacecraft Oceanography (SPOC) Program Office, intended to push forward the field of oceanographic remote sensing. John W. Sherman took the helm of SPOC in 1967 and became a dedicated advocate and sponsor of research in the field.[54] In its early years, SPOC deployed instruments on aircraft and conducted feasibility studies for remote-sensing oceanography. NOAA took over SPOC in 1972. SPOC and Sherman were influential in developing research that supported the deployment of oceanographic remote sensing instruments, both for the microwave range (used for temperature, sea ice, sea surface height, and sea surface winds) and in the visual range.

Dennis Clark initially entered the field of ocean color remote sensing with research funded by SPOC. He started working on the detection of dye tracers in the ocean, then collaborated with the Environmental Research Institute of Michigan (ERIM), who were fabricating instrumentation (particularly infrared scanners) for military applications on

[53] Ray Smith, comments recorded at the Ocean Color Collaborative Historical Workshop, January 13-14, 2009, St. Petersburg, Florida.

[54]Ray Smith, comments recorded at the Ocean Color Collaborative Historical Workshop.

aircraft.[55]

In 1968, Sherman and L. Cheney published "Spacecraft oceanography – its scientific and economic implications for the next decade" at a United Nations conference in Vienna, Austria, dedicated to the exploration and peaceful uses of outer space.[56]

Other Activities

Gemini 5, which carried astronauts Gordon Cooper and Charles "Pete" Conrad into space on August 21, 1965, was an eight-day mission, ending on August 27[57]. During the mission, one of the new crew activities was taking color photographs of the Earth. One picture taken by the Gemini 5 crew was over the northern Gulf of California. Don Ross of International Imaging Systems attempted to determine water depth for this region by analyzing the color density in the photograph. It turned out that the effort was futile; in 1972, Roswell "Ros" Austin showed that the heavy sediment load in this region, maintained by strong tidal cycles, prevented actual observations of the ocean bottom.[58]

Scripps Visibility Laboratory

In the late 1960s and early 1970s, the Scripps VisLab continued to carry out experiments in which aircraft observations were combined with measurements of spectral radiometric properties in the water, under the management leadership of S.Q. Duntley. Dale Kiefer described Duntley as "a scientist from the old school- gentlemanly, generous, humorous, and fully devoted to the field of marine optics."[59] Kiefer also recalled that one reason the VisLab fomented important advances in the field was that there were "at least 3 engineers for every scientist – key members besides Duntley were Ted Petzold, Ros Austin, Ray Smith, and Rudy Preisendorfer."[60]

One of the experimental areas investigated by the VisLab was in the western Atlantic Ocean, working in conjunction with Ewing and Clarke. Another experiment took

---

[55] Dennis Clark interview notes transcript.

[56] John Sherman and L. Cheney, "Spacecraft Oceanography – its Scientific and Economic Implications for the Next Decade", UN PAPER 68-95878, Conference on the Exploration and Peaceful Uses of Outer Space, August 14-27, Vienna, Austria, http://ntrs.nasa.gov/search.jsp?R=815572&id=3&qs=N%3D4294954015%26Ns%3DHarvestDate|1 (accessed 31 January 2009)

[57] Gemini-V (5), http://science.ksc.nasa.gov/history/gemini/gemini-v/gemini-v.html (accessed 31 January 2009)

[58] Roswell Austin, "Remote Sensing of the Oceans: BC (Before CZCS) and AC (After CZCS)"; Roswell Austin, "Surface Truth Measurements of Optical Properties of the Waters in the Northern Gulf of California", *Fourth Annual Earth Resources Program Review*, V.IV NOAA and NRL Programs, Section 106, 21 pages.

[59] Dale Kiefer, "Ocean Optics Papers", unpublished manuscript, received 22 November 2011.

[60] Dale Kiefer, "Ocean Optics Papers", unpublished manuscript.

place off the coast of California in 1971. This experiment was called the High Altitude Ocean Color Experiment (HAOCE) – presumably the researchers had learned from Ewing and Yentsch that flying at 1000 feet was a bit risky. HAOCE was particularly notable for two reasons. The first noteworthy aspect was the combination of observations taken by a spectroradiometer operated by Goddard Space Flight Center engineer Warren Hovis, flying on a Lear jet over VisLab crew members performing optical, chlorophyll and atmospheric measurements. The second aspect of note was that Austin could utilize this combined suite of measurements to precisely describe the optical light path, i.e., from the Sun through the atmosphere, into the ocean, out of the ocean, and back to the airborne instrument – the first time this had ever been done.[61]

According to Dennis Clark, HAOCE was also one of the first times that a calibrated spectroradiometer was used, which utilized an integrating radiance sphere built by Hovis – and dubbed the "Hovisphere." The Hovisphere became well-known in the field (which will be further discussed in Chapter 4), but during HAOCE the researchers were still getting used to it, and they ran into a perplexing problem making the conversion from radiances measured while at altitude to irradiances measured with the Hovisphere. Clark discovered that a necessary calculation step had been omitted; the researchers had forgotten to divide by π. Clark tracked the problem, discovered the error, and made the correction, likely relieving a source of anxiety.[62]

One of the critical aspects at this time was that early difficulties made Warren Hovis – who was flying the spectroradiometer with a satellite instrument in mind – leery of having a "blue" spectral band on the eventual satellite instrument. In Clark's recollection, regarding the blue band, Hovis "didn't want it, didn't think it would work." However, researchers did dye-log tests off Sandy Hook NY, using two Sikorsky helicopters and boats, and they measured particle size distributions and chlorophyll. A key researcher on this project was the previously-mentioned Dale Kiefer (son of legendary Olympic swimming champion Adolph Kiefer).[63] Kiefer determined that in this region, nutrients would get trapped and enriched, and chlorophyll would have a detectable increase, indicating that the blue band was working to detect chlorophyll. The problem was the dreaded signal-to-noise ratio (SNR) – the signal is small in the blue region of the spectrum, where 90% of the signal is from the atmosphere – when conditions are at their best. When moving from the deep blue ocean into coastal waters, there is more red, and less blue, in the spectral hues of the water, and backscatter also increases, so eventually scattering and reflection dominate over absorption. This early work here was important to the eventual inclusion of a blue band on the first

---

[61]Roswell Austin, "Remote Sensing of the Oceans: BC (Before CZCS) and AC (After CZCS)"; Roswell Austin, "Remote sensing of spectral radiance from below the ocean surface", in Nils Jerlov and Einer Steemann Nielsen, editors, *Optical Aspects of Oceanography* (Academic Press, New York, 1974), pp. 317-344.

[62] Dennis Clark, interview notes transcript.

[63] Kari Lydersen, "Where are they now? Adolph Kiefer", *Swimming World and Junior Swimmer*, 44(2), (February 2003), http://findarticles.com/p/articles/mi_qa3883/is_200302/ai_n9168822/pg_2/ (accessed 11 March 2010).

ocean color instrument in space.[64]

One of the primary scientists at the Scripps VisLab in the 1960s was Rudolf W. Preisendorfer, who was primarily a mathematician with a fascination for the behavior of light in ocean waters.[65] Preisendorfer worked on a six-volume series entitled "Hydrologic Optics" that was ultimately published in 1976 by NOAA.[66] Preisendorfer's research interests evolved from the optics field to tsunami forecasting and climate modeling when he moved to the Pacific Marine Environmental Laboratory.[67]

Researchers at the Johns Hopkins University Physics Department, led by Bill Fastie, were important collaborators with the Scripps VisLab. Fastie and his group at Hopkins, including Ray Lee, actually built some of the spectroradiometers (which measured L_u and E_d as a function of wavelength) used by the laboratory. Fastie pioneered the double spectroradiometer design that helped to eliminate stray light. John Tyler and Ray Smith were participants in this development. Another spectroradiometer built at JHU went to faculty biologist Howard Seliger; Wayne Esaias was one of the operators of this early instrument. Harold Yates, Warren Hovis, Jack Sherman, and John Apel were all researchers at JHU during this foundational period for in-water optics research.[68]

André Morel noted that the Scripps spectroradiometer was "…an instrument allowing the spectral composition of the downward as well as upward irradiance to be determined *in situ*." This measurement allowed determination of the spectral reflectance, the ratio of the upward to downward spectral irradiances, and this was critical because "Spectral reflectance at null depth is the fundamental quantity in ocean color remote sensing."[69] Morel indicated that John Tyler and Ray Smith were the first to use the spectroradiometer systematically over the period 1967-1969, and the spectroradiometer was also successfully used during the SCOR Working Group 15 cruise.[70] The Scripps radiometer and the French prototype were deployed side-by-side during the cruise and achieved comparable results.[71]

The Scripps VisLab was also a crucial site for the development of in-water optical instrumentation necessary to analyze the ocean's light field. The VisLab pioneered the

---

[64] Dale A. Kiefer, R. Olson and Wayne H. Wilson, "Reflectance spectroscopy of marine phytoplankton. Part I. Optical properties as related to age and growth rate" *Limnology and Oceanography* 24(4), 664-672, (1979); Warren H. Wilson, and Dale A. Kiefer, "Reflectance spectroscopy of marine phytoplankton. Part II. A simple model of ocean color", *Limnology and Oceanography*, 24(4), 673-682, (1979); Dennis Clark, interview notes transcript.

[65] Curtis Mobley, "Biography of Rudolf William Preisendorfer", http://misclab.umeoce.maine.edu/education/HydroOptics/D:/PAPER/bio.pdf, (accessed 10 November 2009).

[66] Rudolf W. Preisendorfer, *Hydrologic Optics*, published in 1976, U.S. Dept. of Commerce, National Oceanic and Atmospheric Administration, Environmental Research Laboratories, Pacific Marine Environmental Laboratory (Honolulu).

[67] Curtis Mobley, "Biography of Rudolf William Preisendorfer."

[68] Dennis Clark, interview notes transcript.

[69] André Morel, manuscript comments on Chapter 2, received December 2009.

[70] John E. Tyler and Raymond C. Smith, *Measurements of spectral irradiance underwater* (Gordon and Breach Science Publishers, New York, 1970); John Tyler, "Report of SCOR Working Group 15."

[71] John E. Tyler, "Data Report, SCOR Discoverer Expedition, May 1970," University of California, Scripps Institute of Oceanography, SIO Reference 73-16, (June 1973); André Morel, manuscript comments on Chapter 2, received December 2009.

development of transmissometers, instruments that measure the extinction coefficient (*c*) in various media (obviously in the field of ocean optics, the medium is seawater). The VisLab first described their transmissometers in 1959, and continued to develop more advanced designs and models through the 1990s.[72]

Deployment and use of these instruments at sea was never easy, and in those early days, could resembly a jury rig. During the Fresnel II cruise in March 1971 in the Gulf of California, Dale Kiefer mounted his homemade fluorometer next to Ros Austin's state-of-the-art VisLab transmissometer. Kiefer connected his "pump, garden hose, and taped electrical wire" to the expensive transmissometer. By compensating for the 40 seconds it took for the water to flow upward from the sampling depth to the instrument (because there was no depth sensor on the fluorometer), the researchers were able to demonstrate that chlorophyll concentrations varied in tandem with the beam attenuation coefficient. This observation provided "key evidence that there existed a tight quantitative relationship between phytoplankton concentration and optical properties in the upper water column."[73]

Another type of instrument that the VisLab constructed was designed to measure light scattering. In order to describe the transmission of light in the oceans, precise determination of the volume scattering function is necessary. The VisLab's first volume scattering instrument was built in 1958. An instrument built in 1964 was designed for extreme depths, and was intended to be mounted on the U.S. Navy bathyscaphe *Trieste.*[74]

Many of the VisLab's instruments were built by skilled mechanical engineer Ted Petzold, and the collaboration of Petzold with the VisLab ocean optics scientists is an example of the vital interaction between scientists and engineers necessary to produce high-quality instrumentation. One of the best examples of this collaboration was the GASM (General Angle Scattering Meter) instrument, which Petzold employed to write the oft-cited paper "Volume Scattering Functions for Selected Ocean Waters."[75] Great instrumentation keeps on ticking; the GASM was used for ocean research as recently as 2002.[76]

The VisLab also created instruments to measure irradiance, specifically the Scripps spectroradiometer and radiance, notably the Scripps radiance photometer.[77] This instrument measured the radiance propagating toward the surface from all directions, not just that traveling straight up toward the zenith. The radiance photometer actually relied on photographic techniques and required densitometry analysis of the film; subsequent

---

[72] Kenneth Voss and Roswell Austin, "An instrumental history of the Scripps Visibility Laboratory."

[73] Dale Kiefer, "Ocean Optics Papers"; Dale A. Kiefer and Roswell W. Austin, "The effect of varying phytoplankton concentration on submarine light transmission in the Gulf of California," *Limnology and Oceanography*, 19(1), 55-64. (1974).

[74] Kenneth Voss and Roswell Austin, "An instrumental history of the Scripps Visibility Laboratory."

[75] Theodore Petzold, "Volume Scattering Functions for Selected Ocean Waters,", Scripps Institute of Oceanography Technical Report, Reference Number 72-78, http://misclab.umeoce.maine.edu/education/VisibilityLab/reports/SIO_72-78.pdf.

[76] Kenneth Voss and D.A. Phinney, "Light scattering measurements in the coastal zone, Gulf of Mexico," Abstract, OS21A-12, American Geophysical Union Ocean Sciences Meeting, Honolulu, Hawaii, (February 2002).

[77] John Tyler and Raymond Smith, "Submersible Spectroradiometer," *Journal of the Optical Society of America*, 56, 1390-1396, (1966); S.Q. Duntley, R. J. Uhl, R. W. Austin, A. R. Boileau, and J. E. Tyler, "An underwater photometer," *Journal of the Optical Society of America*, **45**, 904, (1955).

instruments have been developed by Ken Voss that utilize electronic digitization.[78]

Early International Activities

SCOR Working Group 15, which included Nils Jerlov, demonstrates the early international flavor of ocean optics research. Another noteworthy member of the group was Einer Steeman Nielsen, also of Denmark. Steeman Nielsen was a major name in the field of biological oceanography, primarily because he pioneered the carbon-14 isotope uptake method of measuring oceanic primary productivity, first described in 1951.[79] Nielsen also had a close association with Jerlov, and they collaborated as editors on the book "Optical Aspects of Oceanography", published in 1974, which contained 19 proceedings papers from a meeting held in Copenhagen in 1972.[80] The book provides a cross-section of the state of the science in 1972, and also indicates the names of several researchers who would be integral to the NASA missions in coming decades. In this book, Nielsen himself was the author of the paper "Light and Primary Production", describing the basic mechanisms of this crucial relationship. The book also features a chapter entitled "Optical properties of water and pure seawater" authored by French scientist André Morel.[81]

Morel's scientific career initiated with studies of light penetration in Mediterranean waters.[82] His work began (and is still located) in the picturesque coastal town of Villefranche-sur-Mer, adjacent to Nice and the peninsula of Cap Ferrat. Morel's extensive work examined many aspects of the oceanic light field, including how light interacts with the individual cells of phytoplankton, and the relationship of light in the ocean to photosynthesis. He collaborated in the 1970s with members of SCOR Working Group 15, and with Raymond Smith of the Scripps VisLab.

Another longtime Morel collaborator was Louis Prieur. Morel and Prieur published together in 1971, but their most important contribution, which has been cited hundreds of times in other research papers, was produced in 1977, entitled simply "Analysis of variations in ocean color." [83] The paper demonstrates the light spectra of different water types, ranging

---

[78]Kenneth J. Voss and Y. Liu, *"Polarized radiance distribution measurements of skylight: I. system description and characterization," Applied Optics*, 36, 6083-6094, (1997).

[79] Einer Steeman Nielsen, "Measurement of the production of organic matter in the sea by means of carbon-14," *Nature* 167, 684-685, (1951); Einer Steeman Nielsen, "The use of radioactive carbon (14C) for measuring organic production in the sea,"*J. Cons. Int. Explor. Mer.* 18, 117-140, (1952).

[80] Eugene Lafond, review of "Optical Aspects of Oceanography," *Limnology and Oceanography*, 20(2), 302, (1974).

[81] Einer Steeman Nielsen, *"Light and primary production."* in Nils Jerlov and Einer Steeman Nielsen, *Optical Aspects of Oceanography* (Academic Press, New York, 1974) , pp. 361–388; André Morel, "Optical properties of pure water and pure sea water," in Nils Jerlov and Einer Steeman Nielsen, editors, *Optical Aspects of Oceanography* (Academic Press, New York, 1974) , pp. 1-24.

[82] André Morel, "Resultats experimentaux concernant la penetration de la lumiere du jour dans les eaux Mediterrane'ennes." *Cahier Oceanogr.* 17, 177-184, (1965).

[83] André Morel and Louis Prieur, "Analysis of variations in ocean color," *Limnology and Oceanography*, 22, 709–722, (1977).

from the deep blue of offshore pelagic ocean waters to the greens of highly productive waters to the shades of tan and brown that can be found where the waters contain turbid coastal sediments. Morel and Prieur's paper also defines two extreme "cases" – cases which are frequently referred to in ocean color remote sensing. Case 1 is the "easy" case. Case 1 waters are those waters where the dominant factor controlling the optical characteristics of the water is the concentration of phytoplankton, so that the main light absorption is caused by chlorophyll. The only other important factor in this case is the interaction of light with water itself. Case 2 waters, on the other hand, are the "hard" case. In Case 2 waters, the dominant factors controlling the optical characteristics of the water are something other than phytoplankton. It could be suspended sediments, or colored organic matter, or *Gelbstoff* ("yellow substance") which is organic matter but not quite the same thing as colored organic matter. Very commonly, Case 2 waters contain a contribution from "all of the above", and thus analysis of their optical constituents is much more difficult.

Summary: Early ocean optics, from its beginnings with investigations of water color and clarity, to early systematic observations from ships, and then to the more rigorous instrumental investigations from ships and aircraft platforms, set the stage for NASA to design a mission that would provide remote sensing observations of the oceans. The dramatic increase in ocean optics research that followed World War II provided scientific impetus for such a mission. The research conducted in the 1950s, 1960s, and 1970s also established a base of expertise from which NASA could draw when scientific input was needed for their pioneering mission.

2

STEPS TOWARD SPACE

In the 1960s, while the public's attention with regard to space was largely focused on manned spaceflight – for the United States, the Mercury, Gemini, and Apollo programs leading to landings on the moon – NASA was also exploring other uses of satellites in space, and one of the primary uses was deemed to be observations of Earth. In the 1960s, this initially meant satellites that could be used for weather observations. NASA's Office of Space Science and Applications (OSSA) partnered with the Weather Bureau to push forward this practical use of the orbital environment.[1]

The idea of using high-altitude observations of the Earth for improvement of weather forecasting, and the science of meteorology in general, dates back to the 1940s, when rockets were used to launch observational instruments, generally in a military context.[2] NASA got involved when it assumed management of the TIROS program in 1958, due to the initial NASA organizational guidelines.[3] Despite the fact that the nation's first satellite (Explorer 1) had only launched in January 1958, NASA managed to launch the nation's first weather satellite, TIROS 1, on April 1, 1960.[4] Whether or not this was considered foolish by anyone, NASA continued with the TIROS program through TIROS-10 in 1965.[5]

Concurrently, NASA initiated another program, the Nimbus program, in the 1960s. While TIROS was primarily an observational platform, i.e., to see what was coming, Nimbus was the research platform, satellites that would try out new instruments and collect new data, allowing increased scientific insight.[6] One downside of this particular approach was that new technology was inherently riskier, and several Nimbus instruments failed rather quickly in orbit. Warren Hovis was Principal Scientist for an instrument on Nimbus 4 that

---

[1] W. Henry Lambright, "NASA and the Environment: Science in a Political Context", in Steven J. Dick and Roger D. Launius, editors, *Societal Impact of Spaceflight*, NASA SP-2007-4801, page 313.
[2] Pamela Mack and Ray A. Williamson, "Observing the Earth from Space", Chapter 2 in John D. Logsdon, Roger D. Launius, David H. Onkst, and Stephen J. Garber, editors, *Exploring the Unknown, Selected Documents in the History of the U.S. Civil Space Program, Volume 3: Using Space*, NASA SP-4407, page 157.
[3] Pamela Mack and Ray A. Williamson, "Observing the Earth from Space", page 158.
[4] "Milestones of Flight: Explorer 1 (backup)", http://www.nasm.si.edu/exhibitions/gal100/exp1.html, accessed 22 October 2010.
[5] "TIROS: Television Infrared Observation Satellite Program", http://science.nasa.gov/missions/tiros/, accessed 22 October 2010.
[6] NSSDC, "Nimbus Program", http://nssdc.gsfc.nasa.gov/earth/nimbus.html, accessed 15 March 2011.

didn't return useful data, and an instrument on Nimbus 5 that only worked for four weeks after launch![7]

The late 1960s were also a period of increasing environmental awareness. Books such as Rachel Carson's *Silent Spring*, published in 1962, and Paul Ehrlich's *The Population Bomb* in 1968, motivated the environmentally-minded public.[8] Salient incidents, notably the Cuyahoga River fire in Cleveland, Ohio in June 1969, honed public opinion and also precipitated legislative action. As Jonathan Adler said, "The fire did contribute a huge amount to the new environmental movement and it put the issue in front of everyone else, too," adding , "Water pollution became a tangible, vivid thing -- like it had never been on a national level.[9]

While the Cuyahoga River fire focused the attention of Americans on the issue of water pollution, in parallel with efforts to clean up the atmosphere (the Clean Air Act had been significantly amended in 1970, and the Clean Water Act was passed in 1972), an incident two years earlier marked a significant increase in the international public's awareness of the potential of man's activities to damage the ocean environment. This incident was the wreck of the *Torrey Canyon* oil tanker on the Scilly Isles on March 18, 1967. While not large compared to the much bigger *Amoco Cadiz* and *Exxon Valdez* spills, or the Ixtoc I oil well blowout in the southern Gulf of Mexico, the *Torrey Canyon* was the first large tanker oil spill to get international publicity. In fact, "The Torrey Canyon disaster did have some beneficial consequences. International maritime regulations on pollution were created. A charismatic young botanist called David Bellamy was asked to comment on the disaster and became a television star; he, and the oil slick, helped raise awareness of pollution." [10] NASA's Nimbus program of Earth observations therefore took place against a backdrop of increased environmental awareness – a factor that program managers and administrators could potentially use to argue for new types of observations, such as ocean color.

Thus, in the late 1960s and early 1970s, a convergence of thought and action began to occur. Clarke, Ewing, and Lorenzen published their paper on the correlation of backscattered light measured from an aircraft with chlorophyll concentrations in 1970, and Hovis operated the TRW spectroradiometer in an airplane during HAOCE in 1971. [11] Those initial campaigns – taken together with Yentsch's paper suggesting that the perceived color of the ocean should vary with different population levels of phytoplankton – therefore suggested to NASA engineers looking for a challenge that it might be possible to detect

---

[7] See Appendix 2: "Setting the Stage: Nimbus 1-5".
[8] Paul R. Ehrlich and Anne H. Ehrlich, "The Population Bomb Revisited", *The Electronic Journal of Sustainable Development*, 1(3), 63-70 (2009).
[8] Michael Scott, "Cuyahoga River fire galvanized clean water and the environment as a public issue", http://blog.cleveland.com/metro/2009/04/cuyahoga_river_fire_galvanized.html, accessed 23 March 2011.
[10] Patrick Clark, "Oil Spills: Legacy of the Torrey Canyon", http://www.guardian.co.uk/environment/2010/jun/24/torrey-canyon-oil-spill-deepwater-bp, accessed 23 March 2011.
[11] George Clarke, Gifford Ewing, and Carl Lorenzen, "Spectra of backscattered light from the sea obtained from aircraft as a measure of chlorophyll concentration"; Roswell Austin, "Remote Sensing of the Oceans: BC (Before CZCS) and AC (After CZCS)".

different levels of light emerging from the ocean surface from a satellite-borne instrument.[12] The data collected by the instrument in space could then be used to conduct the same type of analysis Clarke et al. had done – only with a much broader, large-scale view of the Earth's oceans than could be obtained from an airplane.

There was one particular problem that had already become apparent to the researchers attempting these observations – and that problem was air (really, the atmosphere).

We don't necessarily consider air much of a problem, considering that breathing is very important to our ability to continue our lifestyle in the way to which we have become accustomed. We generally consider air to be a rather transparent factor in our daily lives, and indeed we tend more to notice when it isn't readily available, such as when we dive underwater in a swimming pool. And while we're relaxing on the side of the pool, we don't generally remark on the fact that the pool water appears blue and on a nice sunny clear day, the air doesn't have a noticeable color at all.

But the sky does, of course – it's blue, too. And the reason is the same (see Appendix 1) – scattering and absorption of the light from the Sun by the molecules of air or the molecules of water.

Yet there is another, less obvious effect that light scattering has, unless you are submerged at a considerable depth in water or if it is a particularly hazy day. The scattering of light by water or air attenuates the light intensity – that is, over longer distances, the intensity of the light is reduced.

The early aircraft campaigns conducted at low altitude – with low-sensitivity radiometers – indicated very quickly that one of the main problems with detecting the changes in the light emanating from the ocean surface due to the main absorbing and reflecting components (phytoplankton, sediments, and optically active organic matter) was the scattering of light by the atmosphere. The oceans are a dark target (Appendix 1, page 25), meaning that only about 6-10% of the sunlight that enters the ocean surface comes back out.[13] Compare that to sea ice or snow, which reflects 85-90% of the sunlight that hits it – clouds are similar, though thin clouds don't reflect as much – or even to land surfaces with considerable vegetation, which reflect about 10-20% of the light, and it's immediately clear that this is not a very large signal to work with! [14] Furthermore, only a percentage of that small amount of reflected or scattered light is directed in a sufficiently upwards direction to be detected by an instrument mounted on an aircraft, or on a satellite.

And then there's the atmospheric factor. Scattering is responsible for lovely blue skies or hazy brown or gray skies (depending on what substances, and how much of them, are floating in the air), and this scattering process further reduces the amount of light heading skyward.[15] Thus, only a few dogged and persistent photons that have emerged from the ocean surface will ultimately be captured by the instrument detectors high overhead, as is

---

[12] Charles Yentsch, "Distribution of chlorophyll and phaeophytin in the open ocean"

[13] National Snow and Ice Data Center, "All About Sea Ice: Processes: Thermodynamics: Albedo", http://nsidc.org/seaice/processes/albedo.html, (accessed 17 March 2009).

[14] "Albedo", http://www.absoluteastronomy.com/topics/Albedo, (accessed 17 March 2009).

[15] "Blue Sky", http://hyperphysics.phy-astr.gsu.edu/hbase/atmos/blusky.html, (accessed 10 February 2009).

described in Appendix 1.

Even though preliminary research campaigns had indicated it was possible to detect alteration of the surface ocean light field caused by variable chlorophyll concentrations (and other stuff), the atmospheric problem had to be addressed. The basic question was: if you put a radiometer at very high altitude, or in space, is a sufficient amount of light going to get to that radiometer to allow quantitative determination of chlorophyll concentrations?

The way to answer that question, as a step toward putting an instrument on a satellite, was to build a better radiometer, and then fly it on an aircraft at very high altitude. Fortunately, NASA had already provided a way to do that.

For those who recall the 1950s, the simple phrase "U-2" is synonymous with some very significant events. On May 1, 1960, "U-2" burst into the public consciousness when a U-2 high-altitude surveillance aircraft piloted by Gary Powers was shot down over the Soviet Union, precipitating a Cold War crisis.[16] But that crisis was somewhat overshadowed by the next one, when a U-2 photographed Soviet missile bases being built in Cuba in October 1962.[17]

While the military role of the U-2 was recognized and remembered due to these events, the civilian aerospace role of the U-2C and its scientific heir, the ER-2, is not nearly as well-known. The high-altitude aircraft campaign, however, provided a very important test-bed to prove the capabilities of new instruments seeking a coveted and highly-competitive spot on a NASA satellite.[18]

So, at about the same time it became necessary to move the airborne ocean color radiometer from Lear jet altitude to higher altitude, NASA commenced its high altitude aircraft program in 1971.[19] Part of the reason for the initiation of high-altitude Earth observations may have been the guidance of NASA administrator James Fletcher, who wanted NASA to be considered "an environmental agency".[20] (Interestingly, though the environmental movement was ascendant at this time and the Environmental Protection Agency was established in 1970, NASA didn't ally itself in general with the environmental movement because of its perceived anti-technology bent.[21])

In January 1972, Warren Hovis published a short report in a Goddard Space Flight Center research summary entitled "Measurements of ocean color" describing initial aircraft

---

[16] Thomas Bogharde, "Traitor or Patriot? The Story of U-2 Pilot Francis Gary Powers", 2007, http://www.spymuseum.org/programs/educate/pdfs/AIRSHO2007_Powers.pdf, (accessed 1 April 2009).
[17] "The Cuban Missile Crisis, 1962: The Photographs", http://www.gwu.edu/~nsarchiv/nsa/cuba_mis_cri/photos.htm, (accessed 1 April 2009).
[18] NASA, "ER-2 Program History", http://www.nasa.gov/centers/dryden/research/AirSci/ER-2/history.html, (accessed 1 April 2009).
[19] NASA, "ER-2 Program History".
[20] Roger Launius, "A Western Mormon in Washington, DC: James C. Fletcher, NASA, and the Final Frontier," *Pacific Historical Review* (1995), p. 236, cited in W.L. Lambright, "NASA and the Environment: Science in a Political Context", page 315.
[21] Jack Lewis, "The birth of EPA", *EPA Journal*, November 1985; Roger Launius, "A Western Mormon in Washington, DC: James C. Fletcher, NASA, and the Final Frontier."

instrumentation measurements.[22] Just a couple of weeks after that, NASA received a report from Siebert Duntley entitled "Detection of ocean chlorophyll from earth orbit", which was an evaluation of the feasibility of detecting phytoplankton chlorophyll at satellite orbital altitudes for both clear and hazy atmospheric conditions.[23] The report also examined the effect of solar altitude to determine the best orientation for an ocean color sensor to optimize chlorophyll observations. According to Dale Kiefer, this report included "…a comprehensive treatment of the physics of orbital measurement of spectral radiance leaving the sea surface … issues such as appropriate viewing angle, sun glitter, spatial resolution, temporal and spatial coverage as a function of the sensor's field of view and the satellite's orbit." [24]

Soon after that, in 1973, the Goddard-built Ocean Color Scanner (OCS) was mounted on the ER-2 aircraft and commenced an observational campaign intended to *prove* that it would be indeed possible to measure ocean chlorophyll concentrations from space.[25]

The first results of this campaign were reported in 1973, which described the instrument deployments and observations on the Lear jet and the Convair.[26] They reported that chlorophyll concentrations could indeed be measured at high altitude and return similar results to those from low altitude, though the "intervening atmosphere reduced color contrast", which is a rather wordy way to say that there was less signal to be measured at high altitude compared to low altitude. The instrument used for this research was a precursor to the OCS. Hovis (and presumably a few other instrument engineers) built the OCS as a "breadboard system" for the instrument that was intended to fly on Nimbus G. 1973 also marks the year when the Coastal Zone Color Scanner (CZCS) program became an approved program.[27]

Garnering approval for the CZCS program intensified the need to both gather more data from aircraft observation campaigns, and to start to get a handle on the atmospheric scattering problem. Since there is usually some lag time between when data is collected and when research papers are published, the results of the aircraft observation campaigns supporting the CZCS program were published just before the launch of Nimbus G, and for a few years afterward.

The critical results of the OCS observations were published by Hovis and K.C. Leung. They first collaborated on a paper presented at the Oceans '76 meeting in Washington, DC, entitled "Remote Sensing of Chlorophyll Concentration from High Altitude" .[28] This paper

---

[22] Warren Hovis, "Measurements of Ocean Color", Significant Accomplishments in Science, p 24-29 (1971). Publication date was 1 January 1972.
[23] Siebert Q. Duntley, "Detection of ocean chlorophyll from earth orbit", *NASA Manned Spacecraft Center 4th Annual Earth Resources Program Review*, Volume 4, 25 pages.
[24] Dale Kiefer, "Ocean Color Papers".
[25] Roswell Austin, "Remote Sensing of the Oceans: BC (Before CZCS) and AC (After CZCS)".
[26] Warren Hovis, Michael Forman, and Lambdon Blaine, "Detection of ocean color changes from high altitude", *NASA Technical Memorandum X-70559*, NASA Goddard Space Flight Center, (November 1973), 25 pages.
[27] Roswell Austin, "Remote Sensing of the Oceans: BC (Before CZCS) and AC (After CZCS)".
[28] K.C. Leung and Warren Hovis, "Remote sensing of chlorophyll concentration from high altitude" in *Oceans '76; Proceedings of the Second Annual Combined Conference*, Washington, D.C., September 13-15,

reports the results of a collaboration between the Goddard scientists and the NASA pilots with the Department of Oceanography and Remote Sensing Center of Texas A&M University. (Though it is not stated in this paper, it's very likely that one of the Texas A&M scientists was Dr. Sayed Z. El-Sayed.) The campaign in November 1974 observed three separate areas: coastal waters off of Tampa Bay and Panama City (Florida), and Timbalier Bay, Lousiana. Timbalier Bay was covered by clouds and the waters off of Tampa Bay were too clear, but Panama Bay provided a range of chlorophyll concentrations.

Then the NASA scientists had the temerity to make aircraft observations in the New York Bight, in collaboration with NOAA. Speaking in terms that were hopeful, Leung and Hovis reported that the results were "reasonable" – the spectral shape attributed to chlorophyll in the Gulf of Mexico was similar to that in the New York Bight, but they didn't agree quantitatively.[29] They suspected that the difference was due to a higher concentration of suspended sediments in the New York Bight. (This might have been the first time that the interfering aspects of suspended sediments on ocean color data were reported; this is a theme that would be repeated many times in subsequent years.)

The other "small" thing that Leung and Hovis mention is that they detected a very strong signal from an acid waste dump off of New Jersey.[30] These observations, and observations from the CZCS a couple of years later, would introduce ocean remote-sensing into the public eye with definite overtones of environmental concern.

The next significant contribution from Hovis and Leung was a published paper entitled simply "Remote sensing of ocean color", published in Optical Engineering in 1977.[31] This paper included a remarkable color swath from the OCS obtained over the southwestern coast of Florida – a plume of sediment can even be seen emerging from Johns Pass between the St. Petersburg Beach and Madeira Beach barrier islands north of the mouth of Tampa Bay. The paper also showed images of the New York Bight, and detected a signal from the ocean dumping site off of Sandy Hook. Underscoring how difficult it was to extract meaningful data from the ocean surface, this paper estimated that the "oceanographic information" was "contained in 20% or less of the total detectable signal". The final section of this paper previews the CZCS and notes the importance of the tilt mechanism to reduce the effects of sun glint.

In 1980, Hovis provided a brief programmatic summary of the CZCS program at a meeting entitled "Oceanography from Space" – essentially a followup to the 1964 Woods Hole conference organized by Ewing.[32] Although Ewing was unable to attend the meeting (which was held in Venice, Italy) he provided an introduction to the book containing the published proceedings, and the proceedings book was later dedicated to him. Ewing noted

---

1976. (A78-12827 02-48) New York, Institute of Electrical and Electronics Engineers, Inc.; Washington, D.C., Marine Technology Society, 1976, p. 15E-1 to 15E-4.

[29] K.C. Leung and Warren Hovis, "Remote sensing of chlorophyll concentration from high altitude".

[30] K.C. Leung and Warren Hovis, "Remote sensing of chlorophyll concentration from high altitude".

[31] Warren Hovis and K.C. Leung, "Remote sensing of ocean color", *Optical Engineering,* 16, 158- 166 , (1977).

[32] Warren Hovis, "The Nimbus-7 Coastal Zone Color Scanner (CZCS) Program" in J.F.R. Gower, editor, *Oceanography from Space*, Plenum Press, New York, 213-225, (1981).

in his introduction that "there is good reason to predict that the accomplishments of the next 16 years will surpass those of the previous 16". That 16-year interval would figure prominently in the history of ocean color remote sensing.

In his paper, Hovis gave a short description of the work that led to the CZCS mission concept, crediting Clarke et al. and the ongoing research by Yentsch as providing the necessary theoretical basis.[33] His description of the aircraft campaign is particularly intriguing. After noting the first results of Hovis et al. 1973, he describes the results of the U-2C missions.[34] The following quote acknowledges where some of the difficulty would definitely be expected:

"Although the aircraft information indicated that atmospheric correction would be a formidable problem, especially the correction for the highly spatially variable aerosols in the atmosphere, the data was encouraging enough that it was felt the such corrections could be carried out with spacecraft data, and the Nimbus-7 CZCS was conceived and proposed to NASA for flight."

Hovis mentions another aspect of the aircraft observations that would guide the design and deployment of the CZCS on Nimbus 7: *sun glint.*[35] As Appendix 1 describes, sun glint is the direct reflection of sunlight off the surface to the optics of the satellite, a problem that happens with a direct downward (nadir) view of the ocean surface. The aircraft data confirmed that sun glint would be a "serious" problem; the solution to the problem would be a significant aspect in the design of the CZCS and subsequent ocean color satellite instruments.

Thus, while the OCS provided sufficient data to get the CZCS mission off the ground (at least in the programmatic sense) during the 1971-1973 period, that was certainly not the only role for the OCS as the CZCS was being designed and launched, and even while it was in orbit. The OCS continued to be deployed around the world to improve the scientific understanding of ocean color remote sensing data and to support oceanographic research programs. Table 1 lists some of the significant observational campaigns that included the OCS. In 1975, a study enlisted the help of world-famous oceanographer Jacques Cousteau and his nearly-as-famous ship, the *Calypso.*[36] Cousteau's own appeal to the environmental consciousness of the 1960s and 1970s, particularly highlighting the oceans, stemmed in part from the TV show "The Undersea World of Jacques Cousteau", which aired from 1966 to 1976.[37] Working with Cousteau likely burnished the environmental credentials of the OCS scientific crew.

The OCS, as noted earlier, took part in the evaluations of the ocean dumping sites off of New Jersey prior to the launch of Nimbus 7, and confirmed that it wasn't difficult to

---

[33] Warren Hovis, "The Nimbus-7 Coastal Zone Color Scanner (CZCS) Program".

[34] Warren Hovis, Michael Forman, and Lambdon Blaine, "Detection of ocean color changes from high altitude"

[35] Warren Hovis, "The Nimbus-7 Coastal Zone Color Scanner (CZCS) Program".

[36] J.C. Harlan, J.M. Hill, H.A. El-Reheim, and C. Bohn, "A biological and physical oceanographic remote sensing study aboard the *Calypso*", in *Proceedings of the 10th International Symposium on Remote Sensing of the Environment,* Volume 1, Environmental Research Institute of Michigan, Ann Arbor, 661-670, (1975).

[37] David Pickering, "Cousteau, Jacques", Museum of Broadcast Communications, http://www.museum.tv/eotvsection.php?entrycode=cousteaujac, accessed 23 October 2010.

detect these operations.[38] The OCS was also deployed to Europe, flying over the North Sea and examining coastal process off of Belgium, Holland, and Germany (an area that would turn out to be one of the regions where suspended sediments presented a near-constant problem) and also over the southern Adriatic Sea[39].

The OCS took part in studies of San Francisco Bay concerned with evaluating water quality in the urbanized estuary.[40] In the Bay, the OCS successfully located areas of high biological activity, in contrast to standard aerial photography, which was unable to locate such areas. The OCS data was also somewhat correlated with salinity, probably because in an estuary there is a large range of salinity, in contrast to the much more constant salinity in the open ocean. The San Francisco Bay study included Dr. Rita Colwell, who would continue in a distinguished scientific career by examining the connection between zooplankton (specifically copepods) and cholera outbreaks. Colwell would also serve as the Director of the National Science Foundation from 1998-2004.

One other OCS study (Kim et al. 1980) is worthy of mention.[41] The OCS was deployed over both the Pacific and Atlantic Oceans to evaluate atmospheric correction methods and to test how well chlorophyll maps could be generated. This study included young researchers Charles McClain and James Yoder, who continued in this field of work into the SeaWiFS and MODIS era.

The European Association of Scientists for Experiments on Pollution (EURASEP) was the primary participant in a campaign of OCS flights that took place over locations in Europe in 1977, several of which are discussed above. A conference which convened on October 30, 1979 in Ispra, Italy was devoted to the various OCS observational missions for the 1977 observational campaign.[42]

Summary: NASA's Earth observation program allowed the concept of a satellite ocean color mission, an engineering challenge that was inspired by the research of several individuals, notably Gifford Ewing and Charles Yentsch. Warren Hovis of Goddard Space Flight Center

[38] J.B. Hall and A. Pearson, "Results from the National Aeronautics and Space Administration remote sensing experiments in the New York Bight, 7-17 April 1975", *NASA Technical Memorandum, NASA-TM-X-74032*, (April 1977), 190 pages. (Hall and Pearson are listed as "compilers".)

[39] P. Lohmann and H. van der Piepen, "Evaluation of ocean bottom features from Ocean Color Scanner imagery", *Photogrammetria*, 36, 81-89, (April 1981); L. Giannini, "Differential spectroscopy for the coastal water quality identification by remote sensing", in J.F.R. Gower, editor, *Oceanography from Space*, Plenum Press, New York, 395-402, (1981).

[40] R.N. Colwell, A.W. Knight, and S. Khorram, "Remote sensing analysis of water quality and the entrapment zone in the San Francisco Bay and delta: Final Report", *NASA Technical Report, NASA-CR-163169*, (1979), 40 pages.

[41] H.H. Kim, C.R. McClain, L.R. Blaine, W.D. Hart, L.P. Atkinson, and J.A. Yoder, "Ocean chlorophyll studies from a U-2 aircraft platform", *Journal of Geophysical Research*, 85, 3982-3990, (July 20 1980).

[42] Benny Møller Sorensen, editor, *Proceedings of Workshop on the EURASEP Ocean Color Scanner Experiments*, Commission of the European Communities, Joint Research Centre, Ispra, Italy, 198 pages, (1979).

was the technological leader of this effort, and his work with aircraft instrument observations created the necessary technical background for a proposed satellite instrument. This work also clearly indicated the research areas that would be the most difficult to address. The early collaborations with airborne instruments also established international collegiality that would be crucial to the success of NASA's first ocean color mission.

Figure 2.1. NASA ER-2 aircraft. Photo from NASA Dryden Flight Research Center Photo Collection.

Table 1. OCS observational campaigns

Observation campaign or publication year	Purpose	Location	Reference	Notes
Publ. 1975	Observation/measurement correlations	Gulf of Mexico and Caribbean	Harlan et al. [43]	Conducted off the R/V *Calypso*, Jacques Cousteau's ship
April 7-17, 1975	Ocean dumping evaluation	New York Bight	Hall and Pearson [44]	
October 29, 1975	Data evaluation	Santa Barbara Channel	Kraus et al. [45]	
June 27, 1977	North Sea OCS Experiment	North Sea, Tay Estuary	Cracknell et al. [46]	

[43] J.C. Harlan, J.M. Hill, H.A. El-Reheim, and C. Bohn, "A biological and physical oceanographic remote sensing study aboard the *Calypso*".

[44] J.B. Hall and A. Pearson, "Results from the National Aeronautics and Space Administration remote sensing experiments in the New York Bight, 7-17 April 1975".

[45] S.P. Kraus, J.E. Estes, M.R. Kronenberg, and E.J. Hajic,"Evaluation of Ocean Color Scanner (OCS) photographic and digital data: Santa Barbara Channel test site, 29 October 1975 overflight". *NASA Technical Report, NASA-CR-156822*, (18 October 1977).

[46] A.P. Cracknell, S.M. Singh, and N. Macfarlane, "Remote sensing of the North Sea using Landsat-2 MSS and Nimbus-7 CZCS data", in *Proceedings of the International Symposium on Remote Sensing of Environment*, Volume 3, San Jose, Costa Rica, April 23-30, Ann Arbor, Mich., Environmental Research Institute of Michigan, 1643-1651, (1980).

Observation campaign or publication year	**Purpose**	**Location**	**Reference**	**Notes**
Sep. 25, 1977	Coastal applications	Adriatic Sea	Giannini[47], Frassetto[48]	
Summer 1979	Water quality applications	Lake Ontario	Zwick et al. [49]	Study included Nimbus 7 (CZCS) overflight data
Publ. 1979	Water quality applications	San Francisco Bay	Colwell, Knight, and Khorram, [50] Khorram[51]	Colwell went on to study the phytoplankton – cholera connection and became head of NSF
1980	Superflux III	Chesapeake Bay	Ohlhorst[52]	

[47] R.F.L. Giannini, "Spectral analysis of Ocean Color Scanner data applied to the North Adriatic Sea", in Benny Møller Sorensen, editor, *Proceedings of Workshop on the EURASEP Ocean Color Scanner Experiments*, Commission of the European Communities, Joint Research Centre, Ispra, Italy, 103-114, (30 October 1979).

[48] R. Frassetto, "Pre-launch experiments for coastal pollution studies with the satellite Nimbus G", in *Proceedings of the Workshop on pollution of the Mediterranean*, Antalya, Turkey, International Commission for the Scientific Exploration of the Mediterranean Sea, Monaco. United Nations Environment Programme, Nairobi, 645-646, November 24-27, (1978).

[49] H.H. Zwick, S.C. Jain, and R.P. Bukata, "A satellite/airborne and in-situ water quality experiment in Lake Ontario, , in Benny Møller Sorensen, editor, *Proceedings of Workshop on the EURASEP Ocean Color Scanner Experiments*, Commission of the European Communities, Joint Research Centre, Ispra, Italy, 181-190, (30 October 1979).

[50] R.N. Colwell, A.W. Knight, and S. Khorram, "Remote sensing analysis of water quality and the entrapment zone in the San Francisco Bay and delta: Final Report".

[51] S. Khorram, "Use of ocean color scanner data in water quality mapping**"**, *Photogrammetric Engineering and Remote Sensing*, 47, 667-676, (May 1981).

[52] C.W. Ohlhorst, "Analysis of ocean color scanner data from the Superflux III Experiment", *NASA Technical Memorandum NASA-TM-83290*, 30 pages, (1 March 1982).

Observation campaign or publication year	Purpose	Location	Reference	Notes
Publ. 1980	Testing, atmospheric correction	Pacific and Atlantic sites	Kim et al. [53]	Co-authors include McClain and Yoder
Publ. 1981	EURASEP	Tay River estuary, Belgian and Dutch coasts	Cracknell and Singh [54]	CZCS calibration test
Publ. 1981	Bottom detection	German Bight	Lohmann and Van der Piepen[55]	

[53] H.H. Kim, C.R. McClain, L.R. Blaine, W.D. Hart, L.P. Atkinson, and J.A. Yoder, "Ocean chlorophyll studies from a U-2 aircraft platform".

[54] A.P. Cracknell and S.M. Singh, "Coastal zone research using remote sensing techniques; calibration of coastal zone color scanner", *Proceedings, ESA Applications of Remote Sensing Data on the Continental Shelf*, 157-168, (1 July 1981).

[55] P. Lohmann and H. van der Piepen, "Evaluation of ocean bottom features from Ocean Color Scanner imagery".

3

THE PIONEER – CZCS

Vandenberg Air Force Base is located in southern California, between Lompoc and Santa Maria, about 100 kilometers northwest of Santa Barbara, and almost due north of San Miguel Island, the westernmost of the Channel Islands.[1] Vandenberg lies on the relatively flat coastal plain of the Santa Ynez River, which provides irrigation for agriculture, which in this area for many years was devoted to flowers to produce seeds for gardeners.[2] The coast of California turns to the north from Point Conception up to Point Arguello, which lies just southwest of Vandenberg. It is a good location to launch missiles and rockets from, because the Pacific Ocean provides a fairly safe place for a failed launch to splash into.

Vandenberg, as its name implies, is an Air Force base, and it is also the location of the Western Launch and Test Range (WLTR).[3] The Western Launch and Test Range was originally used to test intercontinental ballistic missiles, and was soon developed into a satellite launch facility. In 1972, it was also selected to be the western launch site for the Space Shuttle, but this expensive capability was never utilized.[4]

Vandenberg was the site of all the Nimbus satellite program launches.[5] On October 24, 1978, a five-year process of instrument design and fabrication, satellite construction, and a Delta launch vehicle converged for the launch of the Nimbus G mission.[6] Attached to the Nimbus satellite perched on top of the Delta launch vehicle – one of the seven instrument packages on the instrument platform – was a "proof-of-concept" instrument called the Coastal Zone Color Scanner, the CZCS[7]. Either watching intently, or waiting for news of the launch, were Warren Hovis and the members of the CZCS Nimbus Experiment Team (CZCS NET).

It is very likely that both the engineers who built the instrument and the scientists who helped plan the mission all entertained a recurring question that day: would it really work?

---

[1] "Vandenberg Air Force Base – Fact Sheet", http://www.vandenberg.af.mil/library/factsheets/factsheet_print.asp?fsID=4606&page=1, (accessed 10 April 1979).

[2] "History of Lompoc", http://www.cityoflompoc.com/government/history.htm, (accessed 10 April 1979).

[3] Joseph F. Wambolt and Jimmy F. Kephart, "A Complete Range of Launch Activities", http://www.aero.org/publications/crosslink/winter2003/04.html, (accessed 10 April 2009).

[4] Joseph F. Wambolt and Jimmy F. Kephart, "A Complete Range of Launch Activities".

[5] "Nimbus Program History". National Aeronautics and Space Administration Goddard Space Flight Center, NP-2004-10-672-GSFC, October 26, 2004, 35 pages, http://atmospheres.gsfc.nasa.gov/nimbus/pdf/Nimbus_History.pdf.

[6] "Nimbus Program History". NASA GSFC.

[7] "Nimbus Program History". NASA GSFC.

Of course, anyone watching a satellite launch might also be thinking – will it actually get into orbit? With regard to satellite launches, this is never a sure thing.

They had certainly tried hard enough to reach that point. As Hovis described in 1973, the high-altitude remote sensing observations with a Goddard radiometer and then the OCS provided enough initially-promising data that the CZCS could be proposed for flight – a decidedly aeronautical way of saying that it could be proposed to be an instrument on a satellite mission.[8] The proposal apparently was directed to William Nordberg (who had moved from the Nimbus Project Scientist position to be the Director of the Earth and Atmospheric Sciences Directorate at Goddard Space Flight Center) as well as Project Manager Charles Mackenzie and Project Scientist Al Fleig.[9]

CZCS design and fabrication commenced in 1973, with Goddard selecting Ball Brothers Aerospace Corporation (now Ball Aerospace) as the instrument fabricator.[10] Reports provided by Robert Woodruff indicate that work on the optical design commenced by December 1973.[11] Woodruff was the chief optical designer, working with Jim Darnell, the optics design engineer. Some of the significant personnel at Ball Brothers included Program Manager Bill Wallschlaeger, system engineer Randy McConaughey and production engineer Bill Flaherty.[12] Other personnel at Ball who worked on the CZCS were Art Ray, who wrote testing software and was also responsible for creating the phase lock loop design for the primary mirror. Neal Zaun worked on the instrument detectors, Bob Loomis was the chief mechanical engineer, and the quality analysis engineer was Gavin Noble, a "classic Scotsman".[13] [Other Ball personnel will be mentioned in subsequent discussion of the fabrication of the CZCS.]

Steve Varlese recalled: "CZCS was the first flight instrument I worked on at Ball, some time early in 1976. I was assisting Eloy "Bill" Roybal, who had been the optical production engineer for the Skylab coronagraphs Ball built. He was extremely knowledgeable about optics fabrication, inspection, assembly, and test, and a demanding teacher. We did assembly and test of the engineering model of the CZCS instrument together, then I did most [of the] assembly and alignment of the protoflight model that actually flew [on Nimbus 7],

---

[8] Warren Hovis, Michael Forman, and Lambdon Blaine, "Detection of ocean color changes from high altitude".

[9] Nicholas M. Short, "Meteorology – Weather and Climate: A Condensed Primer", Section 14, *Remote Sensing Tutorial*, http://www.fas.org/irp/imint/docs/rst/Sect14/Sect14_1a.html, (accessed 10 April 2009); "Nimbus 7". National Space Science Data Center, **NSSDC ID:** 1978-098A, http://nssdc.gsfc.nasa.gov/nmc/spacecraftDisplay.do?id=1978-098A, (accessed 10 April 2009).

[10] Warren Hovis, Dennis Clark, Frank Anderson, Roswell Austin, W. Wilson, Edward Baker, D. Ball, Howard Gordon, James Mueller, Sayed El-Sayed, Boris Sturm, Robert Wrigley, and Charles Yentsch, "Nimbus-7 Coastal Zone Color Scanner: System Description and Initial Imagery", *Science*, 210 (4465), 60-63, (1980).

[11] Ball Brothers Research Corporation, System Engineering Report, dated 10 April 1974. Copy provided by Robert Woodruff.

[12] Woodruff later worked on the Skylab Backscatter Ultraviolet (SBUV) mission, the Hubble Space Telescope, and is currently working on optical design for the James Webb Space Telescope.

[13] Recorded interview, Bill Wallschlaeger and Randy McConaughey, April 3, 2009. Information also provided by S. Varlese.

continuing through integration, test, and support to the spacecraft integration." Varlese is now associated with the Gemini telescopes in Hawaii.[14]

Varlese added that: "the "flight" unit was cancelled just after the telescope and spectrometer optics had been assembled, so put in a box and shipped to some government warehouse."[15] (There will be more about this flight model, and its actual location – no, it is not necessary to consult with Indiana Jones to find it – in Chapter 5).

One particular aspect of the CZCS deserves mention. CZCS was certainly not a pinball machine, but it has one thing in common with those antiquated relics of life-before-video games: it could tilt. Tilt may have been the bane of pinball wizards, but the tilt of the CZCS was designed to significantly reduce one of the banes of ocean color remote sensing, which is sun glint. The CZCS was designed such that its viewing angle was not directly downward (the nadir view), but it could view forward of the satellite's position above the Earth, or behind it. By doing this, the amount of direct reflection from the sunlit surface of the ocean was curtailed. CZCS could tilt in 2° increments backward or forward, up to 20°. The CZCS tilt capability was a major first for satellite Earth observations.[16] The actual degree of tilt was commanded by ground controllers. The tilt capability is extraordinarily significant for ocean color observations, either for instruments that have the capability – or for instruments that do not – as will be described in subsequent chapters.

Nimbus 6

While the CZCS was being designed in 1973 and 1974, the next satellite in the Nimbus series was being prepared for flight. This era marked NASA's increasing efforts in Earth Observation; the first Landsat satellite, pushed by James Fletcher, launched in 1972.[17] Nimbus F on the ground – to be designated Nimbus 6 when in orbit – was designated to carry an ambitious payload of seven instruments.[18] These instruments included new versions of the ESMR and THIR; the new and groundbreaking Earth Radiation Budget (ERB) instrument; the High Resolution Infrared Radiation Sounder; the Limb Radiance Inversion Radiometer; the Pressure Modulated Radiometer; and the Scanning Microwave Spectrometer. In addition, Nimbus 6 conducted a Tracking and Data Relay experiment testing satellite data transmission to ground stations, and the Tropical Wind Energy Conversion and Reference Level (TWERLE) experiment. TWERLE used the Nimbus 6 random access measurements system (RAMS) to determine the location and movement of

---

[14] Email message received from Steve Varlese, May 6, 2008. Steve Varlese supplied valuable information about the CZCS program at Ball, including the names of other Ball personnel who worked on the CZCS, and pictures of CZCS fabrication and launch. Soon after moving to work on the Gemini Telescope project, Mr. Varlese passed away suddenly in late November 2008.
[15] Email message received from Steve Varlese, 6 May 2008.
[16] Warren Hovis, "The Nimbus-7 Coastal Zone Color Scanner (CZCS) Program".
[17] W. Henry Lambright, "NASA and the Environment: Science in a Political Context".
[18] "Nimbus 6: NSSDC ID 1975-052A", http://nssdc.gsfc.nasa.gov/nmc/spacecraftDisplay.do?id=1975-052A, (accessed 16 April 2009).

weather balloons. This information could be used to determine upper atmospheric wind speeds and other dynamic atmospheric variables. The description of TWERLE indicates that RAMS was also utilized to collect oceanographic and atmospheric data from ocean buoys.[19]

The Nimbus history notes that TWERLE became useful beyond its research intentions by demonstrating that a satellite could be used for search-and-rescue operations on the Earth.[20] Balloonists Ben Abruzzo and Max Anderson attempted to cross the Atlantic Ocean in September 1977 in a balloon named "Double Eagle". Luckily for them, they brought along a TWERLE platform. Strong storms off Newfoundland pushed them off course, and they had to ditch Double Eagle near Iceland. They were located by the Nimbus Control Center using their TWERLE platform. Just about a year later, Abruzzo, Anderson and Larry Newman made the first successful Atlantic balloon crossing in August 1978. The narrative of this flight does not mention if they brought along a TWERLE platform, but given what happened to Double Eagle I, it would be surprising if they didn't.

Nimbus 6 was launched on June 12, 1975, from Vandenberg on a Delta 2910 launch vehicle.[21] The ERB instrument, which was intended to measure the radiation emitted by, or reflected from, the Earth to quantify Earth's radiative balance (i.e., how much radiation enters the Earth climate system from the Sun versus how much is radiated back into space) operated successfully, but developed mechanical scan problems two months into the mission, and so it could only collect data viewing directly down (nadir) after March 1976.[22]

Caught in the NET

While OCS observational campaigns continued, NASA continued to push forward with both CZCS engineering and with the establishment of the Nimbus Experiment Teams (NET). Nimbus 7, in a departure from the way that the prior Nimbus experiments had operated, determined to significantly increase the utilization of the data collected by the instruments by research scientists.[23] In order to do this, the management of the Nimbus program and each instrument package had to evolve. Prior to Nimbus 7, the primary "owner" of the data from the instrument was the Principal Scientist for the instrument mission. Responsibility for converting the instrument data into meaningful parameters for Earth science (also called "geophysical parameters") was thus part of the Principal Scientist's duties.[24]

---

[19] "Tropical Wind Energy Conversion and Reference Level (TWERLE)", http://nssdc.gsfc.nasa.gov/nmc/experimentDisplay.do?id=1975-052A-01, (accessed 16 April 2009).
[20] "Nimbus Program History", NASA GSFC.
[21] "Nimbus 6: NSSDC ID 1975-052A", NSSDC.
[22] "Earth Radiation Budget (ERB)", http://nssdc.gsfc.nasa.gov/nmc/experimentDisplay.do?id=1975-052A-05, accessed 16 April 2009.
[23] "Nimbus Program History", NASA GSFC, page 28.
[24] "Nimbus Program History", NASA GSFC, and additional information on page 28.

For Nimbus 7, the role of the Principal Scientist was transferred to the members of the NET for each instrument. The NET members were selected from responses received to a NASA Announcement of Opportunity (AO) for Nimbus 7. The ultimate responsibility for managing this menagerie of outstanding scientific talent fell to Nimbus 7 Project Scientist Al Fleig.[25]

Fleig provided considerable insight into the transition between the Nimbus "Principal Scientist" era and the science team concept for Nimbus 7. The Principal Scientist model extended back to the nation's first satellite, Explorer 1, which carried the instrument that discovered the Van Allen radiation belts. Essentially, the Principal Investigator provided the instrument, and he got the data back. There was no requirement or contract to give the data to anybody else!

Part of the reason for this, Fleig noted, was that back in the early days, much of the Nimbus instrumentation consisted of "gee-whiz" science, as in "gee whiz, can we actually do that?" The main goals at this state were to make the instruments better.

That began to change with the Backscatter Ultraviolet Instrument (BUV) on Nimbus 4. There were 30-40 ground stations collecting ozone data, but there was no global coverage. Because of early concerns about the ozone layer (predating the concerns about chlorofluorocarbons), scientists wanted answers from the BUV data. But the data was not being processed or released for general distribution. So an Ozone Processing Team (OPT) was created that operated from 1970-1973, which created an ozone data set. This was a change in style; rather than just a Principal Scientist on an instrument, ozone researchers were given funding to create an ozone data set.

While Nimbus 5 and Nimbus 6 basically followed the Principal Scientist concept, it was becoming "more and more obvious" that the instruments were no longer just for gee-whiz science, they were actually making useful measurements. So when the decision was made to fly the Solar Backscatter Ultraviolet and Total Ozone Mapping Spectrometer (SBUV/TOMS) on Nimbus 7, this decision came with the recognition that the OPT should also do the data processing for this instrument. A science data team had already been selected, but this recognition established the concept of having a team dedicated to creating the data set. Thus, as the Nimbus 7 launch approached, data processing teams were established for other Nimbus 7 instruments, which are described subsequently. The CZCS had a slightly different mode, as the University of Miami had already become the *de facto* processing group for much of the data, partly based on the fact that according to Fleig, they had "astonishingly good computer equipment".[26]

Ultimately, Fleig noted, he was able to get the concept of data processing team and science team synergy into NASA's thinking about Earth remote sensing. He pushed the idea that funding to do science with data didn't insure a long-term data set, which was more expensive to do. However, old concepts die hard, creating "clear, and at times, bitter demarcations"; because eager young scientists were hired for the processing teams, they were

---

[25] "Nimbus Program History", NASA GSFC, page 28.

[26] During the CZCS mission, the University of Miami system was used primarily for research; the next chapter will describe the role of the University of Miami in CZCS data processing.

also capable of doing science with data that they understood intimately. But when they did this, and published research findings, the science teams wondered why they were being "allowed" to do science, because their funding was supposed to be only for processing the data. Some of the most vociferous griping about this stemmed from scientists who had not been selected for the instrument science teams.

Fleig also noted that being a scientist on a data processing team didn't establish scientific credentials in the way that publishing papers did. It is a slow, and still on-going evolution, to give scientific weight to the creation of a data set used widely by other scientists that is equivalent to an extensive publication record.

1975 was the year that the NET members were selected for CZCS. One of the members of the team was not too much of a surprise: Charles Yentsch. The other team members were Warren Hovis (GSFC), Dennis Clark of the National Environmental Satellite Service (NESS) at NOAA; Frank Anderson of the National Research Institute for Oceanology, South Africa; Roswell Austin of the Scripps Visibility Laboratory; Howard Gordon of the University of Miami; James Mueller, GSFC; Sayed Z. El-Sayed from Texas A&M University; Boris Sturm from the Joint Research Center in Ispra, Italy (located on the scenic shores of Lago Maggiore); Robert Wrigley from the NASA Ames Research Center; and John Apel and Edward Baker, both affiliated with NOAA's Pacific Marine Environmental Laboratory (PMEL) in Seattle. Charlie Yentsch complemented Hovis on his clever selections for the CZCS NET; Sayed El-Sayed noted that one reason that the team worked was due to the fact that Hovis selected members of the team with "noteworthy expertise in their respective fields".[27] Clark noted that when he found out about the CZCS call for proposals, he asked Ros Austin about it, and Ros Austin had not known about it, so Clark added him to his proposal, which fortuitously resulted in placing both of them on the NET.[28] This particular science team was going to provide its scientific expertise to Jim Mueller, who would direct the data processing effort.

Clark recalled that one of the first problems the NET grappled with was whether the regions of the world's oceans could be divided up into seven different optical provinces. Alternatively, the optical water mass classes that had been described by Nils Jerlov might do the trick.[29]

Howard Gordon also recalled with some surprise that when he joined the team, he found out that Hovis had already designed the instrument with the Ball Brothers engineering team. Dennis Clark echoed this comment, saying that Hovis told him "the instrument had been designed, it was in the can, and he wanted to get rid of the blue band if he could." What Clark only realized later was that Hovis and John Sherman "were having a running argument about the blue band." When it was finally proven that the blue band actually

---

[27] James G. Acker, "The Heritage of SeaWiFS: A Retrospective on the CZCS NIMBUS Experiment Team (NET) Program" in S.B. Hooker and E.R. Firestone, editors, *NASA Technical Memorandum 104566,* Vol. 21, NASA Goddard Space Flight Center, Greenbelt, Maryland, 44 pp, (1994).

[28] Dennis Clark, comments recorded at the Ocean Color Collaborative Historical Workshop, January 13-14, 2009, St. Petersburg, Florida.

[29] Dennis Clark, interview notes transcript.

worked, "Hovis presented Sherman with a blue ribbon." [30]

With the team selected, the scientific support program began in earnest. At-sea data collection commenced even before the CZCS NET members had their first meeting. Charlie Yentsch took some time off from guiding the Bigelow Laboratory for Ocean Sciences in its first year of existence (Yentsch had already become fairly experienced at establishing new oceanographic research institutions, because the Bigelow Laboratory was his third) to board the R/V *Pacific Clipper* with Ros Austin. Yentsch and Austin collected bio-optical data at seven stations off of San Diego and the southern California coast. This short cruise took place from October 26 to November 3, 1975.[31]

To Boldly See What No Instrument Had Seen Before: Building the CZCS

Back in Boulder, Colorado, the Ball Brothers engineers were contending with the issue of polarization (see Appendix 1). Measurements of the ocean light field had to be monochromatic, i.e., the instrument had to measure the brightness of the radiance independent of polarization. However, light leaving the ocean surface is strongly polarized. Polarization occurs just about any time that light passes through a different medium or reflects off of something – this is the basic reason that polarized sunglasses reduce glare by reducing the amount of polarized light that passes through the lens.[32] So if the satellite instrument received the light directly from the ocean surface, it would be sensitive to the viewing angle of the instrument; in some directions, polarization would reduce the signal to nearly zero, and in other directions, the signal would be increased.

The design of the CZCS made depolarization more important, and even more difficult. The CZCS design team decided to create the wavelength bands that the instrument would measure by using a diffraction grating (called the "polychromator"). Diffraction gratings used to be common devices in a high school physics laboratory, because a simple cardboard frame holding a diffraction grating generates a visible spectrum when sunlight is passed through it. The fine lines of the diffraction grating separate the light from the Sun (or in the case of the CZCS, the light from the ocean) into the visible spectrum, and then light filters capture the light at specific wavelengths.[33] The problem with the diffraction grating approach was that diffraction gratings are quite sensitive to polarization. The problem at Ball Brothers was that no one at that time had thought of a way to depolarize the light beam,

---

[30] Howard Gordon, comments recorded at the Ocean Color Collaborative Historical Workshop, January 13-14, 2009, St. Petersburg, Florida; Dennis Clark, comments recorded at the Ocean Color Collaborative Historical Workshop, January 13-14, 2009, St. Petersburg, Florida.

[31] James G. Acker, "The Heritage of SeaWiFS: A Retrospective on the CZCS NIMBUS Experiment Team (NET) Program"; Dennis K. Clark, "Phytoplankton algorithms for the NIMBUS-7 CZCS" in J.F.R. Gower, editor, *Oceanography from Space*, Plenum Press, New York, 227-238, (1981).

[32] Sönke Johnsen and Heidi Sosik, "Shedding Light on Light in the Ocean", *Oceanus*, 43(2), http://oceanusmag.whoi.edu/v43n2/johnsen/sosik.html (accessed 17 April 2009).

[33] Rod Nave, "Diffraction Grating", http://hyperphysics.phy-astr.gsu.edu/hbase/phyopt/grating.html, (accessed 17 April 2009).

to essentially "scramble" the polarization.

In order to meet design specifications, the degree of polarization of the light utilized by the CZCS had to be less than 2% – the most stringent requirement of the CZCS optical requirements to achieve.[34] And it was also important to reduce the polarization while preserving as much of the precious energy in the light beam as possible; some of the available depolarizers reduced the energy in the light beam by half.

Supposedly, Newton conceived the theory of gravity sitting in an apple orchard and getting conked on the noggin by a well-timed ripe apple. The word "Eureka" is attributed to the philosopher Archimedes, who interrupted his leisurely bath by conceiving the principle of buoyancy and subsequently entertained the citizenry of Syracuse with a naked dash through the streets. The invention of Teflon stems from a fortuitous refrigeration of gaseous tetrafluoroethylene.

The success of the CZCS may be owed in part to Robert Woodruff's walking route to work; particularly his careful avoidance of horse manure. While placing his feet carefully during one of his daily commutes, which took him through a horse pasture, Woodruff conceived the design of the CZCS depolarizer, which could also be called a quartz-quartz wedge depolarizer. The depolarizer works by taking two birifringent quartz wedges and aligning them so that the optical axes of the quartz are different by 45°. Two wedges are used to maintain the beam direction. While depolarizers had been in existence before, Woodruff's horse pasture inspiration was invented for the CZCS.[35]

After conceiving the depolarizer, Ball had to construct the optics system and then test it, under the direction of Jim Darnell. Because of the polarization requirement, the standard reflective material that would have been preferable, aluminum, couldn't be used. The alternative was a mirror with a reflective surface made of silver, but silver oxidizes when exposed to reactive oxygen atoms in the rarefied environment of space. Normally, a silver mirror would be protected with a layer of sputtered quartz, but that couldn't be done for the CZCS because of infrared absorption (as the CZCS was also equipped with a thermal IR band). So Ball had a coatings company put a protective coating on the mirror. Because the formulation of the coating was proprietary, even the Ball engineers didn't know what it was; but it worked.[36]

They also had to correct for variations in the CZCS scan mirror. The CZCS system included a dichroic beamsplitter that split the beam into visible and infrared beams. The IR beam was intended to be used by Channel 6 of the instrument to measure sea surface temperature. The beam splitter utilized a thin gold coating to direct the beam to the visible range detectors or the infrared detector.[37]

---

[34] Barton Howell, "Measurement of the polarization effects of an instrument using partially polarized light", *Applied Optics*, 8(6), 806-812, (15 March 1979); "Coastal Zone Color Scanner for Nimbus G", Ball Brothers Research Corporation, Part 2, Technical Proposal (copy provided by R. Woodruff).
[35] Recorded interview, Robert Woodruff, July 3, 2008.
[36] Recorded interview, Bill Wallschlaeger and Randy McConaughey, April 3, 2009.
[37] Recorded interview, Robert Woodruff, July 3, 2008.

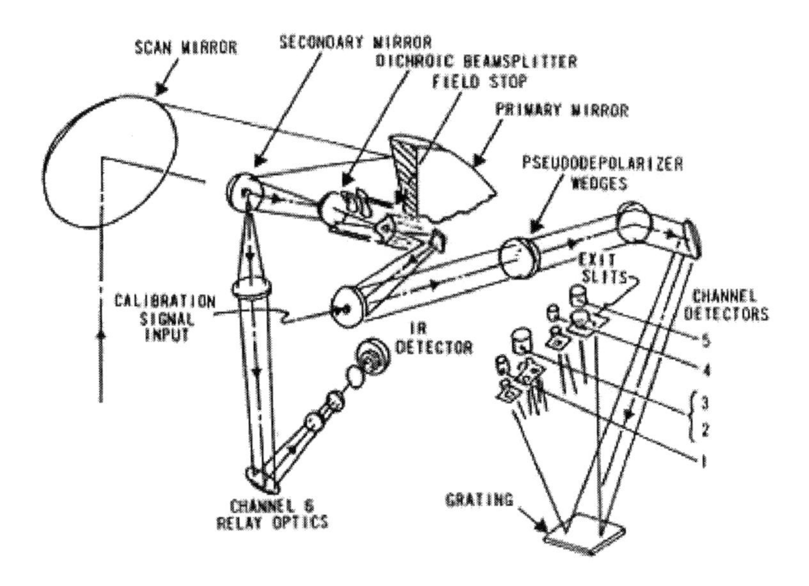

Figure 3.1. CZCS optical design.

When the optics package was complete, Ball Brothers subjected it to a "detailed and massive" analysis of the polarization sensitivity. The depolarizer worked; the tests indicated that the degree of polarization was less than 1%.[38]

Having overcome the polarization problem, Ball Brothers continued to work on the engineering model, and then the protoflight model of the CZCS. One groundbreaking aspect of the CZCS was the use of solid state electronics (photodiodes) instead of photomultiplier tubes. Wallschlaeger said "I knew that ITT [who was competing on the CZCS proposal] would use PMTs". [39] In the CZCS, the photodiodes in conjunction with the spectral diffraction grating established the bandwidths for each band.

McConaughey recalled that the CZCS used very large feedback resistors, and they saw a drop-off in the frequency response. This required him to put in a peaking amplifier after the pre-amplifier to boost the high frequency response. This actually had to be tested before Ball submitted the CZCS proposal, because they had to make sure that they didn't add random noise.[40]

McConaughey also recalled that he put in the 55° compensation mirror, which is called the "secondary mirror" in the CZCS diagram (Figure 3.1) which also addressed the instrument polarization.[41] The optical design and gratings were a team effort; Darnell did the design and Roybal did the specs, because "Darnell didn't know how to do specs". Woodruff brought in optical design ideas and worked part-time on the project; the main

---

[38] Recorded interview, Robert Woodruff, July 3, 2008.

[39] Recorded interview, Bill Wallschlaeger and Randy McConaughey, 3 April 2009.

[40] Recorded interview, Bill Wallschlaeger and Randy McConaughey, 3 April 2009.

[41] From the Coastal Zone Color Scanner Instrument Guide, http://disc.sci.gsfc.nasa.gov/oceancolor/documentation/scientific-documentation/CZCS_Sensor.gd.shtml.

thing he contributed was the depolarizer. "That was a key thing to make this work," Wallschlaeger confirmed.[42] Roybal also designed the CZCS optical detector area to allow easy access to adjust the optics for the purpose of achieving the right spectral response.

Sometimes, vexing problems occurred that contributed to sleepless nights for the CZCS team. Ball built the drive systems and bearings in-house, lubricated them and ran them for testing. After about a month, they noticed one drive unit was slowing down and drawing more current. They took the drive apart and analyzed everything, and discovered cornstarch on the bearings. Wallschlaeger explained that "normally, we used latex gloves with no lubrication for work, special gloves." It turned out that "some genius in purchasing decided to save money and bought gloves with cornstarch lubricant, and it came off when the gloves were reversed" – and some of it got on the bearings and gummed up the works.

Roybal was also a stickler for cleanliness in the clean room – he obviously was not the person responsible for purchasing the wrong gloves! A small area around the CZCS radiative cooler was covered with Mylar, and they kept the Mylar clean. One time, a representative from Goddard [Bob Lambeck] was observing the "manufacturing, assembly, and test" of the CZCS, and saw a speck of lint on the Mylar. He conscientiously tried to flick it off – leaving a small oily fingerprint on the Mylar.[43] Roybal was livid, but the fingerprint was successfully removed with alcohol.

Other problems were not so easy to address. When the CZCS was in Thermal Vacuum testing, one of the capacitors was in backwards and shorted out. The CZCS electronics box proceeded to overheat "sky high", and the team realized there had not anticipated all of the heating from the components and that a heat sink would be necessary. Ingenious Rube Goldbergian improvisation was required; the fix for this problem was a braided copper wire attached to the electronics box and connected to the outside of the instrument (Figure 3.2). The excellent heat conducting characteristics of copper were able to shunt the excess heat away from the vital electronics.

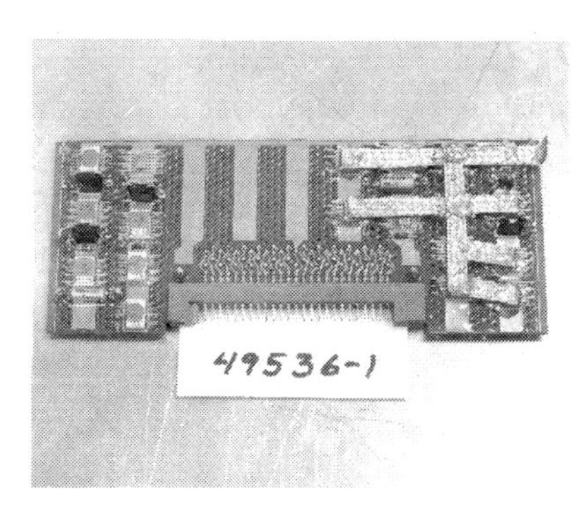

Figure 3.2. CZCS electronics board with braided copper wire for heat dissipation. Photo provided by Randy McConaughey.

---

[42] Recorded interview, Bill Wallschlaeger and Randy McConaughey, 3 April 2009.

[43] Email message from Randy McConaughey, 5 March 2009; Recorded interview, Bill Wallschlaeger and Randy McConaughey, 3 April 2009.

According to McConaughey and Wallschlaeger, GSFC ran 12-hour shifts for Thermal Vac testing under the direction of Herb Richards. They also recalled that Joe Arlauskas, Herb Richards' boss, was a stickler on testing: "he was demanding, but it paid off". [44]

Wallschlaeger also remembered that the testing phase was a bit stressful for Bill Roybal and his chief assistant, Steve Varlese. "They had terrific arguments, a Hispanic versus an Italian. We'd just leave them alone to work it out. At the end of the day they'd go out and have a beer together." [45] Roybal was invaluable on the instrument and subsystems, both men agreed.

Steve Varlese said that he began working on the CZCS early in 1976. Ball Brothers constructed an engineering model of the CZCS and then commenced work on the protoflight model, which is the subject of the fabrication process described above. [46] "Protoflight" is likely a NASA-coined term for an instrument that is built and tested, and then flown on the satellite. In this conflation of Latin and English, it essentially means "first to fly". So the protoflight CZCS was the CZCS. As Varlese noted, there was also a flight model of the CZCS that never flew.

According to Wallschlaeger, he had actually first known Hovis when he was at ITT, and they built the HRIR on the earlier Nimbus satellites. Hovis actually talked to Ball before the CZCS RFP, and he also built the integrating sphere, the "Hovisphere", which was used for the CZCS testing phase. McConaughey added that Wallschlaeger and Hovis had a working relationship – Bill "could translate theory into engineering". Wallschlaeger said that was because he was the only one that did radiometric calculations.[47]

Wallschlaeger noted one anecdote from Hovis: "Back when they flew their stuff [i.e., the OCS] on the U-2, they flew over Los Angeles and the adjacent ocean. Hovis knew there was a pipe that carried sewage offshore, but they didn't see it. Hovis said the people of Los Angeles were not cooperating!" [48]

Pre-launch CZCS NET Activities

At GSFC, the Nimbus project was settling on the final instrument payload – Warren Hovis had proposed both the CZCS and the SSTR (Sea Surface Temperature Radiometer, the "Sister") for Nimbus 7, but the SSTR was dropped, partly because the CZCS included an IR channel.[49] With a bit of bio-optical data in hand, it was time for the CZCS NET to meet face-to-face; somewhat logically and logistically, the first meeting was held at GSFC. Though he was already the Principal Scientist for the instrument, the CZCS NET selected Hovis as the leader of the team. The work taking place at Ball Brothers was a hot topic for

[44] Recorded interview, Bill Wallschlaeger and Randy McConaughey, 3 April 2009.
[45] Recorded interview, Bill Wallschlaeger and Randy McConaughey, 3 April 2009.
[46] Email message from Steve Varlese, 20 September 2008.
[47] Email message from Steve Varlese, 20 September 2008.
[48] Email message from Steve Varlese, 20 September 2008.
[49] Recorded interview, William Barnes, 3 July 2008.

this meeting, as was the topic of "oceanographic usage of the data". [50] Whether or not this means the team was concerned about how useful this data would actually be is unknown.

Preparing a satellite for launch, especially with a new plan for data product development and data distribution that had never been tried before, can produce a hectic schedule. The next CZCS NET meeting followed a month after the first, also at GSFC, with a much more packed agenda than the first meeting. A fly on the wall at this meeting might have been stunned by the variety of concerns that were discussed. An important subject was the actual function and construction of the Nimbus G satellite, particularly with regard to where the CZCS would be located, and how much power would be allocated to its operation. Nimbus G power, like political power in a coalition government, would be shared among all the instruments on the satellite. CZCS was initially given a power allocation of two hours of observation a day. One somewhat convenient scheduling aspect of CZCS was that it would only observe on the lighted side of the Earth; other instruments could make observations on the dark side.[51]

Discussion of the critical element of atmospheric correction was led by Howard Gordon. Gordon had come to the University of Miami as a physics professor, sharing his time at the University of Miami Physics Department and on Rickenbacker Causeway at the Rosenstiel School of Marine and Atmospheric Sciences (RSMAS). He was expecting to work on atmospheric optics, but there was a bit of funding available for ocean optics work, so that is where he began his academic research career. He initially worked with Otis Brown, measuring the AOPs and IOPs of the ocean surface reflectance.[52] His emetic experiences with at-sea research made the remote sensing field even more appealing. When one of Gordon's graduate students, W. Ross McLuney, told Gordon about the CZCS Announcement of Opportunity, Gordon applied to be on the team.[53] Also at the meeting, the "Surface Truth" subcommittee was formed, and a large number of task assignments were doled out.

Ball Brothers hosted the CZCS NET on May 24-25, 1976. Charlie Yentsch noted that Ball Brothers gave the members of the team a set of salt and pepper shakers. Gordon remembered that the salt and pepper shakers were miniature Ball jars. Randy McConaughey explained that the souvenirs were mementos from the Ball Corporation original business – canning and bottling fruits and vegetables in Muncie, Indiana. The "Ball" on the side of collectible pickling jars represents the same business origination as the present-day Ball Aerospace Corporation.[54]

---

[50] James G. Acker, "The Heritage of SeaWiFS: A Retrospective on the CZCS NIMBUS Experiment Team (NET) Program".

[51] James G. Acker, "The Heritage of SeaWiFS: A Retrospective on the CZCS NIMBUS Experiment Team (NET) Program".

[52] Apparent Optical Properties and Inherent Optical Properties, described in Appendix 1.

[53] Email message from Howard Gordon, 19 September 2008; Howard Gordon comments recorded at the Ocean Color Collaborative Historical Workshop, January 13-14, 2009, St. Petersburg, Florida.

[54] Recorded interview with Charles Yentsch, 16 July 2008; Howard Gordon comments recorded at the Ocean Color Collaborative Historical Workshop, January 13-14, 2009, St. Petersburg, Florida; Email message from Randy McConaughey, 5 May 2009.

The scientific media in 1976 apparently received some information on the instrument complement of Nimbus G. An article in *Science News* discussed the aircraft observational campaign and the corresponding surface oceanographic effort, and said that Scripps researchers were "analyzing the data to see how well the characteristic spectral 'signature" of chlorophyll-a is likely to hold up when observed from satellite heights".[55] The article notes that Dennis Clark of the NOAA team developing the "coastal zone color scanner" (marking perhaps the first time it was referred to in the popular media) said that there were "very encouraging signs" that the chlorophyll concentration detected at the surface gave a "a representative indication of the mean concentration in the water column beneath".

Because of the cessation of manned space flight activities following the moon landing, even NASA itself found it more difficult during this era to get attention from the media, scientific or otherwise. Skylab had been launched and repaired in space, hosting three crews in 1973 and 1974.[56] Following Skylab, the last mission of the Apollo program was the "détente in space" Apollo-Soyuz mission, which took place in 1975.[57] And in 1976, the main focus of public attention for NASA was on Mars, due to the two Viking mission landings that took place on July 20, 1976 (which had originally been scheduled for July 4 to coincide with the 200th bicentennial of the United States) and on September 3, 1976.[58] So NASA's unmanned earth observation activities were relatively unknown, garnering only a few lines of coverage in newspapers and magazines.

One of the goals of the CZCS NET was to get the word out about this mission to the oceanographic community, and the NET took a step in that direction by attending, *en masse*, the European Association of Scientists in Environmental Pollution (EURASEP) meeting in Brussels, Belgium, in late September 1976 (just a few days after the second Viking landing on Mars). The NET presented a comprehensive overview of activities, and set up plans for a prelaunch calibration cruise with an OCS overflight in September-October 1977, utilizing the Texas A&M workhorse research vessel R.V. *Gyre* in the Gulf of Mexico (Figure 3.3).[59] Clearly the NET members were not considering hurricane frequency statistics when they came up with the initial dates for this cruise.[60]

During this same period, there was still a "surprising amount of political wind blowing against the CZCS being launched", according to Ray Smith. Smith recalled a meeting about the sensor in the office of William Nierenberg, the director of the Scripps Institute of Oceanography, which Siebert Duntley asked Smith to attend. Smith faced noted physical

---

[55] "Will Nimbus G be Fishsat A?" *Science News*, 109 (21) , page 331, (22 May 1976).
[56] NASA, "Part 1: The History of Skylab", http://www.nasa.gov/missions/shuttle/f_skylab1.html, accessed 15 January 2011.
[57] NASA, "The Flight of Apollo-Soyuz", http://history.nasa.gov/apollo/apsoyhist.html, accessed 15 January 2011.
[58] NASA, "Viking: Trailblazer for All Mars Research", http://www.nasa.gov/mission_pages/viking/viking30_fs.html, accessed 15 January 2011.
[59] R/V *Gyre* and R/V *Researcher* photographs from Norman L. Guinasso, Jr., http://tabs.gerg.tamu.edu/~norman/cruisehistory.html, (accessed 8 January 2010).
[60] James G. Acker, "The Heritage of SeaWiFS: A Retrospective on the CZCS NIMBUS Experiment Team (NET) Program".

oceanographers Walt Munk from Scripps, Carl Wunsch from the Massachusetts Institute of Technology, and Francis Bretherton, director of the National Center for Atmospheric Research. Smith said that both Munk and Wunsch were against it, but Bretherton stuck up for it.[61]

An incident in late 1976 served to raise awareness in the United States of pollution threats to the ocean environment. The oil tanker *Argo Merchant* ran aground on Middle Rip Shoal near Nantucket Island on December 15, and broke up on December 21, resulting in the loss of all its cargo. Fortunately, favorable winds and currents kept the oil offshore. Nevertheless, the incident "generated concern among the public and members of Congress about the condition of tankers that were increasingly being used to import oil into the United States", a concern motivated in part by the damage that oil spills could do to the coastal environment.[62] The wreck of the *Argo Merchant* marked the first major coordinated oil spill response by NOAA.[63]

1977

The calendar turn in 1977 found the CZCS NET immersed in a critical year. But they did choose their meeting locations well – the first meeting of the year moved to the friendly environs of the Scripps Institution of Oceanography in La Jolla, California. Someone on the team had noted that the Gulf of Mexico in September might possibly entertain a hurricane or two, so the calibration cruise and OCS overflight was rescheduled for mid-October. An interesting aspect of this meeting was the discussion of in-water instruments, an effort led by Hovis, spearheading an effort that would dramatically increase in importance for the ocean color community. The algorithm development team faced a long list of things to do at the end of this meeting.[64]

The algorithm developers were still busy in April during the next meeting, which was back at GSFC. The data system engineers made a polite request for at least a preliminary sketch of the algorithms, and asked the amenable Howard Gordon to work on the important algorithms for chlorophyll and sediments. The team set up a delivery schedule, with the preliminary algorithms to be delivered in early June, and the final algorithms due on March 1, 1978. Facing up to the two-hour limit on observations, the Nimbus project also requested

---

[61] Ray Smith, comments recorded at the Ocean Color Collaborative Historical Workshop, January 13-14, 2009, St. Petersburg, Florida.
[62] Dennis Bryant, "Loss of the Argo Merchant", *Maritime Professional*, http://www.maritimeprofessional.com/Blogs/Maritime-Musings/December-2009/Loss-of-the-Argo-Merchant.aspx, accessed 10 January 2011.
[63] NOAA,"Responding to Environmental Catastrophes: An Evolving History of NOAA's Involvement in Oil Spill Response", http://celebrating200years.noaa.gov/transformations/spill_response/welcome.html, accessed 10 January 2011.
[64] James G. Acker, "The Heritage of SeaWiFS: A Retrospective on the CZCS NIMBUS Experiment Team (NET) Program".

observational targets. Presumably "water" was considered too general.[65]

The June meeting of the CZCS NET this year, where the preliminary algorithms were delivered, contains a truly critical ascertainment: the atmospheric correction process required an unacceptably long amount of processing time, as determined by Jim Mueller. So the first plan for image processing did not include atmospheric correction.[66] Howard Gordon was in the process of discovering that the OCS did not have a radiometric calibration good enough to test his atmospheric correction algorithm, and he was also finding out his algorithm would challenge the fastest and biggest computers on the market at the time – if indeed anyone was actually selling computers that were as fast and big as necessary.[67]

The 1977 hurricane season turned out to be mild; only the first two hurricanes of the season, Anita and Babe, affected the Gulf of Mexico in late August and early September. Anita was nonetheless powerful, forming right in the center of the Gulf of Mexico and reaching Category 5 strength before storming ashore on September 2 in northeastern Mexico, responsible for 10 deaths. Babe was only a Category 1 that went ashore in western Louisiana.[68] So the weather was relatively good for the calibration cruise, which had to have been good news for Howard Gordon's emetic propensity. In addition to the *Gyre*, the services of the NOAA R/V *Researcher* had also been obtained (Figure 3.3), so the two ships conducted simultaneous observations in the Gulf, meeting up near the mouth of Tampa Bay and proceeding north and west around the northeast Gulf of Mexico, with stations near the Mississippi River delta, and then moving into the blue waters of the central Gulf south of Louisiana and ending near the Yucatan Peninsula.[69] Whether they knew it or not at the time, stations 2-7 of this cruise took place in waters with such optical complexity that they still confound the most advanced ocean color remote sensing algorithms available today. The waters are laden with sediments from the Mississippi River, organic matter from the black water rivers of Florida, notably the Suwannee, and the area also features areas with shallow depth and reflective sediments.

---

[65] James G. Acker, "The Heritage of SeaWiFS: A Retrospective on the CZCS NIMBUS Experiment Team (NET) Program".
[66] James G. Acker, "The Heritage of SeaWiFS: A Retrospective on the CZCS NIMBUS Experiment Team (NET) Program".
[67] Howard Gordon, "Removal of atmospheric effects from satellite imagery of the oceans". *Applied Optics*, 17, 1631-1636, (1978).
[68] "1977 Hurricane Anita", http://www.tropmet.com/gallery/hurricane/gal_1977_anita.htm, accessed 17 April 2009; "Hurricane Babe, September 3-9, 1977", http://www.hpc.ncep.noaa.gov/tropical/rain/babe1977.html, (accessed 17 April 2009).
[69] James G. Acker, "The Heritage of SeaWiFS: A Retrospective on the CZCS NIMBUS Experiment Team (NET) Program".

Figure 3.3. R/V *Gyre* (left) and R/V *Researcher* (right).

Along with Gordon, the *Gyre* carried Sayed El-Sayed, Dennis Clark, and Ed Baker, as well as the young and untested Charles "Chuck" Trees. The *Researcher* hosted Ros Austin.[70]

Where in the World is Mill Creek?

Sometime the best science takes place in your backyard; or the nearest creek. Needing a site with muddy water, shallow bottom, and surrounded by land, Dennis Clark chose Mill Creek, a dendritic branch of the Chesapeake Bay on Maryland's Eastern Shore (most notably located near Wye Oak State Park), and made some in-water radiometric observations off a dock. These close-to-home observations were a valuable end-member for the initial CZCS data product algorithms.[71]

The Fellowship of Nimbus 7

Approximately September 1977, Ball Brothers delivered the CZCS to General Electric in King of Prussia, Pennsylvania for spacecraft integration. Fabrication of the protoflight CZCS, including calibration and testing by Ball Brothers, was completed earlier in the year. One of the final tests was a requirement; show that the instrument was actually detecting chlorophyll. To accomplish this requirement, Ball had to take the CZCS out of the clean room and set it up outside, turn on the scanner, and make a simulated scan image by turning on the tilt drive and tilting through the full tilt range, 40°. Unfortunately there is no record of any of the Ball engineers saying about this test, "Let's go full tilt," but that is what they

[70] James G. Acker, "The Heritage of SeaWiFS: A Retrospective on the CZCS NIMBUS Experiment Team (NET) Program".
[71] James G. Acker, "The Heritage of SeaWiFS: A Retrospective on the CZCS NIMBUS Experiment Team (NET) Program"; Warren Hovis et al., "Nimbus-7 Coastal Zone Color Scanner: System Description and Initial Imagery"; Dennis K. Clark, "Phytoplankton algorithms for the NIMBUS-7 CZCS".

did – in the parking lot, taking an image of the Continental Divide (Figure 3.4). Even though, as Steve Varlese describes it, it was a "beautiful fall day", both the scan mirror and calibrator received a coating of dust and had to be cleaned. The CZCS did collect an acceptable scan, which was used as the cover image of the Ball Brothers 1978 calendar. According to Varlese, the instrument had been subjected to calibration and radiative transfer analysis using the integrating sphere built by Warren Hovis, but the test drive in the parking lot was the final test before the trip to Pennsylvania.[72]

Figure 3.4 CZCS test image of the Continental Divide. (More pictures of the CZCS team are found at the end of the chapter.) Photo provided by Steve Varlese.

The CZCS had company on the instrument platform; Nimbus G carried eight instruments, seven of which were packed onto the circular base of the satellite (Figure 3.5). The eighth instrument, the Scanning Multichannel Microwave Radiometer (SMMR), was perched under the solar array and attitude control structure.[73]

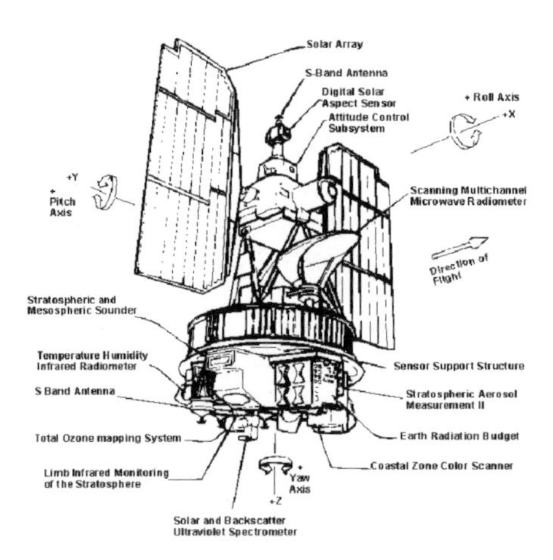

Figure 3.5 Nimbus 7.

[72] Email message from Steve Varlese, received 20 September 2008.
[73] National Aeronautics and Space Administration. 1978. *The Nimbus 7 Users' Guide.* C. R. Madrid, editor. Goddard Space Flight Center. Nimbus 7 image (Figure 4.5) from the Total Ozone Mapping Spectrometer Web site, http://ozoneaq.gsfc.nasa.gov/nimbus7tech.md.

The august company of instruments accompanying the CZCS:

Scanning Multichannel Microwave Radiometer (SMMR): The SMMR measured microwave brightness temperatures in 10 bands, intended to determine sea surface temperature and near-surface winds. SMMR measurements paved the way for numerous microwave radiometers on subsequent missions. SMMR turned out to be a particularly valuable instrument for observations of sea ice in the polar regions.[74]

Stratospheric and Mesopheric Sounder (SAMS): SAMS was a "pressure-broadening" spectral radiometer, constructed for the purpose of measuring the concentrations of chemical species in the atmosphere, notably water vapor, methane, nitrogen oxides (N_20 and NO), and carbon monoxide. SAMS was also a limb sounder, conducting its measurements by scanning through Earth's "limb", the thin line of atmosphere at the edge of the Earth's sphere as viewed by a satellite in space, which allowed determination of the vertical distribution of these species.[75]

Limb Infrared Monitor of the Stratosphere (LIMS): LIMS was another limb sounder, concentration on the concentrations of ozone, water vapor, nitrogen oxide, and nitric acid. LIMS measured infrared radiation, utilizing the spectrum of carbon dioxide to determine temperature and thus allowing the production of temperature vs. concentration profiles.[76]

Stratosopheric Aerosol Measurement II (SAM-II): SAM-II examined the vertical distribution of aerosols over the Arctic and Antarctica. The instrument accomplished this by measuring the extinction coefficient of the atmosphere as the Sun came into view over the poles.[77]

Temperature Humidity Infrared Radiometer (THIR): A heritage instrument, THIR had a predecessor instrument on Nimbus 4, 5, and 6. As its name indicated, THIR detected infrared radiation, used for both measurements of atmospheric moisture content and the temperatures of cloud tops, the terrestrial surface, and the surface of the ocean.[78]

Earth Radiation Budget (ERB): ERB was another heritage instrument, with a

---

[74] NSSDC, "Scanning Multichannel Microwave Radiometer (SMMR)". http://nssdc.gsfc.nasa.gov/nmc/experimentDisplay.do?id=1978-098A-08, (accessed 17 April 2009).
[75] NSSDC "Stratospheric and Mesopheric Sounder (SAMS)". http://nssdc.gsfc.nasa.gov/nmc/experimentDisplay.do?id=1978-098A-02, (accessed 17 April 2009).
[76] NSSDC, "Limb Infrared Monitor of the Stratosphere (LIMS)". http://nssdc.gsfc.nasa.gov/nmc/experimentDisplay.do?id=1978-098A-01, (accessed 17 April 2009).
[77] NSSDC,"Stratosopheric Aerosol Measurement II (SAM-II)". http://nssdc.gsfc.nasa.gov/nmc/experimentDisplay.do?id=1978-098A-06, (accessed 17 April 2009).
[78] NSSDC, "Temperature Humidity Infrared Radiometer (THIR)". http://nssdc.gsfc.nasa.gov/nmc/experimentDisplay.do?id=1978-098A-10, (accessed 17 April 2009).

predecessor on Nimbus 6. The two ERBs established an important observational data set that provided impetus for a satellite devoted entirely to measurement of Earth's radiation budget, appropriately entitled the Earth Radiation Budget Experiment (ERBE). Three ERBE experiments were actually launched; the Earth Radiation Budget Satellite (ERBS) was deployed by Sally Ride from the Space Shuttle, and two other ERBE instruments were carried on NOAA polar-orbiting satellites.[79]

Solar Backscatter Ultraviolet and Total Ozone Mapping Spectrometer (SBUV/TOMS): Considered to be one experiment, SBUV/TOMS actually had two components located next to each other. The intention of SBUV/TOMS was to measure the concentration of ozone in the atmosphere, as well as determining the amount of ultraviolet radiation impinging on the Earth from the Sun and backscattered in the Earth's atmosphere. SBUV/TOMS accomplished these observations quite well, arguably becoming the most famous instrument on Nimbus 7 when TOMS observations indicated the extent of the Antarctic "ozone hole" (ozone depletion was first observed in British Antarctic ground station data) where reduction of atmospheric ozone concentrations by chemical species related to chlorofluorocarbons is catalytically accelerated by polar ice clouds. The TOMS observations confirmed scientific concerns about ozone depletion and alarmed scientists, the public, and the international community sufficiently to lead to the phasing-out of chlorofluorocarbon use for refrigeration, cleaning of electronics, and many other common functions.[80]

This complement of instruments caused Nimbus 7 to also be referred to as the "Air Pollution and Ocean Observation Satellite", or APOOS.[81] (Fortunately this acronym never entered into wide usage.) In the annals of space exploration, however, Nimbus 7 was noteworthy because it was the first satellite to have a three-axis stabilized attitude control system utilizing gyroscopes and horizon scanners.

The CZCS

The CZCS itself is a multichannel radiometer. The main moving component was a rotating plane mirror, and the reflected light was directed into a Cassegrain telescope. The mirror constantly made full rotations, but the width of the ground swath was determined by

---

[79] NSSDC, "Earth Radiation Budget (ERB)". http://nssdc.gsfc.nasa.gov/nmc/experimentDisplay.do?id=1978-098A-07, (accessed 17 April 2009).
[80] NSSDC, "Solar Backscatter Ultraviolet and Total Ozone Mapping Spectrometer (SBUV/TOMS)". http://nssdc.gsfc.nasa.gov/nmc/experimentDisplay.do?id=1978-098A-09, (accessed 17 April 2009).
[81] "SP-4012 NASA Historical Data Book: Volume III: Programs and Projects, 1969-1978, Chapter Four, Space Applications". http://history.nasa.gov/SP-4012/vol3/ch4.htm, (accessed 17 April 2009).

the allowed scan angle for data collection, ± 40° from nadir. This created a ground swath with a width of 1556 km. As it rotated, the mirror was then exposed to a view of space or the interior of the instrument. The CZCS had six channels, commonly referred to by their center wavelengths at 443, 520, 550, 670, and 750 nm, and the IR channel at 11.5 μm. The combination of orbital altitude and detector elements created a nadir pixel resolution of 825 meters. Note in Figure 3.1 how the diffraction grating separates the visible range light into the specific detector wavelengths, and how the calibration signal from the on-board lamps is shown to enter the optical path.[82] This aspect of the calibration method was an important consideration in the final calibration methods used by the CZCS NET.

The process of instrument integration on the spacecraft continued in King of Prussia during the fall and winter of 1977 (Figure 3.6).[83] Randy McConaughey was one of several Ball Brothers employees who went to GE to assist with spacecraft integration.[84] The pace was sufficiently intense that Steve Varlese was just barely able to get a flight home on Christmas Eve. Varlese notes that the wet December of 1977 was apparently similar to the weather endured by Washington's troops at Valley Forge 200 years earlier – and he was glad to get home to Colorado.[85]

Figure 3.6. Photograph of Nimbus 7 during instrument integration and test, from the TOMS archive.

1978: The Launch Year of Nimbus 7 and the CZCS

With data collected from the Gulf of Mexico cruise and the OCS overflight, and with Nimbus 7 fabrication, instrument integration and testing taking place in Pennsylvania, the

---

[82] NASA GES DISC, "Coastal Zone Color Scanner (CZCS) Instrument Guide", http://disc.gsfc.nasa.gov/guides/GSFC/guide/CZCS_Sensor.gd.shtml, (accessed 17 April 2009).
[83] "Nimbus 7 TOMS Instrument and Satellite Information", http://jwocky.gsfc.nasa.gov/n7toms/n7sat.html, (accessed 17 April 2009).
[84] Recorded interview, Bill Wallschlaeger and Randy McConaughey, 3 April 2009.
[85] Email message received from Steve Varlese, 20 September 2008.

CZCS NET members devoted themselves to the tasks identified in 1977, primarily algorithm development, programming, testing, and wishing that someone would quickly produce much faster computers. The lack of information for the period through August indicates that this was probably a period of intense work on the spacecraft and by the scientific team members. Varlese recalls that he had to make a rush trip to GE in March of 1978 to perform late-stage troubleshooting on the CZCS, which by that time was already mounted on the instrument platform.[86] March 1978 was also the deadline for the final form of the data processing algorithms.[87]

During March 1978, another event focused public attention on the oceans, due to one of the most destructive oil spills in ocean history – a spill which at the time was the largest ever, and which now ranks sixth on the dubiously famous "all-time" list. The wreck of the *Amoco Cadiz* off the coast of Brittany was not blessed with any of the favorable conditions that prevented coastal damage, as had happened for the *Argo Merchant* spill. Instead, the tanker, which ran aground at Portsall on March 16, 1978, immediately began leaking oil. Bad weather deterred recovery efforts, and like the Argo Merchant had, eventually the Amoco Cadiz split in half, spilling its entire cargo of over 1.6 million barrels of oil. A large (321 km) stretch of Brittany's peninsular coast was impacted, affecting shellfish beds. The number of oiled birds killed by the spill was estimated at over 20,000.[88] According to the BBC, "Devastating scenes of marine life dying under a film of oil were broadcast around the world." Though the *Amoco Cadiz* wreck is not mentioned in the CZCS NET deliberations, there is little doubt that these marine scientists were as aware as the public (if not more so) of this environmental catastrophe.

Later that year, the ninth meeting of the CZCS NET provides a brief summary of what had occurred in those fast-paced months, and what had to happen next – in a hurry, as the schedule for the next few months was packed with events. The ninth meeting was hosted by Ed Baker and John Apel at PMEL on August 8th and 9th. One critical path was the scheduling of a post-launch data validation cruise. The ship schedule has to be coordinated with the spacecraft launch schedule, and the intersection of inevitable launch delays with the tight schedules of oceanographic research vessels likely invoked a few headaches. The R/V *Researcher* was originally scheduled as the vessel of choice. Anderson provided a description of cruises that would take place near South Africa. Wrigley reported on the OCS overflight campaign in the Gulf of Mexico. Gordon and Morel jointly presented the preliminary atmospheric correction algorithm. Warren Hovis pressed the team for a full report of these prelaunch cruises.[89]

---

[86] Email message received from Steve Varlese, 20 September 2008.
[87] James G. Acker, "The Heritage of SeaWiFS: A Retrospective on the CZCS NIMBUS Experiment Team (NET) Program".
[88] Green Nature, "Amoco Cadiz Oil Spill", http://greennature.com/article219.html, accessed 10 January 2011; BBC News, "1978: Tanker Amoco Cadiz Splits in Two", http://news.bbc.co.uk/onthisday/hi/dates/stories/march/24/newsid_2531000/2531211.stm, accessed 10 January 2011.
[89] BBC News, "1978: Tanker Amoco Cadiz Splits in Two".

Though it was only a short mention, the NET recommended changing two important items. Originally, the mission had planned to produce a chlorophyll concentration data product and a suspended sediment data product. Growing realization of the limitations of both the instrument and the data it would acquire caused the team to recommend changing the data product name *Chlorophyll* to *Pigment* – meaning that the data product would represent the concentration of all photo-active pigments in ocean waters, including chlorophyll and the related phaeopigments [see Appendix 1]. The daunting task of producing valid sediment concentrations, which is still a daunting task in the modern era, caused the team to recommend changing that product name to the "*Diffuse Absorption Coefficient – K*".[90] This algorithm would return a data product related to the reduction of light intensity with depth in the water column [again, refer to Appendix 1].

During the summer, Nimbus G was shipped to Vandenberg, to be placed on the Delta launch vehicle. Varlese recalls that this process took a couple of months.[91]

There is no list of who attended the Nimbus G launch; Steve Varlese was one of the eyewitnesses, and he captured a photograph of the event (Figure 3.7).[92] The Delta successfully placed the 907 kg (one pound less than a standard ton) Nimbus 7 into a polar orbit with an altitude between 943 and 957 km. Each orbit took 104 minutes to complete.[93]

Figure 3.7 Night launch of Nimbus G. Photo taken by and provided by Steve Varlese.

As Hovis had learned from bitter experience, satellite instruments do not come with even a 90-day warranty. Having received news of the successful launch, the NET scrambled to complete last-minute preparations for the post-launch validation cruise. With the *Researcher* unavailable, an alternate vessel was pressed into the service, the R/V *Athena II* operated by the Office of Naval Research (ONR). The *Athena II* had been converted from a U.S. Navy patrol gunboat into an oceanographic research vessel (Figure 3.8).[94] The ship

90 BBC News, "1978: Tanker Amoco Cadiz Splits in Two", .

91 Email message received from Steve Varlese, 20 September 2008.

92 Email message received from Steve Varlese, 20 October 2008.

93 NASA GES DISC, "Coastal Zone Color Scanner (CZCS) Instrument Guide".

94 Coastal Observing Research and Development Center, "Athena Ship Information", http://cordc.ucsd.edu/projects/hydro/athena_2004/, (accessed 8 January 2010).

was capable of high speeds, up to 35 knots, but fortunately for the members of the NET, most ocean optics measurements are conducted while the ship is relatively stationary. "Relatively stationary" does depend on the local weather conditions; holding station if winds and waves are brisk can make both the observations and the researchers a bit shaky, particularly if the gales of November come early. The *Athena II* put to sea on November 1 in the Gulf of Mexico, and the last hurricane of 1978, Kendra, tracked over the central Atlantic Ocean in late October and early November and never threatened U.S. waters.[95] The *Athena II* completed 10 research stations during the two-week cruise, departing from Panama City, Florida and heading westward past the Mississippi River before making a sharp turn south to the Dry Tortugas, and then finishing outside the mouth of Tampa Bay.[96] There was one notable casualty due to the *Athena II*'s high speed and resulting hull vibration: Dennis Clark recalled that Dale Kiefer's spectrometer was destroyed in two hours.[97] A few days after the *Athena II* made port, Sayed El-Sayed and the *Gyre* were also back at sea in the Gulf of Mexico for a cruise under the CZCS.[98]

Figure 3.8. The *R/V Athena II.*

William Barnes recalls that the Nimbus Processing System couldn't keep up with the remarkable flow of data from the Nimbus instruments, and was only capable of producing 40 images a day at optimum speed, which didn't usually happen. Therefore, the system required a "man in the loop" – usually Hovis – who would examine the images daily, all morning examining the quick look images and usually selecting 12 mostly cloud-free images for further processing. At the beginning of the mission, each two-minute observational scene acquired by the instrument required approximately an hour of processing. Barnes

---

[95] National Hurricane Center, "North Atlantic Hurricane Tracking Chart: 1978", http://www.nhc.noaa.gov/tracks/1978atl.gif, (accessed 20 April 2009).
[96] James G. Acker, "The Heritage of SeaWiFS: A Retrospective on the CZCS NIMBUS Experiment Team (NET) Program".
[97] Dennis Clark, interview notes transcript.
[98] Dennis Clark, interview notes transcript.

also notes that occasionally the CZCS would be aimed to pick up the sun glint pattern on the sea surface, and the signature of internal waves would be visible in the glint pattern.[99] These observations predate a dedicated internal wave observational effort conducted by Space Shuttle astronauts.[100] CZCS NET member John Apel wrote several papers on remote sensing of internal waves, including one analyzing photographs taken by Apollo/Soyuz astronauts.[101]

1979 – First Year of Operations

With an instrument operating successfully – to a point – in space, NET activities moved to an urgent state. At the same time, the Goddard engineers operating the spacecraft were dealing with a vexing problem. Channel 6, the IR channel intended to provide SST observations to accompany the visible range observations, wasn't working properly. The main problem was identified to be in the cooling system. In order for IR radiation to be detected, the IR sensor, a mercury cadmium telluride detector, had to be below a certain temperature – at least below 121 K (which is 152 degrees C below the freezing point of water). The cooler operated by being exposed to the blackness and cold of outer space and radiating excess heat, so it is called a *radiative cooler*. The problem with the radiative cooler was that it was not cooling the detector to quite the proper temperature. Though it was only a few degrees warmer than 121 K, the detector saturated at that warm temperature, and thus could not detect the IR radiation it was designed to detect.[102] General Electric produced a series of analyses trying to understand and fix the problem early in the mission[103]. Some of the possible causes of the problem were contamination of the external cooler elements, failure of the tape on one of the components, or excess heat generated by sun glint or light scattering. Analyzing the cooler failure was one of the main activities of the spacecraft engineers in the first year of the mission. Bill Wallschlaeger said that their biggest disappointment with the CZCS was the failure of the IR channel. He said that he never felt they had "all the information they needed about the radiation environment – the thermal models were incomplete, they didn't know what was around it." Thermal engineer Bob Poley of Ball was involved with the attempts to fix the radiative cooler.[104]

Even though Channel 6 was having problems, the other channels were working well,

[99] William Barnes, recorded interview, 3 July 2008.

[100] Q. Zheng, Q., X.H.Yan, and V. Klemas, "Statistical and Dynamical Analysis of Internal Waves on the Continental Shelf of the Middle Atlantic Bight from Space Shuttle Photographs," *Journal of Geophysical Research*, 98, 8495–8504, (1993).

[101] John R. Apel, "Observations of Internal-Wave Surface Signatures in ASTP Photographs", *NASA SP 412*, Vol. 2, pp. 505- 509, (1978).

[102] NASA, "The CZCS Instrument", http://oceancolor.gsfc.nasa.gov/CZCS/czcs_instrument.html, (accessed 20 April 2009).

[103] General Electric Flight Evaluation Report (Gene R. Feldman archives).

[104] Recorded interview, Bill Wallschlaeger and Randy McConaughey, 3 April 2009.

although at the next CZCS NET meeting early in January, Hovis reported that there were already signs of degradation in CZCS and other Nimbus 7 instruments. Given the remarkable longevity of some of the Nimbus instruments, this assessment seems rather pessimistic. The CZCS was thirsting for power, utilizing an average of 118 minutes of observational time out of the 120 minutes it was allowed to have each day.[105]

Mixed results were reported for the first post-launch calibration cruise, due to persistent cloud cover. Planning for the next cruises was underway, as were several observational missions. In this context, an observational mission meant that the CZCS would be tasked to acquire as much data as possible for specific regions, according to a specific set of priorities.[106] The U.S. Northeast Coast from Nova Scotia to the Chesapeake Bay was one such area, as was the South African upwelling region on the western coast of southern Africa. During the mission, the U.S. West Coast from California to the Gulf of Alaska and the Mediterranean Sea were frequently targeted.

The next two calibration cruises took place in March and late May – early June. The first cruise utilized the R/V *Velero IV* (Figure 3.9), a University of Southern California ship that was also frequently utilized by staff of the U.S. Geological Survey in Menlo Park, California.[107]

This cruise was apparently inspired by John Steinbeck and his marine biologist friend Ed Ricketts, as it targeted the clear waters of the Pacific Ocean west of Baja California, as well as several stations in the Gulf of California (also known as the Sea of Cortez). Dennis Clark and Ed Baker were identified NET members on this cruise, and a young researcher named B. Greg Mitchell performed many of the ocean optics observations. Due to the clarity of the waters and the largely cloudless conditions, the data from this cruise was a major contribution to the initial CZCS pigment concentration algorithm.[108]

Figure 3.9: R/V *Velero IV.* Picture provided by the U.S. Geological Survey.

---

[105] James G. Acker, "The Heritage of SeaWiFS: A Retrospective on the CZCS NIMBUS Experiment Team (NET) Program".
[106] Warren Hovis, "The Nimbus-7 Coastal Zone Color Scanner (CZCS) Program".
[107] Christina Kellogg, "Microbial ecology of deep-sea corals", http://coastal.er.usgs.gov/coral-microbes/corals.html, (accessed 8 January 2010).
[108] Dennis K. Clark, "Phytoplankton algorithms for the NIMBUS-7 CZCS"; James G. Acker, "The Heritage of SeaWiFS: A Retrospective on the CZCS NIMBUS Experiment Team (NET) Program".

The second cruise of the year, and also the final cruise of the CZCS data calibration campaign, covered a large region of the Atlantic Ocean east of the eastern United States. This cruise also took place on the R/V *Athena II.* Setting out from Charleston, South Carolina, the cruise visited several parts of the Gulf Stream, including an eddy system, and also targeted the northern front of the current. The coastal waters near Cape Cod and Martha's Vineyard, familiar to Yentsch and Ewing, were visited, and then the cruise moved into the blue Atlantic. Apparently daring Davy Jones to open his locker, the stations far offshore were designated "Nowhere", "Bermuda Triangle" and the legendary graveyard of sailing ships, the "Sargasso Sea".

The legendary dangers of the relatively idyllic central Atlantic were not a concern of the crew on the *Athena II*, but the operating characteristics of the ship led to a nearly-disastrous close encounter with the encircling reefs of Bermuda. As noted earlier, the *Athena II* was a converted Navy gunboat, and it could achieve speeds up to 35 knots, by virtue of a 13,300 horsepower gas turbine engine. The high speed capabilities of the *Athena II* allowed it to move quickly between stations, despite the high noise levels generated by the engine, and thus to establish oceanographic stations in favorable viewing locations for the CZCS in time for each noon overpass of the satellite.[109]

During the cruise, it was necessary for the *Athena II* to refuel, which it did at Charleston, South Carolina; Bermuda; Groton, Connecticut; and Norfolk, Virginia. During the cruise, while on approach to Bermuda, the crew elected to use the gas turbine engine as they sped toward the harbor at St. Georges. During the approach, the crew was in radio contact with the Bermuda harbormaster, and when informed that the ship was "on turbine", the harbormaster inquired urgently if the crew had nautical charts for Bermuda. After determining that they did, the harbormaster asked if the crew could see North Rock Light. The crew on the bridge immediately sighted the light and the reef below it, dead ahead, and performed high speed turns to narrowly avoid (by less than a ship's length) wrecking on the reef, as chief scientist Dennis Clark ordered all of the scientific crew to don lifejackets, just in case. After moving safely out to sea and then maneuvering to meet the Bermuda harbor pilot – while still cruising at 40 knots – the *Athena II* finally slowed down and entered the harbor at Bermuda.[110]

Despite this memorable near-miss, the cruise was not lost at sea, and returned to just north of the Bahamas, and a final station in the fast currents of the southern Gulf Stream off of Florida. Once again meeting with apparent success, the stations visited in this cruise contributed data to the development of the CZCS pigment algorithm.[111]

The Ixtoc I Blowout: While the researchers on the *Athena II* were heading into the Atlantic, an event in the Gulf of Mexico off the Yucatan Peninsula burst onto national and

---

[109] William W. Broenkow, "Measuring the Color of the Sea, or Approaching Bermuda at Forty Knots", unpublished manuscript provided by Dennis K. Clark, 12 pages, (October 1991).
[110] William W. Broenkow, "Measuring the Color of the Sea, or Approaching Bermuda at Forty Knots".
[111] Dennis K. Clark, "Phytoplankton pigment algorithms for the NIMBUS-7 CZCS".

international headlines, focusing attention on the oceanic environment just as the wreck of the *Amoco Cadiz* had done in 1978. A Pemex (Petróleos Mexicanos) oil rig called Ixtoc I, drilling in the Gulf of Campeche, blew out on June 3, leaving an uncapped oil well spewing immense volumes of oil into the Gulf of Mexico – an event which would be strikingly and unfortunately paralleled by the British Petroleum Deep Horizon oil rig disaster in 2010.[112] The American public viewed nightly scenes of flames on the sea surface and oil roiling from the uncapped well, spreading "chocolate mousse" slicks that circulated around the Gulf.[113] Eventually, Ixtoc I oil washed up on Texas beaches, and Ixtoc I tar balls were found on the beaches of the Florida Keys and South Florida.[114] Oil flowed from Ixtoc I for 297 days, resulting in the estimated discharge of nearly 3 million barrels of oil, before the well was capped and relief wells were drilled.

With the cruise campaign finished, the CZCS NET met again at Scripps in November. From an engineering standpoint, the instrument operations report from Hovis might sound partially disastrous, as if the mission was teetering on the edge of complete failure. As noted above, the concerns about IR channel 6 dominated the instrument operations discussion for most of the first year, and Hovis reported that the radiation cooler was degrading significantly. As for the rest of the optics, the calibration had changed "precipitously" by two counts in all bands.[115] The CZCS data was called 8-bit data, which meant that the radiance count value was reported between 1 and 256 in the instrument telemetry broadcast to the receiving stations on Earth. (Actually the top 6 count values, 251-256 were reserved for special circumstances, like clouds). So a shift of two counts in the calibration was indeed a cause for concern. Sudden events such as this would require the development of alternate calibration methods as data processing and reprocessing proceeded – in the first year of the mission, the NET was still concerned primarily with getting valid data from the instrument and valid sea-truth data from the research cruises.

Putting them both together – data validation – was still impeded by the slow pace of data processing. To that end, the team wanted to identify both specific "high" priority and "super" priority scenes to be processed that would be used in the validation effort. The actual images were being examined, and the grey scale images showed banding. Not everything that happened at this meeting was dour and glum; Nimbus 7 was still operating, and the Scripps VisLab had recently completed an image processing facility. While at the meeting, the NET members watched data acquisition during a Nimbus 7 overpass.[116]

---

[112] Robert Campbell, "BP's Gulf battle echoes monster '79 Mexico oil spill", http://www.reuters.com/article/2010/05/24/us-oil-rig-mexico-sidebar-idUSTRE64N57U20100524, accessed 1 February 2011.

[113] Arne Jernelöv and Olof Lindén, "Ixtoc I: A Case Study of the World's Largest Oil Spill", *Ambio*, 10(6), 299-306 (1981).

[114] Edward S. Van Vleet, William M. Sackett, Susan B. Reinhardt, and Margarita E. Mangini, "Distribution, sources and fates of floating oil residues in the Eastern Gulf of Mexico", *Marine Pollution Bulletin*, 15(3), 106-110 (March 1984).

[115] James G. Acker, "The Heritage of SeaWiFS: A Retrospective on the CZCS NIMBUS Experiment Team (NET) Program".

[116] James G. Acker, "The Heritage of SeaWiFS: A Retrospective on the CZCS NIMBUS Experiment Team (NET) Program".

Furthermore, even though this meeting took place slightly more than a year after the launch of Nimbus 7 and CZCS, Charlie Yentsch was already contemplating the future, noting that the CZCS had already established that the primary factor which determined the color of the ocean was the level of biological productivity. So he suggested that ocean color remote sensing should become a "significant element" of NASA's oceanographic remote sensing portfolio, because this method of remote sensing could allow estimates of global ocean productivity[117]. Unlike Jack's proverbial beanstalk, this particular seed took years to successfully take root and grow.

1980 – Year of Acceleration

Two months after the West Coast swing, the CZCS NET reconvened at GSFC. The team intended to review a bouquet of quick look images, to identify areas that would get the full processing treatment, i.e., production of geophysical products. James Mueller had conducted a study of how to produce archive data tapes (called CZCS Radiation and Temperature Tapes, or CRTTs) and the production of these tapes would be discussed.[118]

As usual, the thermal channel was not working properly. However, the banding in the images had been reduced. The team discussed a number of scientific issues; Frank Anderson and Sayed El-Sayed were planning on connecting South African coastal investigations to the Antarctic. The team also determined that raw radiance values would be used for CRTT production, because of problems with the information recorded for CZCS calibration, as reported by Mueller.[119]

Mueller was particularly active during this period. He described a number of issues that affected the initial methods used for CZCS data processing and calibration. One issue that is hard to recall (in this era of desktop supercomputers and gigabyte laptops) is how difficult it was to make software changes in the early 1980s for mainframe computer systems, the kind of computer necessary to perform the intensive computational requirements of remote sensing data processing at that time. Changes in the actual code running the machines had to be programmed in, compiled, tested, and then the results reviewed. So the decision to make changes to the existing production and calibration system was not entered into lightly.

Still, it was necessary. The initial CRTT production runs generated random banding. Banding in remote sensing images arises from different causes; in more modern remote sensing instruments, persistent banding has been due to very subtle differences in the

---

[117] James G. Acker, "The Heritage of SeaWiFS: A Retrospective on the CZCS NIMBUS Experiment Team (NET) Program".
[118] James G. Acker, "The Heritage of SeaWiFS: A Retrospective on the CZCS NIMBUS Experiment Team (NET) Program".
[119] James G. Acker, "The Heritage of SeaWiFS: A Retrospective on the CZCS NIMBUS Experiment Team (NET) Program".

reflectivity of the sides of a double-sided mirror.[120] For CZCS, the active calibration method varied by 1 or 2 digital counts. The scan was recalibrated every sixteen lines. In data processing, the variation was amplified by the calibration table to a resulting variation of 4 to 8 counts in the actual data, which was fully unacceptable. On the spacecraft, the voltage delivered to the calibration lamps and the actual radiant output of the lamp shifted in a matter of minutes; and every CZCS scene was two minutes long. So the decision was made to only use the initial voltage and calibration test lamp value for each scene, and to apply the prelaunch calibration coefficients to the raw CZCS counts, because the team believed that the radiance detectors in the instrument would be far more stable than the calibration sources inside the instrument. Detectors would still degrade over time, so new calibration schemes would still be developed.

In April 1980, some of the NET members arrived at GSFC for more algorithm development work. This meeting had some definite bright points; some ideas were floated to speed up processing time, and Austin and Petzold were making progress on the diffuse attenuation coefficient algorithm[121].

At this stage of the CZCS mission, there was still very little public commentary on the CZCS, or indeed any of the data from the other Nimbus 7 instruments. This may have been due in part to the slow trickle of data released and analyzed, and the omnipresent lag time between data analysis and the eventual publication of research results. However, the importance of pollution in the public mind and the relatively fresh memories of the *Amoco Cadiz* spill were evidently connected; an article in the April 17, 1980 issue of *New Scientist* mentioned the "coastal zone colour scanner", stating that the instrument "can identify and track oil spills in 11 major areas used by tankers".[122]

Bienvenue, France: André Morel hosted the NET in Villefranche-sur-Mer on the 21-22 of May. Warren Hovis reported the happy news that 131 validation scenes had been processed, and considerable discussion involved the calibration methodology. Howard Gordon was also happy to report that the processing speed had improved, but they would also have to consider the effects of the instrument tilt.

The scientific aspects of the mission were proceeding at a rapid pace. Ray Smith was already examining features in the California Bight; the South African experiments were underway; and the Europeans were active with cloud-masking (as reported by Boris Sturm) and validation studies using an aircraft instrument similar to the OCS.[123]

The choice of Villefranche-sur-Mer as the venue for the meeting was likely strategic, because right on its heels was the COSPAR/SCOR/IUCRM Symposium on Oceanography

---

[120] X. Xiong, Vincent Salomonson, and William Barnes, "Overview of the EOS/MODIS On-Orbit Calibration and Performance", *Proceedings of the International Geoscience and Remote Sensing Symposium*, Volume 5, 3424-3427, 2005, http://ieeexplore.ieee.org/stamp/stamp.jsp?arnumber=01526578, (accessed 20 April 2009).

[121] James G. Acker, "The Heritage of SeaWiFS: A Retrospective on the CZCS NIMBUS Experiment Team (NET) Program".

[122] Tony Allen and John Latham, "Landsat orbits in a melting pot", *New Scientist*, page 145 (17 April 1980).

[123] Minutes of the 13th CZCS NET Meeting; James G. Acker, "The Heritage of SeaWiFS: A Retrospective on the CZCS NIMBUS Experiment Team (NET) Program".

from Space, held in Venice, Italy on May 26-30, 1980. COSPAR stands for "Committee on Space Research"; SCOR is the "Scientific Committee on Oceanic Research"; and IUCM represents the "Inter-Union Commission on Radio Meteorology". Getting from Villefranche-sur-Mer to Venice looks like a nice jaunt; first a swing around the Italian Riviera through Genova, then cross-country through either Parma (remembering to pick up some *prosciutto*) or Bologna (stopping at a cafe for a *cappuccino* and *pasta bolognese*) and then heading to the Adriatic coast and the storied floating city of Venice. By whatever route and cuisine they chose (Howard Gordon said that he and Jim Mueller took three bottles of wine and traveled there by train), many of the NET members arrived in Venice with eagerly awaited presentations.[124] They also availed themselves of the Venetian atmosphere and the European traditions; Howard Gordon noted that during a keynote address by William Nierenberg, director of the Scripps Institution of Oceanography, "quite a bit of wine was consumed".[125]

The book *Oceanography from Space* contains the proceedings of this meeting, with a forward by Gifford Ewing, who unfortunately was unable to attend. [126] (Gifford Ewing died in La Jolla on December 10, 1987.) In addition to the results from CZCS, this amazing meeting had results from several different pioneering missions for oceanographic remote sensing. The Seasat mission, which had been launched in June 1978 and which failed suddenly on October 10, 1978 – two weeks before the launch of Nimbus 7 – had nonetheless demonstrated several pioneering technologies; scatterometry to measure ocean surface wind speeds; radar altimetry to measure sea surface height, and synthetic aperture radar to observe circulation and wave features on the sea surface. And in addition to CZCS on Nimbus 7, the SMMR instrument was returning significant data on sea ice concentrations. Multi-faceted John Apel, a member of the CZCS NET, led off the sea surface altimetry session.

The session entitled "Water Colour Measurements" hosted a remarkable 26 presentations, and constituted "Opening Night" for the traveling CZCS show. Gordon and Morel led off the session with a general discussion of water colour measurements.[127] They were followed by Warren Hovis, who described the CZCS mission; Dennis Clark, who presented the critical phytoplankton pigment concentration algorithm; and Ros Austin (with instrumentation guru T.J. Petzold as co-author) on the diffuse attenuation coefficient algorithm.[128] Howard Gordon proceeded to clarify the atmospheric correction algorithm,

---

[124] Howard Gordon, comments recorded at the Ocean Color Collaborative Historical Workshop, January 13-14, 2009, St. Petersburg, Florida.

[125] Howard Gordon comments recorded at the Ocean Color Collaborative Historical Workshop, January 13-14, 2009, St. Petersburg, Florida.

[126] J.F.R. Gower, editor, *Oceanography from Space*, Plenum Press, New York, 978 pages, (1981).

[127] Howard Gordon and André Morel, "Water colour measurements – An introduction", in J.F.R. Gower, editor, *Oceanography from Space*, Plenum Press, New York, 207-212, (1981).

[128] W.A. Hovis, "The Nimbus-7 Coastal Zone Color Scanner (CZCS) Program"; Dennis K. Clark, "Phytoplankton pigment algorithms for the NIMBUS-7 CZCS"; Roswell Austin and Theodore Petzold, "The determination of the diffuse attenuation coefficient of sea water using the Coastal Zone Color Scanner", in J.F.R. Gower, editor, *Oceanography from Space*, Plenum Press, New York, 239-256, (1981).

and Boris Sturm discussed how chlorophyll concentrations were retrieved.[129] (It is common to refer to data from remote-sensing satellites as "retrievals".) Ray Smith discussed how surface and satellite observations interacted in the California Bight; Jim Mueller described remote sensing with ocean color and sea surface temperature to observe oceanic fronts; and Charlie Yentsch described investigations at his anchorage in the Gulf of Maine.[130] Later in the day, the South African group described their ocean color experiment.[131] The presentations following the CZCS NET member presentations demonstrated how rapidly the appreciation of the CZCS data, and the anticipation of more, was pervading the oceanographic community. Sufficient data had already been distributed that research had been conducted outside of the CZCS NET group in such areas as the northern Adriatic Sea, the Baltic Sea, and the Ligurian Sea, the lovely oceanic region that makes the shoreline of the Italian and French Riviera such an attractive destination.

Four months later, the peripatetic CZCS NET was back on the familiar grounds of GSFC. Ray Smith had apparently been at sea for quite awhile between the May meeting and this meeting, as he reported results from a trans-Pacific cruise on the *Researcher.* Results from the central Pacific and a cruise related to fisheries applications were presented by Bob Wrigley, and Sayed El-Sayed discussed results from a research cruise in February and March. El-Sayed had also journeyed on the South African vessel *Agulhas* to the Southern Ocean between Africa and Antarctica.[132]

October 1980 marked the two-year anniversary of the CZCS launch. The varied activities of the busy CZCS NET – it is likely that Warren Hovis was enjoying being the head of an active research mission despite the various vexatious vagaries of the instrument itself – show that the process of operating on the cutting edge with a new and untried system was both challenging and arduous. Yet this was also groundbreaking, unprecedented science. The first true public presentation of the success of the CZCS occurred just before the two-year anniversary of the launch, consisting of two papers published in the October 3 issue of *Science.* The first was entitled "Nimbus-7 Coastal Zone Color Scanner: System Description and Initial Imagery" with Hovis as lead author and co-authored by several of the members of the CZCS NET (and apparently a couple of important programmers, Warren

---

[129] Boris Sturm, "Ocean colour remote sensing and quantitative retrieval of surface chlorophyll in coastal waters using Nimbus CZCS data", in J.F.R. Gower, editor, *Oceanography from Space*, Plenum Press, New York, 267-280, (1981).

[130] Raymond Smith and W.C. Wilson, "Ship and satellite bio-optical research in the California Bight", in J.F.R. Gower, editor, *Oceanography from Space*, Plenum Press, New York, 281-294, (1981); James L. Mueller and Paul E. LaViolette, "Color and temperature signatures of ocean fronts observed with the Nimbus-7 CZCS", J.F.R. Gower, editor , *Oceanography from Space*, Plenum Press, New York, (1981); Charles Yentsch and N. Garfield, "Principal areas of vertical mixing in the waters of the Gulf of Maine, with reference to the total productivity of the area", in J.F.R. Gower, editor, *Oceanography from Space*, Plenum Press, New York, 303-312, (1981).

[131] Frank Anderson, L.V. Shannon, S.A. Mostert, N.M. Walters, and O.G. Malan, "A South African Ocean Colour Experiment", in J.F.R. Gower, editor, *Oceanography from Space*, Plenum Press, New York, 381-386, (1981).

[132] Minutes of the 14th CZCS NET Meeting; James G. Acker, "The Heritage of SeaWiFS: A Retrospective on the CZCS NIMBUS Experiment Team (NET) Program".

H. Wilson of Scripps and David Ball of the Computer Sciences Corporation).[133] This paper presented the theory underlying the operation of the instrument to detect the concentration of phytoplankton pigments using the absorption of light by chlorophyll, summarized the important operating characteristics of the instrument, and presented two initial images of the eastern Gulf of Mexico. A particularly noteworthy point made by the authors was to compare the CZCS to the better-known Multi-Spectral Scanner (MSS) on the Landsat satellite; the CZCS was approximately 60 times more sensitive than the MSS. The first scene shown in the paper was a pseudo-true color image, and the second was a phytoplankton pigment image. In the paper, the authors describe how the second image was derived through the application of the critical atmospheric correction algorithm. The paper also describes how the data tapes were to be produced and distributed; the pigment image of the Gulf of Mexico was provided as an example of the geophysical product data that would be distributed to researchers. At the time that the paper was being written, the researchers still had very few actual processed scenes from CZCS to publish.

The second paper was entitled "Phytoplankton Pigments from the Nimbus-7 Coastal Zone Color Scanner: Comparisons with Surface Measurements" authored by Gordon, Clark, Mueller, and Hovis.[134] This paper extends the discussion of the atmospheric correction method and describes the initial form of the CZCS pigment algorithm, noting that the algorithm calculates the combined concentration of chlorophyll and phaeopigments. The paper demonstrates atmospheric correction with grayscale images and shows comparison images of the eastern Gulf of Mexico scene which also appeared in Hovis et al. 1980, utilizing different ratios of the CZCS band data. Dennis Clark noted that the scene used here was from CZCS Orbit 130, and that Jim Mueller had made the tape for these data, which was "remarkable", and was actually the first demonstration of the atmospheric correction algorithm applied to the data.[135] Clark indicated that in contrast to later missions, the algorithms had to be frozen and could not evolve – Hovis didn' t want to get too far behind, and the processing time was very slow.

There is a very interesting section of this paper that describes the difficulty that the CZCS NET was dealing with – the same difficulty that subsequent researchers, and teams and scientists would also struggle with – the atmosphere and everything in it. In fact, the Gordon et al. 1980 paper may even underestimate the magnitude of the problem a bit:

"It can be anticipated then that as much as 80 percent of the radiance detected at satellite altitudes could be due to atmospheric scattering. This atmospheric radiance is difficult to remove from CZCS imagery because of the component arising from aerosol

---

[133] Warren A. Hovis, Dennis K. Clark, Frank Anderson, Roswell W. Austin, W.H. Wilson, Edward T. Baker, D. Ball, Howard R. Gordon, James L. Mueller, Sayed Z. El-Sayed, Boris Sturm, Robert C. Wrigley, and Charles S. Yentsch, "Nimbus-7 Coastal Zone Color Scanner: System Description and Initial Imagery", *Science*, 210 (4465), 60-63, (1980).

[134] Howard R. Gordon, Dennis K. Clark, James L. Mueller, and Warren Hovis, "Phytoplankton pigments derived from the Nimbus-7 CZCS: Initial comparisons with surface measurements", Science, 210 (4465), 63-66, (1980).

[135] Dennis Clark, interview notes transcript.

scattering. The aerosols are highly variable (spatially and temporally) in concentration, composition, and size distribution."

This is one of the clear early statements in the history of ocean color remote sensing that underscores two particular aspects of the science; one, that the scientific analysis (the algorithms that generate the geophysical products) is actually based on a very small signal, and two, that the oceanic radiance signal is significantly smaller than the atmospheric radiance signal. These two factors together are what make the initial success of the CZCS NET so significant, given the constraints of computer speed and radiometer technology at the time of the mission.

The lack of computer capability to process CZCS data was particularly troublesome; getting the CZCS results into print was a significant achievement on several levels. Howard Gordon described this aspect of the mission in detail:

> "A system for processing CZCS imagery was not put in place until several years after launch (mostly because there were no computers fast enough to process the data in a reasonable amount of time). Also, at launch the algorithms were not implemented on any single computer. Under pressure to validate the CZCS data, the NET had to beg and borrow computer time wherever they could find it. In late 1979 and early 1980 Gordon, Clark, and Mueller managed to get computer time on an image processing system at GSFC called AOIPS [Atmospheric and Oceanographic Image Processing System]. It used a PDP 1155 computer coupled to an image display system. The group was allowed to use the system at night from time to time. The computer room was so cold that Gordon often wore his parka hood to keep warm. Using a faster UNIVAC computer in Miami, Gordon carried out the time-consuming computation of the Rayleigh scattering component of the radiance for each 512×512 pixel sub-scene of the 1968×955 pixels in a scene to be examined, and delivered it to GSFC on tape. Jim Mueller's programmer Dave Ball then coded the rest of the atmospheric correction algorithm on the PDP and the correction applied to a few images, line by line. As the correction was applied to each line, the corrected image replaced the uncorrected image on the display system, and it appeared that a veil was being removed from the image at a rate of one line every 2-3 minutes. The water-leaving radiances were then inserted into Clark's algorithm, and the pigment concentration derived, and compared with the *in-situ* data. The agreement was good – and it was decided to announce the CZCS's success in two publications in *Science.*"[136]

[136] Howard R. Gordon, manuscript comments, received 17 December 2009.

Netting a Red Tide

Serendipity has been known to enter into the process of science on various occasions. The choice of the image processed and displayed in Hovis et al. 1980 and Gordon et al. 1980 turned out to be significantly serendipitous, as was the cruise track of the R/V *Athena IV* through this region in November 1978. The CZCS NET researchers all expected that if the instrument functioned as it was designed to, it would detect phytoplankton chlorophyll, and that this would allow detection of phytoplankton blooms – the profuse eruptions of phytoplankton that occur at favorable convergences of nutrient concentrations, sunlight, oceanic circulation, and temperature. The November 1978 image of southwestern Florida happened to capture a particularly disagreeable type of phytoplankton bloom – a harmful algal bloom, in fact the archetypical "red tide". Many different types of harmful and noxious algal blooms are referred to in general as "red tides", but the manifestation that occurs in the Gulf of Mexico is the one to which the name is most commonly applied, particularly in the United States. Gulf of Mexico red tides are caused not by the ubiquitous chlorophyll cells of diatoms, but by a more unusual phytoplankter, a dinoflagellate. All dinoflagellates are characterized by a common element, the flagellum, a whiplike tail that allows them to move through the water under power, unlike the passively floating diatoms. The tail is connected to a cell or cells containing chlorophyll, so dinoflagellates are classified as phytoplankton.

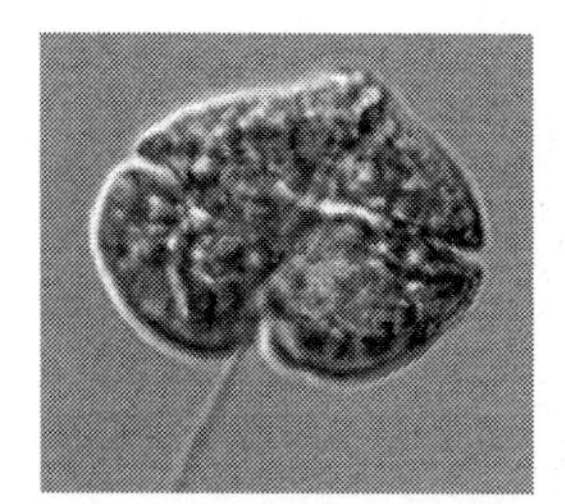

Figure 3.10. *Karenia brevis.*

The dinoflagellate bane of Florida is an organism named *Karenia brevis* (Figure 3.10).[137] It used to be named *Gymnodinium brevis*, but it was renamed in honor of Florida Fish and Wildlife Research Institute scientist Dr. Karen Steidinger, who has studied it extensively.[138] In addition to the chlorophyll, *Karenia brevis* cells contain something else – a potent neurotoxin, a poisonous paralytic agent. When *Karenia brevis* cells bloom, the concentration of this neurotoxin in the water increases markedly. Fish die in great numbers and their rotting carcasses wash up on Florida's fabled beaches. Seabirds that eat affected fish can

[137] "Isolating and Maintaining Cells for FWRI's Culture Collection", http://research.myfwc.com/features/view_article.asp?id=23559, (accessed 20 April 2009).
[138] Christopher Brown, "Detecting blooms of the dinoflagellate *Karenia brevis* from space", http://cics.umd.edu/~chrisb/gbreve.html, (accessed 20 April 2009); Jeff Klinkenberg, "A bloom of her own bloom," St. Petersburg Times, (4 September 2006).

also be affected, exhibiting strange behavior due to paralysis, and may also die. Even the great lumbering manatees and playful dolphins that ingest water containing the dinoflagellate can be poisoned and killed. And *Karenia* is no friend to humans, either; when the waters of the bloom are cast into the air by wind-driven waves and surf, the air can carry the neurotoxin into the lungs of beach-going tourists, causing respiratory distress and asthma-like symptoms. More dangerously, passive filter feeding shellfish like scallops and clams do not discriminate between good and bad phytoplankton, so shellfish that ingest *Karenia brevis* in large quantities concentrate the neurotoxin in their flesh. Eating shellfish that have been infused with the neurotoxin can kill humans. For this reason, red tide alerts and shellfishing bans are a regular activity of the Florida Department of Environmental Protection.[139]

When CZCS captured the image and the *Athena II* visited the area, a red tide was raging off the shore of Sanibel Island, with enormously high cell counts, according to in-water sampling (Figure 3.11).[140] Despite this condition, the CZCS overestimated the concentration of chlorophyll here; this is likely due to other complicating factors of ocean color remote sensing, suspended sediments and shallow waters, and of course the simple fact that the CZCS was the first instrument to ever attempt quantitative remote sensing of phytoplankton chlorophyll and pigments![141] The observation of the red tide bloom so early in the mission indicated one potential area of useful application for this new type of data.

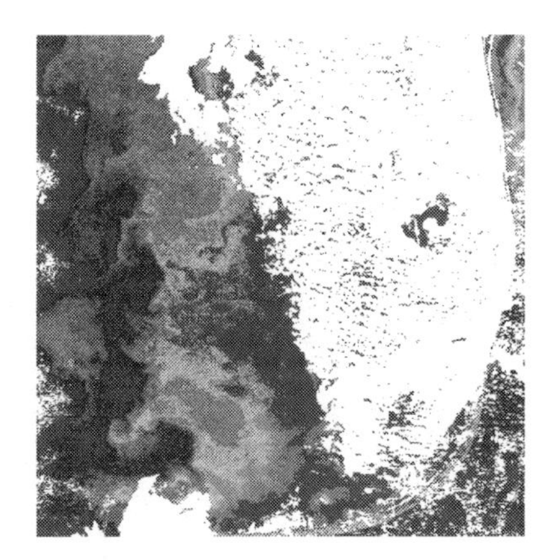

Figure 3.11. CZCS image of Florida and the Gulf of Mexico, November 1978, showing red tide bloom off the West Coast. Image provided by the GES DISC.

---

[139] Jay Abott, "Florida Red Tide Monitoring and Reporting", PDF of Powerpoint presentation, http://www.floridadep.org/coastal/WaterMonitoringCouncil/files/meetings/2008/09-24/Florida_Red_Tide_Monitoring.pdf, (accessed 20 April 2009).

[140] GES DISC, "Chapter 12 – Plankton Blooms – The Good, the Bad, and the Shiny", http://disc.sci.gsfc.nasa.gov/oceancolor/additional/science-focus/classic_scenes/12_classics_blooms.shtml, (accessed 8 January 2010).

[141] Howard R. Gordon, Dennis K. Clark, James L. Mueller, and Warren Hovis, "Phytoplankton pigments derived from the Nimbus-7 CZCS: Initial comparisons with surface measurements".

Beyond the level of serendipity of encountering this massive red tide bloom, this particular area of coastal waters was not even in the original cruise plan. This day of the cruise, however, turned out to coincide with the broadcast of National Football League *Monday Night Football*, and the crew realized that if they maneuvered close enough to Fort Myers, they could pick up the TV broadcast signal and watch the game on the ship. As Howard Gordon described it, "… we were off the Dry Tortugas… the ship track goes in real close, goes through this huge bloom, then goes back off toward Tampa … it basically goes right through this bloom … what a wonderful thing to have done. But the ship was supposed to go straight … the crew wanted to watch Monday Night Football … so they had to swing in toward Fort Myers." Thus, the *Athena IV* cruised close to the southwest Florida shore, the crew watched *Monday Night Football*, and in so doing the ship and scientists encountered a historic red tide bloom.[142]

So according to the traditions of science, by October 1980, two years into the mission, the word on CZCS was officially out. It worked. The scientists who had been working with the mission and with the data already knew this, and they were already reconfiguring their views of the ocean and their former understanding of it. As Dick Barber put it, "I had to rearrange everything I knew… we had been thinking of things on a straight line… now we realized there were lots of lines, with structure." Barber indicated that one surprising aspect was the observation of upwelling "source spots". The CZCS view from space indicated that there was no such thing as alongshore continuity, i.e., samples obtained at one coastal location didn't indicate what was happening only a few kilometers away.[143]

Charlie Yentsch did realize that the system worked as he had hoped, which he had also expected as long as his simplified view of the primary influences on ocean optics was correct. Georges Bank, which he was very familiar with, looked like he had expected it would from space. On the other hand, the coastal phenomena that Barber had noted, variously called "squirts" and "jets" created by variability in coastal currents, surprised him as well. And Yentsch was both particularly gratified and surprised that the CZCS images so effectively captured images of the spinning eddies that are generated by the Gulf Stream in the Atlantic Ocean.[144]

Yentsch also recalled that the first time Warren Hovis showed him a CZCS image of a coccolithophorid bloom, it was so bright he thought that someone had accidentally put a thumbprint on the film.[145] Coccolithophorids are a particular kind of phytoplankton that produce highly-reflective microscopic shells made out of calcium carbonate ($CaCO_3$), and blooms of coccolithophorids cause very bright blue-green waters. (Coccolithophorids will be described in more detail in subsequent chapters.)

Ros Austin of the Scripps VisLab produced a publication of note in 1980, describing the methods used in the at-sea validation cruises conducted in the Gulf of Mexico. His paper was entitled "Gulf of Mexico, ocean color surface truth measurements" and was

[142] Howard Gordon comments recorded at the Ocean Color Collaborative Historical Workshop, January 13-14, 2009, St. Petersburg, Florida.
[143] Recorded interview with Dr. Richard Barber, 1 May 2008.
[144] Recorded interview with Dr. Charles Yentsch, 16 July 2008.
[145] Recorded interview with Dr. Charles Yentsch, 16 July 2008.

published in *Boundary Layer Meteorology*.[146] This is a landmark paper for the field, as it described methods for measuring numerous variables (upwelling radiance and irradiance, downwelling irradiance, vertical profile techniques, integration of in-water measurements with solar irradiance measurements) that are necessary for the data validation effort. While this paper was published in regard to the CZCS data validation efforts, it is a foundational paper for the further development of methods and establishment of measurement protocols that enabled improved data accuracy.

So 1980 closed on a high note for the CZCS and its associated cadre of researchers. The CZCS had been a major topic of interest at the *Oceanography from Space* symposium, and the first research results had been published in major scientific media. The first two years of the mission also produced significantly more data that subsequent years, due to declining power and instrumental operation difficulties.[147]

1981 – The Year of Calibration

A particular subdefinition of *feedback* is "the transmission of evaluative or corrective information about an action, event, or process to the original or controlling source". With regard to the CZCS, feedback was acquired from two different sources; the internal feedback of the operations of the CZCS NET, devoted to the process of refining and improving the data; and the external feedback from the larger oceanographic community, where interest in the proven concept of the CZCS was starting to spread.

The CZCS NET met, according to a fairly consistent schedule, again in January, this time at the Naval Postgraduate School in Monterey, California. There is no record of how many of the NET members were able to work in a round of golf at nearby Pebble Beach. Several team members presented the progress of their work. Despite the emphasis on the coastal zone, Jim Mueller suggested that about 20% of the CZCS observational budget should be devoted to open ocean observations. Charlie Yentsch, still intent on planning for the future, provided an overview of possible research applications for the data.[148]

During the austral summer, El-Sayed's Texas A&M team conducted another cruise in the Southern Ocean spanning February and March. In December of 1981, the Texas A&M group initiated a series of four cruises off the coast of Israel.[149]

The Algorithm Development Team met again at Goddard in March. The next meeting of the NET took place in May, also at Goddard. Instrument operations still appear to have been fairly routine; Warren Hovis only mentioned that one of the calibration lamps was

[146] Roswell Austin, "Gulf of Mexico, ocean color surface truth measurements". *Boundary Layer Meteorology*, 19, 269-285, (1980).
[147] NASA GES DISC, "Coastal Zone Color Scanner (CZCS) 1km Level 1 Calibrated Radiance and Temperature Tape (CRTT) Dataset Guide Document ", (accessed 21 April 2009).
[148] Minutes of the 15th CZCS NET Meeting; James G. Acker, "The Heritage of SeaWiFS: A Retrospective on the CZCS NIMBUS Experiment Team (NET) Program".
[149] James G. Acker, "The Heritage of SeaWiFS: A Retrospective on the CZCS NIMBUS Experiment Team (NET) Program".

degrading and that the demise of the IR channel appeared to be final.[150]

This May 1981 CZCS NET meeting was packed with content. The Level 2 data product development was reviewed; this was the data product eagerly anticipated by the oceanographic world, as it provided the CZCS determinations of chlorophyll and pigment concentration. To this end, Dennis Clark recommended that the NET should choose a single Level 2 algorithm. His own algorithm (Clark 1981, presented at the *Oceanography from Space* symposium) was clearly a prime candidate.[151] Howard Gordon described further development of the atmospheric correction algorithm, and Frank Anderson discussed its application off the coast of South Africa. Jim Mueller's work on sun glint contamination and its removal, and how this affected data processing, was another topic of discussion. Yentsch and El-Sayed described research on Georges Bank and Antarctic waters, respectively. And there was also a point of polite contention; Boris Sturm indicated that the European community was working on its own algorithm, because the NASA algorithm worked fairly well in the Mediterranean but underestimated chlorophyll concentrations in the North Sea.[152]

The Space Shuttle: A very important event took place on April 1981, a month before the May CZCS NET meeting – an event which would prove to be a marked influence on the course of NASA's oceanographic remote sensing mission planning. On April 12, 1981, the first manned Space Shuttle Mission was launched, crewed by John Young and Robert Crippen.[153] The budget devoted to Space Shuttle operations had grown to be a major part of NASA's budget before the launch; the Space Shuttle program became a significant factor in the planning and budgeting for NASA remote sensing missions.

The next CZCS NET meeting, the 17th, was hosted by Charlie's Place, also known as the Bigelow Laboratory for Ocean Sciences in West Boothbay Harbor, Maine, in September. In between lunches of lobster rolls and dinners of fried clams finished off with blueberry pie, the NET members had to face the fact that the CZCS sensors were degrading. Dealing with sensor degradation would become an ongoing theme of the rest of the mission and would influence the data processing effort to a large degree. Research conducted in the Mediterranean, off of Monterey, the U.S. East Coast, and the Gulf of Mexico were all described by various team members. Charlie Yentsch led the discussion on the salient issue of a CZCS sequel; the CZCS "follow-on" mission.[154]

The Clear Water Calibration Method: One other journal publication was a very significant result in 1981. Gordon and Clark published "Clear water radiances for

---

[150] James G. Acker, "The Heritage of SeaWiFS: A Retrospective on the CZCS NIMBUS Experiment Team (NET) Program".
[151] Dennis K. Clark, "Phytoplankton algorithms for the NIMBUS-7 CZCS".
[152] Minutes of the 16th CZCS NET Meeting; James G. Acker, "The Heritage of SeaWiFS: A Retrospective on the CZCS NIMBUS Experiment Team (NET) Program".
[153] NASA Kennedy Space Center Science, Technology, and Engineering, "STS-1". http://science.ksc.nasa.gov/shuttle/missions/sts-1/mission-sts-1.html, accessed 21 April 2009.
[154] Minutes of the 17th CZCS NET Meeting; James G. Acker, "The Heritage of SeaWiFS: A Retrospective on the CZCS NIMBUS Experiment Team (NET) Program".

atmospheric correction of coastal zone color scanner imagery", which appeared in *Applied Optics* in December 1981.[155] In essence, this paper presented a method by which the water-leaving radiances at the sea surface could be determined as a starting point for the atmospheric correction method. I.e., if a pixel in a CZCS scene could be designated as a reference pixel, meaning that the radiances in this pixel were established in advance using *a priori* assumptions, then this pixel can be used to supply the coefficients for the atmospheric correction algorithm to be applied to the entire scene. The **a priori** assumptions allowing this method to work were based on the data acquired on the calibration and validation cruises taking place prior to and after the CZCS launch.[156] In the clear water case, which is where the actual chlorophyll concentrations are less than 0.25 mg m^{-3}, the radiances in the green and yellow CZCS bands (520 and 550 nm) are assumed to be constant and the radiance at 670 nm (red) is assumed to be zero.

The importance of the Gordon and Clark paper, which led to the application of the clear water radiance method in CZCS processing, was very significant. This method allowed the data to be used independent of the calibration of the actual optics of the instrument. Thus, the sensor could vary or the calibration lamps could vary, but the radiance measurements acquired by the instrument could be used by themselves to provide an accurate atmospheric correction for the entire two-minute scene. The authors estimated that application of the method would lead to errors less than 30% in the estimate of pigment concentration. The Clear Water Radiance method (which also came to be known as the "dark pixel" method) allowed a consistent atmospheric correction to be applied to the entire CZCS data set.

Tasmania – On November 27, 1981, the CZCS captured what may be the most striking single image of the entire mission, an image of Tasmania surrounded by swirls and eddies of current-driven phytoplankton pigment concentrations (Figure 3.12). This image demonstrates quite clearly the advantages of the view from space, with the atmospheric veil corrected away; over a distance of only a few kilometers, phytoplankton pigment concentrations vary by factors of 100 or more, and classical ship-based research would easily miss small features of high productivity, such as the tight eddy spiral off the northeast coast.

[155] Howard R. Gordon and Dennis K. Clark, "Clear water radiances for atmospheric correction of coastal zone color scanner imagery. *Applied Optics,* 20**,** 4,175-4,180, (1981).
[156] Dennis K. Clark, "Phytoplankton algorithms for the NIMBUS-7 CZCS".

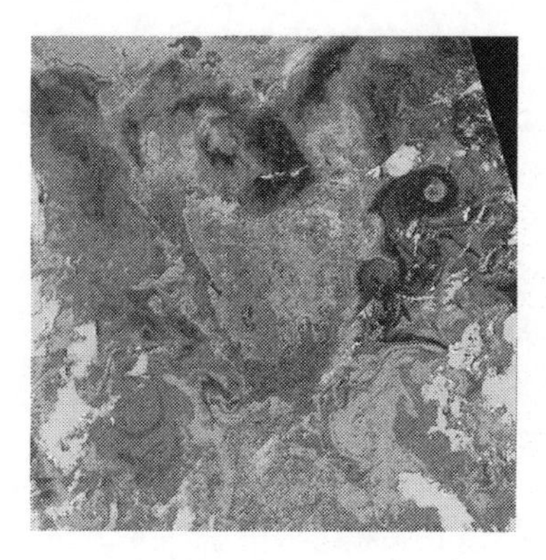

Figure 3.12. CZCS image of Tasmania, acquired on November 27, 1981.

1982 – The Year of El Chichón

At the first meeting of the CZCS NET in January 1982, the instrument status is not discussed. One of the main items that Warren Hovis presented was a memorandum to be signed by the CZCS NET members to push for more open ocean coverage. These developments indicate that the success of the CZCS was evolving the perception of the mission from a coastal focus to a global ocean focus. At the same time, all of the instruments were facing the continuing aspect of a declining satellite power budget. There were less instruments working on the platform now to share the power; LIMS had been turned off in June 1979 after its supply of cryogenic coolant had been exhausted, and the ERB narrow-angle scanner failed in June 1980, though the instrument continued to collect solar radiation data at least through 1983.[157] The request for increased open ocean coverage was seen as important for the issue of a CZCS follow-on mission. The issue was not just power; data on Nimbus 7 was stored on tape recorders between opportunities for telemetry downlink, and if CZCS data took up more and more space on the tape recorder, the recording of data from other instruments would be reduced.

Charlie Yentsch led the discussion of the Level 2 processing, and he was seeking scenes that had a high degree of oceanographic interest. Howard Gordon, as would be expected, was still working on atmospheric correction.

Boris Sturm described his research in the northern Adriatic Sea, measuring chlorophyll and suspended sediment concentrations. (This area is particularly noted for sediments, considering that the city of Venice is built on coastal sediment deposits.) The South African group had conducted research in Lambert's Bay, a few miles north of Cape Town, according to Frank Anderson.[158]

---

[157] NSSDC, "Limb Infrared Monitor of the Stratosphere (LIMS)"; NSSDC, "Earth Radiation Budget (ERB)".

[158] Minutes of the 18th CZCS NET Meeting; James G. Acker, "The Heritage of SeaWiFS: A Retrospective on the CZCS NIMBUS Experiment Team (NET) Program".

El Chichón: It was one of the most unexpected and powerful volcanic eruptions in human history – and it was hardly noticed by most of the world's population. With very little warning, a small bump on the landscape of the Mexican province of Chiapas – previously little more than a forested hill that was hard to recognize as volcanic – suddenly exploded into a brief period of devastating activity between March 29 and April 4th, 1982. The powerful eruption generated fierce glowing avalanches of gas and rock that killed over 2,000 people in the surrounding region. The collapse of the small summit created a kilometer-wide caldera hosting a steaming, bubbling crater lake.[159] Yet most of the power of El Chichón's eruption was directed straight into the sky. The explosive column blasted into the stratosphere, and the peculiar geological background of this volcano caused the eruptive gases to be laden with a very high concentration of sulfur dioxide, which creates a sunlight-blocking aerosol that was tracked by the Nimbus 7 TOMS[160]. The cloud was so acidic that it attacked the windows of commercial airliners over the next year, causing millions of dollars of damage.[161] The effects of the El Chichón eruption on CZCS were immediate; the El Chichón sulfur aerosol soon became distributed globally, attenuating the light from the sea surface that CZCS was attempting to collect.

While El Chichón was noted for its effect on the atmosphere, another eruption four months earlier actually set the stage. The December 1981 eruption of Nyamuragira volcano in Tanzania produced a "mystery cloud" that was detected by the Nimbus 7 TOMS, as was the El Chichón cloud.[162] While its observations of stratospheric ozone depletion were its most famous discovery, TOMS was also adept at tracking the sulfur aerosol clouds from volcanic eruptions. Because the Nyamuragira eruption was not explosive or devastating, it took researchers some work to track down the source of this particular aerosol cloud.

Gulf Stream Rings

The National Science Foundation funded an oceanographic program called the "Warm Core Rings Program" in the early 1980s, which was based out of Woods Hole and directed by Peter H. Wiebe. This program, which investigated the physical and biological dynamics

---

[159] S. De la Cruz-Reyna and A.L. Martin del Pozzo, "The 1982 Eruption of El Chichón volcano, Mexico: Eyewitness of the disaster", *Geofísica Internacional*, 48 (1), 21-31, (2009).
[160] Jørgen S. Aabech, "El Chichón, Mexico", http://www.vulkaner.no/v/volcan/latinam/chicon-e.html, (accessed April 21, 2009); Arlin J. Krueger, "Sighting of El Chichón sulfur dioxide clouds with the Nimbus 7 Total Ozone Mapping Spectrometer", *Science*, 220, 1377-1379, (1983).
[161] A. Bernard and W.I. Rose, Jr., "The injection of sulfuric acid aerosols in the stratosphere by the El Chichón volcano and its related hazards to the international air traffic", *Natural Hazards*, 3, 59-67, (1990).
[162] Arlin J. Krueger, C.C. Schnetzler, and L.S. Walter, "The December 1981 eruption of Nyamuragira Volcano (Zaire), and the origin of the ``mystery cloud'' of early 1982", *Journal of Geophysical Research*, 101 (D10), 15191-15196, (1996).

of warm-core rings birthed by the Gulf Stream current system, resulted in a collection of publications in the journal *Deep-Sea Research* in 1986.[163] Early in 1980, Stan Wilson, head of the newly-created oceanography program at NASA HQ, selected Kendall Carder of the University of South Florida to be the first 2-year appointee to head NASA's Ocean Productivity program. Just a few months earlier, Wilson had been recruited to come to NASA by Bill Bishop, deputy to Ron Greenwood in the Environmental Observations Division of the Applications Office at NASA HQ. Wilson said that this recruitment created somewhat of a dilemma, because he had been "bad mouthing NASA's wild-ass claims" after the failure of SeaSat. Another reason for the creation of the oceanography program, according to Dixon Butler, was that the NASA administrator at the time was Robert Frosch, an oceanographer – the only time that NASA was headed by a scientist. According to Butler, Frosch said, "We need an oceanography program!", followed by "Go make me an oceanography program." Thus, when the oceanography program was made, Wilson agreed to take it on.[164]

While still settling in to the job, Carder realized that getting CZCS data out to researchers on ships would be an excellent demonstration of NASA's new commitment to oceanographic science. So he helped to arrange the deployment of radio receivers on ships going to sea to investigate the rings in April of 1982. Bob Evans recalled that the antennas had to be bolted onto the bridge of the ship. CZCS data was downlinked, processed, and an image was radioed out to the ships. Ray Smith indicated that this novel access to near-real-time data was important both for guiding ship operations and tasking the CZCS observational region selections for subsequent days.[165] This effort marked the first time images of ocean biological activity viewed from space had been sent to ships at sea to assist in oceanographic research.

Texas A&M got its chance to host the CZCS NET in May 1982. Ros Austin described a particularly busy set of activities at Scripps, covering atmospheric aerosol effects, fishery applications, and studies of sensor degradation and calibration. Howard Gordon addressed sensor drift corrections. Charlie Yentsch was now enamored of the CZCS capabilities to observe cold and warm core Gulf Stream eddies.[166] The eddies and a remarkable image of the Gulf Stream from the CZCS had just been featured months before in an article in the December 1981 issue of *National Geographic* (Figure 3.13).[167] (The image which appeared in *National Geographic* may be the most familiar CZCS image ever published.) The memorandum advocating open ocean coverage had been received by the Nimbus Project management –

[163] Peter H. Wiebe, and T.J. McDougall, guest editors, "Warm-Core Rings. Studies of their physics, chemistry, and biology". *Deep-Sea Research*, 33 (11/12), 467 pages, (1986).

[164] Stan Wilson, recorded interview, 18 March 2009; Dixon H. Butler, oral history, http://www.jsc.nasa.gov/history/oral_histories/NASA.../ButlerDM_6-25-09.pdf, *accessed 26 May 2011.*

[165] Raymond Smith, Robert Evans, and Kendall Carder, comments recorded at the Ocean Color Collaborative Historical Workshop, January 13-14, 2009, St. Petersburg, Florida.

[166] Minutes of the 19th CZCS NET Meeting; James G. Acker, "The Heritage of SeaWiFS: A Retrospective on the CZCS NIMBUS Experiment Team (NET) Program".

[167] Samuel W. Mathews, "New World of the Ocean", *National Geographic*, 160 (6), 792-932, (December 1981).

their concern was the availability of tape recorder space.

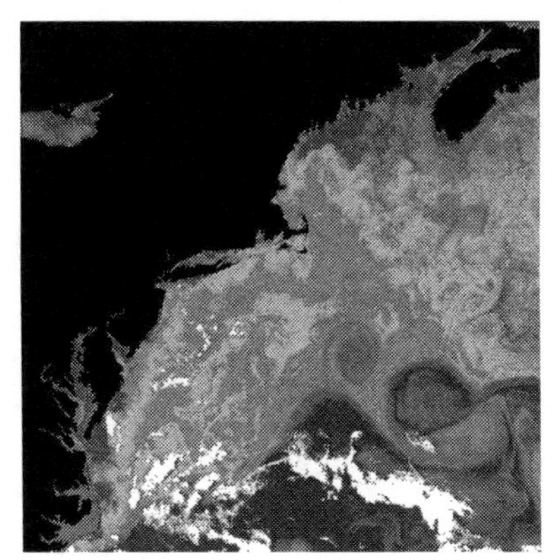

Figure 3.13. CZCS image of Gulf Stream eddies, published in National Geographic. Image provided by Howard Gordon.

Following fairly quickly after this meeting was the next NET meeting, at the University of Rhode Island, in late July. The focus of this meeting was highly scientific; an atlas of CZCS observations was being produced in collaboration with the Walter A. Bohan Company in Park Ridge, Illinois. The team heard from Dale Kiefer and B. Greg Mitchell regarding light levels and primary productivity, and Charlie Yentsch presented seven (yes, seven) papers, one regarding productivity on Georges Bank.[168]

1983 – The Year of El Niño

Following its customary schedule, the CZCS NET held another January meeting, selecting the home center of Robert Wrigley, the NASA Ames Research Center (Sunnyvale, California) as the venue. This meeting had both highlights and low-lights; one of the highlights was that the goal of the Level 2 production target was to be 500 CZCS scenes. One of the low-lights was the degradation of the solar power cells on Nimbus 7, reducing the available power budget on the satellite. The data production team was also working on a global geolocation system to properly place the CZCS scenes on the Earth's surface, a development leading to partially global Level 3 data products covering larger areas of the ocean.[169]

The effects of El Chichón entered into the conversation (as far as records of the CZCS NET indicate) at this meeting. The evaluated results indicated that the CZCS atmospheric correction algorithm had "capably" handled the atmospheric optical effects caused by the

[168] Minutes of the 20th CZCS NET Meeting; James G. Acker, "The Heritage of SeaWiFS: A Retrospective on the CZCS NIMBUS Experiment Team (NET) Program".
[169] Minutes of the 21st CZCS NET Meeting; James G. Acker, "The Heritage of SeaWiFS: A Retrospective on the CZCS NIMBUS Experiment Team (NET) Program".

dramatic eruption of the little-known Mexican volcano. Howard Gordon recalled that "coherency [in the time-series of CZCS data] was important to validate atmospheric corrections and anomalies that were introduced by the El Chichón event". Gene Feldman in conducting his thesis research had to look at images for both the El Niño event of 1982-1983 (described subsequently) and the El Chichón eruption, which he called the "big events" during the CZCS program.[170] The AVHRR on the NOAA polar-orbiting satellites did not fare as well as the CZCS after El Chichón; the sea surface temperature data record has a gap of several months following the El Chichón eruption due to the inadequate capability of the atmospheric correction algorithm to deal with the volcanic aerosols.[171]

The 1982-1983 El Niño: Even as the CZCS NET members were convening in California to discuss the events of the preceding year, an event which had been brewing for months in the Pacific ascended into its full, most frightening potential. The recurrent alteration of the normal climate of the Pacific Ocean known as El Niño (or as ENSO, which stands for El Niño/Southern Oscillation) was well-known to oceanographers. The genesis of the name, which represents the Christ Child, is attributed to the occurrence of rainfall along the parched desert coast of Peru, indicating a shift of ocean currents and winds resulting in warmer offshore water. The warmer offshore waters suppress the upwelling of deep ocean waters that brings nutrients to the ocean surface at the coast, nourishing legions of phytoplankton which in turn had fed millions of anchovies and sardines. When unusual December rains fell in Peru, fishermen recognized that fishing would suffer, and named the phenomenon El Niño because its effects were generally noted around the time of Christmas.[172]

Previous ENSO events, as well as they could be characterized without the advantages of the broad view of the oceanic basins provided by satellites, had followed a general pattern of effect and impact. The effects had been noted as an alteration of the equatorial currents and upwelling, warming of eastern Pacific waters, and the reduction of fishery catch, most notably off South America. In general, the effects of El Niño had been primarily felt in the tropical Pacific and both Central and South America. On the western side of the Pacific, El Niño could result in reduced rainfall and drought in both Indonesia and Australia. The "normal" El Niño was expected to persist through the winter months and fade back to the normal flow of winds and ocean currents by late spring and summer.

The El Niño of 1982-1983 was not normal. It was a record-breaker, a massive sea change, bringing unprecedented effects across wide areas of the Pacific basin, stretching

---

[170] Howard Gordon comments recorded at the Ocean Color Collaborative Historical Workshop, January 13-14, 2009, St. Petersburg, Florida; Gene Feldman, comments recorded at the Ocean Color Collaborative Historical Workshop, January 13-14, 2009, St. Petersburg, Florida.

[171] A.E. Strong, E.J. Kearns, and K.K. Gjovig, "Sea surface temperature signals from satellites - An update". *Geophysical Research Letters*, 27, 1667-1670, (2000).

[172] Curt Suplee, "El Niño/La Niña: Nature's Vicious Cycle", http://www.nationalgeographic.com/elnino/mainpage.html, (accessed 21 April 2009); NASA, "El Nino – and what is the Southern Oscillation anyway?", http://kids.earth.nasa.gov/archive/nino/intro.html, (accessed 21 April 2009).

across the entire equatorial band and from the coasts of Chile and Peru to the Straits of Juan de Fuca in Washington State. In numerous descriptions, the 1982-1983 El Niño is referred to as the El Niño of the century or a 100-year event, an event so strong it was unlikely to occur more often than once a century.[173] (Of course, in 1983 there were still 17 years left in the 20th Century.)

The wind-shift harbingers of the event were detected in June 1982; though largely unrecognized then, this timing later led to some scientists wondering if the eruption of El Chichón the previous month may have somehow triggered the El Niño event.[174] This potential connection has been largely discounted, but El Chichón nonetheless had an important and confounding influence on the effects of the 1982-1983 El Niño. Climate models predicted that the El Niño would have a considerable warming effect on climate, but the cooling from El Chichón's stratospheric sulfur cloud delayed the onset of the warming effects of the massive El Niño for about six months.[175] After the strongest effects of the El Chichón aerosols had dissipated, global atmospheric temperatures shot up at the beginning of 1983 due to El Niño.

By August 1982, current meters on buoys in the Pacific were detecting a dimunition of the normal westward flow of the surface equatorial current, which is created by the predominant direction of the trade winds.[176] The normal sea level rose in the east and dropped in the west, leading to prolonged exposure and damage to coral atolls and reefs. The thermocline marking the demarcation of warm surface water and cold deep water deepened by tens of meters over much of the Pacific, causing surface fish to abandon their normal feeding locations. Seabirds, unable to locate fish, abandoned rookeries and essentially disappeared, confounding ornithologists.[177] Coastal flooding in Ecuador and Peru caused the flowering of deserts where it is normally abnormal to receive an inch of rain in a year; instead, 100 inches of rain fell in six months.[178] Finch populations on the Galapagos Islands changed markedly due to the enhanced availability of seeds from

---

173 Jan Null, "El Niño, La Niña, and California flooding" http://ggweather.com/nino/calif_flood.html, accessed 21 April 2009; W.H. Quinn, and D. O. Zopf, "The unusual intensity of the 1982-1983 ENSO Event," in David Halpern, editor, *Tropical Ocean-Atmosphere Newsletter*, 26, NOAA Pacific Marine Environmental Laboratories, Seattle, WA., 17-20, (1984); R. S. Quiroz, "The climate of the El Niño winter of 1982-1983. A season of extraordinary climatic anomalies," *Monthly Weather Review*, 111, 1685-1706, (1983); E.M. Rasmusson, and J. M. Hall, August 1983. "El Niño: The Great Equatorial Pacific Ocean Warming Event of 1982-1983," *Weatherwise*, 36, p. 166-175, (August 1983).

174 A. Robock, K.E. Taylor, G.L. Stenchikov, and Y. Liu, "GCM evaluation of a mechanism for El Niño triggering by the El Chichón eruption cloud," *Geophysical Research Letters*, 22(17), 2369-2372, (1995).

175 James K. Angell, "Tropospheric temperature variations adjusted for El Niño, 1958-1998," *Journal of Geophysical Research*, 105(D9), 11,841-11,849, (16 May 2000).

176 Eric Firing, Roger Lukas, James Sadler, and Klaus Wyrtki, "Equatorial undercurrent disappears during 1982-1983 El Niño," *Science*, 222, 1121-1123, (9 December 1983).

177 Richard T. Barber and Francisco P. Chavez, "Biological consequences of El Niño," *Science*, 222, 1203-1210, (16 December 1983).

178 NASA EOSDIS Data Sampler, "El Nino Events", http://outreach.eos.nasa.gov/EOSDIS_CD-03/docs/hotevent.htm, (accessed 21 April 2009).

flowering plants.[179]

As the El Niño progressed, Indonesia and Australia experienced massive wildfires due to the spreading drought conditions.[180] El Niño-driven storms caused billions of dollars in damage in California.[181] Coastal erosion from a succession of battering winter storms was widespread. Homes on the coastal cliffs were severely damaged, some threatening to fall off the cliffs to the ocean surf below.[182]

Not all the effects were as damaging. One of the humorous events was the appearance of a bright red open ocean tropical crab (*Pleuroncodes planipes*) on the shores of the Monterey Peninsula.[183] Normally found no further north than San Diego, the warm waters of El Niño flowing north along the coast carried *Pleuroncodes* to regions where the coastal waters are normally much too cold for it. Mass strandings of the crimson crab allowed some seabirds to gorge beyond the capabilities of flightworthiness.

CZCS was perfectly situated to observe the effects of the 1982-1983 El Niño on the phytoplankton populations of the Pacific Ocean. Although the CZCS did not provide views of the global ocean, it could observe particular areas, and during the winter of 1982-1983 it focused its detectors on the remote archipelago that Darwin made famous, the Galapagos Islands.

Under normal conditions, the Galapagos archipelago waves a flag of phytoplankton that indicates the direction of the predominant equatorial currents. Caused both by enhanced upwelling and a bit of iron from the volcanic sands, this productive plume usually extends to the west for more than 100 kilometers, providing an excellent and reliable location for fish and squid to find plankton – and for seabirds and seals to find fish and squid! But as El Niño peaked in power in January and February 1983, this flag fluttered as the currents faded. Instead of the reliable plume flowing to the west, phytoplankton populations became scattered through the island realm, even concentrating in knots and eddies of anomalous flow to the east of the islands (Figure 3.14).[184] The dramatic changes caused seals to lose marked amounts of weight, and seabirds to abandon their nests. The dramatic effects of the El Niño as viewed by the CZCS were described in "Satellite color observations of the phytoplankton distribution in the eastern equatorial Pacific during the

---

[179] Jonathan Weiner, *The Beak of the Finch: A Story of Evolution in Our Time*, Vintage Books, Random House, New York, 332 pages, (1994).

[180] Thomas W. Swetnam and Julio L. Betancourt, "Fire-Southern Oscillation Relations in the southwestern United States," *Science*, 249, 1017-1020, (31 August 1990).

[181] Jan Null, "El Niño, La Niña, and California flooding".

[182] Kenneth R. Lajoie and Scott A. Mathiesen, "1982-1983 Coastal Erosion: San Mateo County, California", http://elnino.usgs.gov/SMCO-coast-erosion/introtext.html, 1998, (accessed 21 April 2009).

[183] Sanctuary Integrated Monitoring Network (SIMON), "Pelagic red crabs enter Monterey Bay (1982-1983), http://sanctuarysimon.org/monterey/sections/other/sporadic_crabs82.php, (accessed 21 April 2009).

[184] Gene C. Feldman, Dennis Clark, and David Halpern, "Satellite color observations of the phytoplankton distribution in the eastern Equatorial Pacific during the 1982--1983 El Niño", Science, 226, 1,069-1,071 (1984).

1982-1983 El Niño", published in *Science*, the November 30, 1984 issue.[185] This publication resulted in the first cover page of a major scientific journal featuring a CZCS image. (In order to conduct his research, Feldman would regularly come down to Maryland from the State University of New York at Stony Brook and select scenes at the NOAA Census building in Suitland, Maryland, where Warren Hovis held the data. Feldman normally worked from 6 PM to 4 AM for that process. He noted that because he started working on the project just as the El Niño commenced, he had to go back earlier in the mission to characterize the normal state of the tropical Pacific Ocean.)[186]

[Footnote: In a strange convergence of publications, the November 30, 1984 issue also featured a paper by Betzer et al., entitled "The Oceanic Carbonate System: A Reassessment of Biogenic Controls".[187] This paper discussed the potential influence of the mineral aragonite, created as the shells of some types of oceanic zooplankton, on the carbon chemistry of the Pacific and global ocean. One of the co-authors on this paper was graduate student James Acker, who at the time knew very little about the CZCS. While on a research cruise in May 1982 collecting the data which resulted in this publication, the daily radio news brief – this was an era long before the availability of the Internet by satellite on research vessels – had reports about observations of a mysterious stratospheric cloud. Even though the El Chichón eruption was known, the connection between the eruption and the cloud had not apparently been widely publicized a month later.]

As the El Niño finally began to fade in the Pacific, the CZCS NET decided to move its familiar Maryland meeting from Goddard in Greenbelt to the U.S. Naval Academy in Annapolis in July. Publications were proliferating; one of the most important at the time was Gordon's analysis of the degradation of the CZCS sensitivity. At the meeting, Dennis Clark amplified Gordon's analysis, showing that without corrections, the data in some of the CZCS scenes was off by a factor of 10. Charlie Yentsch had written an overview of biological remote sensing of the oceans, and the South African scientists had several publications. The instrument itself was continuing to be affected by the declining power availability on Nimbus 7.[188]

---

[185] Gene C. Feldman, Dennis Clark, and David Halpern, "Satellite color observations of the phytoplankton distribution in the eastern Equatorial Pacific during the 1982--1983 El Niño".

[186] Gene Feldman, comments recorded at the Ocean Color Collaborative Historical Workshop, January 13-14, 2009, St. Petersburg, Florida.

[187] Peter R. Betzer, Robert H. Byrne, James G. Acker, Carolyn S. Lewis, Robert R. Jolley, and Richard A. Feely, "The Oceanic Carbonate System: A Reassessment of Biogenic Controls", Science, 226 (4678), (1984).

[188] James G. Acker, "The Heritage of SeaWiFS: A Retrospective on the CZCS NIMBUS Experiment Team (NET) Program".

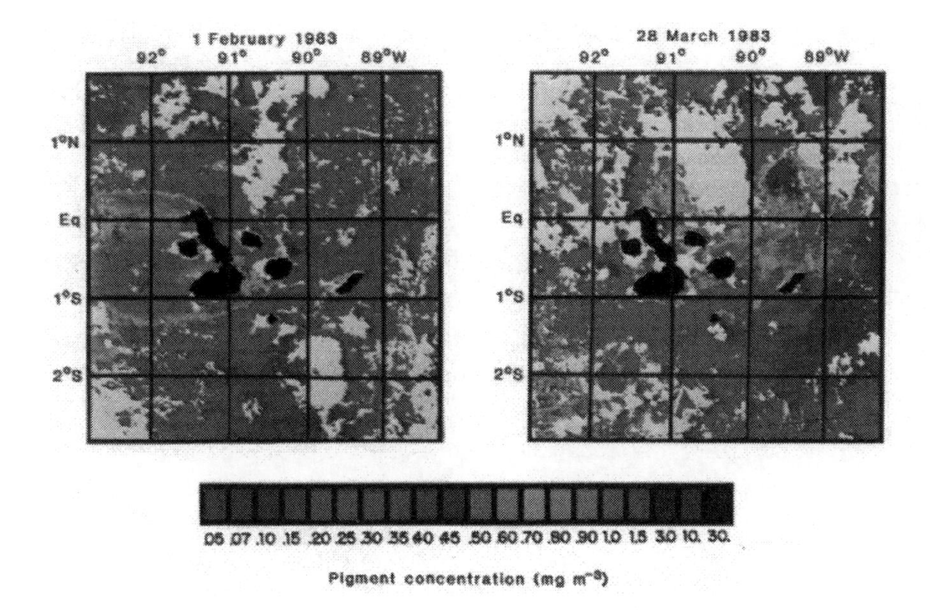

Figure 3.14. Figure from Feldman et al. 1984, showing the change in the Galapagos pigment plume under the influence of El Niño.

1984 – The Year of the On/Off Switch

For satellite remote sensing of the Earth, for any of its multiple aspects, algorithm development amounts to the homework that nobody ever wants to do, but if it wasn't done, you wouldn't pass the class. The algorithms that underly the colorful images and stunning scientific discoveries are intended to produce accurate results. When they do that, their effective functioning is widely ignored. When they don't do that, they are widely critiqued, criticized, and corrected. Early in 1984, the CZCS algorithm development team met at GSFC, examining how the Clear Water Radiance calibration method was being implemented. Because the process of science is rarely complete, the algorithm development team asked for all of the validation reports where CZCS data products were compared to sea surface measurements.[189]

The second manager of the Ocean Productivity Program at NASA Headquarters, Wayne Esaias (who had arrived at that position for a two-year stint in 1982 after being a professor at the State University of New York at Stony Brook and a previous posting at the Langley Research Center) wrote a letter to the algorithm development team indicating that there was ongoing concern with the accuracy of the CZCS derived products.[190] Given what

[189] James G. Acker, "The Heritage of SeaWiFS: A Retrospective on the CZCS NIMBUS Experiment Team (NET) Program".

[190] James G. Acker, "The Heritage of SeaWiFS: A Retrospective on the CZCS NIMBUS Experiment Team (NET) Program".

Clark and Gordon had discussed in Annapolis, this was understandable. The letter did note that the CZCS project had definitely shown the potential usefulness of satellite ocean color data.

The CZCS NET journeyed back to Maine in August for another meeting at the Bigelow Laboratory, the 23rd meeting of the instrument team.[191]

Up in space, the CZCS had started to experience an occasional operational anomaly.[192] As has been noted, Nimbus 7 shared power generated by its solar wings amongst all the instruments it carried. For that reason, all the instruments had to be commanded when they would turn ON and when they would turn OFF. Unfortunately, CZCS, like a recalcitrant tot at bedtime, had sometimes refused to turn ON when asked to do so, and started to require repeated telemetric entreaties before spinning up and commencing its scanning operations. The problem was traced to the mechanism which compensated for the momentum of the spinning scan mirror. If the scan mirror spin was not compensated perfectly, the imbalance would cause the satellite to pitch and yaw excessively, requiring thruster correction so that the sensors were properly pointed. According to Randy McConaughey, "As I recall, the scan mechanism consisted of the scan mirror, driven by a brushless dc motor/resolver, a momentum compensator (just a mass) also driven by a brushless dc motor/resolver in the opposite direction, and an encoder. The encoder was for controlling the speed of the scan mirror, not for positional information as is usually thought of with encoders. We built the encoder in-house with purchased parts – a glass encoder disc with fine lines deposited (I believe) and a LED and photo diode. The frequency of the output from the photo diode was fed into a phase locked loop to control the speed of the scan mirror. I do not remember what drove or controlled the momentum compensator, but maybe a separate track on the encoder disc. During turn-on, the speed of the scan mirror and the momentum compensator were controlled and slaved together to minimize the net momentum imparted to the spacecraft."[193]

As the satellite aged, the power going to the LED diminished, especially during the low power season when the sunlight reaching the solar wings was at low ebb. And the LED was also degrading, producing less light. So sometimes when turning on the instrument, even when the lines on the encoder disc were aligned perfectly, the detector did not receive enough light to allow the instrument to power up, and this caused the instrument start-up process to be shut down. The shutdown command was automatic, and there was no capability of overriding it with a manual command.[194]

So the spacecraft engineers had to come up with ways to get the LED to shine bright enough to convince the detector everything was working properly. The first way they did this was simply to increase the voltage to the LED, causing it to brighten when the ON

[191] James G. Acker, "The Heritage of SeaWiFS: A Retrospective on the CZCS NIMBUS Experiment Team (NET) Program".
[192] James G. Acker, "The Heritage of SeaWiFS: A Retrospective on the CZCS NIMBUS Experiment Team (NET) Program".
[193] Randy McConaughey, email message, received 23 May 2009. Gene Feldman initially described watching telemetry of a switch-on attempt late in the mission.
[194] Gene Feldman, personal anecdote, telephone conversation, 16 September 2008.

command was sent. This procedure initially worked quite well.[195]

In December 1984, the article "A Multidisciplinary Oceanography Program on the Southeastern U.S. Continental Shelf" appeared in the American Geophysical Union weekly newspaper *Eos*.[196] This article featured a CZCS image of the Gulf Stream and Charleston Bump. So at the end of 1984, more CZCS images were starting to appear in print. The use of CZCS data for oceanographic research was commencing.

Charles McClain provided more context about this image. It was acquired in conjunction with a cruise out of the Skidaway Institute of Oceanography, and McClain noted "there was a perfect CZCS scene over the cruise". They applied the Rayleigh scattering correction and Gordon's algorithm, and in McClain's words "we processed that scene and this Gulf Stream frontal feature jumped out … the chlorophylls went from 0.1 to 7 or 8 [milligram per cubic meter] with three or four kilometers along the front … we read off the chlorophyll values and they matched almost exactly what the ship had recorded … I just couldn't believe it." [197]

1985-1986: The Years of Crossed Fingers and the Open Ocean

As the mission moved into its seventh year, every ON command to the CZCS became a balancing act worthy of the Great Wallendas. The spacecraft telemetry indicated when the momentum wheels were spinning synchronously, so that the ON command could be sent – and the satellite operators then waited to see if the detector would see the LED and allow the CZCS to power up. Sometimes the commands would have to be sent several times, with the engineers increasing the voltage to the LED and even warming up the satellite itself to up the power a few more critical millivolts. If the CZCS refused the command, the process would have to be restarted. The instrument also began to give the controllers additional heartburn when it started to spontaneously shut down.

While the instrument operations became problematic, the CZCS NET met for a final time at Goddard in May 1985. Item One on the agenda was the operation of the instrument. Brighter notes were sounded on the subjects of data product production, the CZCS Atlas, and archive plans.[198]

Because of the behavior of the instrument, the Open Ocean coverage effort had been green-lighted, with John Sissala guiding the data acquisition. The goal of this effort was to acquire at least one image over the entire global ocean. It was already known that large

---

[195] Gene Feldman, personal anecdote, telephone conversation, 16 September 2008.

[196] J. O. Blanton, James A. Yoder, Lawrence P. Atkinson, Thomas N. Lee, Charles R. McClain, D. W. Menzel, G. A. Paffenhofer, L. J. Pietrafesa, L. R. Pomeroy, and H. L. Windom, "A Multidisciplinary Oceanography Program on the Southeastern U.S. Continental Shelf", *Eos, Transactions, American Geophysical Union*, 65: 1202-1203, (December 1984).

[197] Charles McClain, comments recorded at the Ocean Color Collaborative Historical Workshop, January 13-14, 2009, St. Petersburg, Florida.

[198] James G. Acker, "The Heritage of SeaWiFS: A Retrospective on the CZCS NIMBUS Experiment Team (NET) Program".

expanses of the oligotrophic ocean basins didn't have much chlorophyll to detect – that's one of the characteristics of oligotrophic waters – but having a full global image would put the areas of higher productivity in global context. The data processing for this effort would be accomplished utilizing software from the University of Miami, according to Wayne Esaias. And despite the publicity and acclaim accorded to the chlorophyll + pigment product, it was also recommended that the K(490) measurements be continued as well. At the end of the meeting, the CZCS NET voted in Esaias and Charles McClain as associate members.[199]

[Dennis Clark described his experience with the NET by saying that "it was the best experience I ever had… it was competitive, there were egos… there was a free exchange of views, it functioned as a team. … Dealing with a problem, problems would come up, they'd try to fix them right away". He also noted that the CZCS NET resources in the United States were primarily devoted to making the instrument work. The one problem with this devotion to instrumental duty was that the scientific endeavor was secondary. In contrast, a consortium of sorts was formed in Europe featuring some members of the NET, and this team, which Clark indicated was well-funded, could be more dedicated directly to the science. The NET meeting in Brussels in 1976, which was coordinated with EURASEP, featured this European CZCS science group.][200]

The instrument behaviors persisted into 1986. Due to the increasing difficulty of turning on the instrument, on March 9 the CZCS was given the highest observational priority on Nimbus 7, attempting to finish the Open Ocean coverage effort, and operated continuously into June, when the low power season forced a shutdown. The Nimbus 7 project planned to attempt to turn the CZCS back on in December when the power budget was better, but the attempts failed, and the CZCS mission was declared over on December 18, 1986.[201]

Even though the CZCS stopped functioning in June 1986, its success was gaining attention. Mary Jane Perry described the potential use of satellite ocean color data to assess marine primary productivity in *Bioscience* in the summer of 1986.[202] An article in *Weatherwise* published in October 1986 described the extremely accurate measurement and images of ocean color provided by the CZCS.[203]

1987: Not an End, But a Beginning

The observational phase of the CZCS experience ended in December 1986. Warren

---

[199] James G. Acker, "The Heritage of SeaWiFS: A Retrospective on the CZCS NIMBUS Experiment Team (NET) Program".
[200] Dennis Clark, interview transcript notes.
[201] NASA GES DISC, "Coastal Zone Color Scanner (CZCS) Instrument Guide".
[202] Mary Jane Perry, "Assessing Marine Primary Production from Space," *BioScience*, 36(7), 461-467, (July-August 1986).
[203] Patrick Hughes, "Ocean Color—A Key to Climate Change," *Weatherwise*, 39(5), 267-270, (October 1986).

Hovis had led an unprecedented eight-year mission of ocean observation; certainly a personal triumph after his previous two missions had ended shortly after launch, and a dramatic success for a "proof-of-concept" mission which was only planned to function for a single year. As the mission ended, the data distribution process was being significantly improved, such that researchers could now obtain the data in some form of storage media and start processing it at their home institutions. Janet Campbell recalled that an entire room at the Bigelow Laboratory was filled with CZCS magnetic tapes.[204]

On the West Coast, Mark Abbott had commenced the creation of the West Coast Time Series data processing effort at the Jet Propulsion Laboratory, using the DSP software developed at the University of Miami, using both CZCS ocean color and AVHRR sea surface temperature data collected by the Scripps satellite data receiving facility. Stan Wilson, the head of NASA's Oceanic Processes program, said that Nierenberg (the Scripps director) and Buzz Bernstein "wanted to get something going", so they utilized the Scripps downlink. While at ONR prior to coming to NASA, Wilson had helped fund the establishment of this downlink in 1978, and he contributed to it further once at NASA. The West Coast Time-Series allowed an early opportunity to bypass the NET and get CZCS data directly into the hands of users; it was also not bad politics to have that facility, and most of its users, located at one of the nation's leading oceanographic research institutions.[205] This program allowed rapid progress on understanding data processing, and both algorithms and processing were aided by the Scripps VisLab. Wayne Esaias remarked that creating the West Coast Time Series took "guts" on Wilson's part, and there were personality clashes, but Wilson clearly "didn't like data piling up".[206] When Abbott took approximately 50 data tapes with him from Scripps to JPL in the trunk of his car, this journey constituted one of the fastest satellite remote-sensing data transfer rates to that date.[207] Abbott calculated that he transported more than 2,000 9-track data tapes during his weekly drives from La Jolla to Pasadena. He also noted that the original data, collected by Ray Smith and Karen Baker, was easier to work with, as it was a 10-12 minute swath corresponding to a satellite overpass, not the 2-minute scenes produced by the Nimbus data processing system. Abbott said that the West Coast Time Series had data that was never in the Nimbus archive.[208]

The minutes of the final CZCS NET meeting in 1985 indicate that the CZCS Atlas was close to printing. This atlas, entitled "Nimbus-7 CZCS: Coastal Zone Color Scanner Imagery for Selected Coastal Regions", contains many different color and grayscale images

---

[204] Janet Campbell, recorded interview, 14 July 2009.

[205] Stan Wilson, email comments received 7 January 2010.

[206] Wayne E. Esaias comments recorded at the Ocean Color Collaborative Historical Workshop, January 13-14, 2009, St. Petersburg, Florida; Stan Wilson recorded interview, 18 March 2009.

[207] Mark Abbott, recorded interview, 26 November 2008. Donald Collins of the JPL Physical Oceanography Distributed Active Archive Center (PODAAC) first told the author about this high data transfer rate in the trunk of Abbott's car at an American Geophysical Union Ocean Sciences meeting in New Orleans, Louisiana in February 1992.

[208] Mark Abbott, recorded interview, 26 November 2008.

with accompanying short text descriptions of what is being observed in the images.[209] The subjects cover Gulf Stream eddy systems, the California Current, island effects around Vancouver Island, the Amazon River outflow, and the Agulhas Current off of South Africa. The Agulhas current images are somewhat unique as they show both the sea surface temperature and suspended solids data products from CZCS, products which were never widely utilized. Fittingly, the atlas featured a Foreword from Gifford C. Ewing,

So as the mission ended, the scientific pressure was increasing, consisting of data and images awaiting analysis. That pressure would soon break through the dual barriers of data processing and the preparation and publication process of scientific research, resulting in a onrush of scientific results from the pioneering CZCS mission.

Summary: The success of the CZCS was not unexpected, but it was unprecedented. Not only were the scientific cognoscenti uncertain that the instrument would actually provide data of value, they were uncertain if the investment was worth it, even close to the time of its launch on Nimbus 7. Scientists and engineers that were part of the CZCS mission recognize now that what they were doing was truly pioneering; at the time, they were motivated by the scientific and technical challenges, and not by some greater sense of purpose. The CZCS was part of a larger NASA tradition of undertaking difficult and risky challenges, and the mission provided much more return than even its participants expected, even though the public was only marginally aware of its significance. The CZCS proved the worth of ocean color remote sensing from space: the view of scientific history is that the CZCS was a somewhat serendipitous intersection of technological achievement and remarkable scientific discovery. One of the main lessons of the CZCS was that part-time views of the ocean, while providing a much better view of large oceanic regions from the vantage point of space, were insufficient to truly capture the dynamics interplay of physical forces and biological response – so the next step was to discover all that could be learned from the data, while at the same time conceiving and launching a full-time global ocean color instrument.

[209] NASA Goddard Space Flight Center, *Nimbus-7 CZCS: Coastal Zone Color Scanner Imagery for Selected Coastal Regions*, prepared for NASA GSFC by the Walter A. Bohan Company, Park Ridge, Illinois under contract NAS 5-29079, 99 pages and front matter, no publication date.

Additional CZCS pictures

Dr. Warren Hovis is partially obscured, second from left in this picture, at a 1971 teacher's workshop. This picture is from the September 1971 issue of the Goddard News, http://library01.gsfc.nasa.gov/goddardnews/September_1971.pdf, accessed January 6, 2010.

CZCS test set-up for the Foothills image, from Steve Varlese.

Ball Brothers CZCS engineering staff: (left to right) Bill Frank, Project Manager Bill Wallschlaeger, John Lohman (contracts), standing: Howard Vogt (Configuration and Data Management). Picture provided by Randy McConaughey.

Ball Brothers CZCS engineering staff: (left to right) Rich Wolfkiel, production technician; George Hicks, lead production engineer; Jerry McCoy, production engineer for electronics; Ken Lersen, machinist. Standing is Randy McConaughey, system engineer. Picture provided by Randy McConaughey.

Ball Brothers CZCS engineering staff: (left to right) Bob Loomis, lead mechanical design; Ken Hegy, data management/program administration; the third person could not be identified. Picture provided by Randy McConaughey.

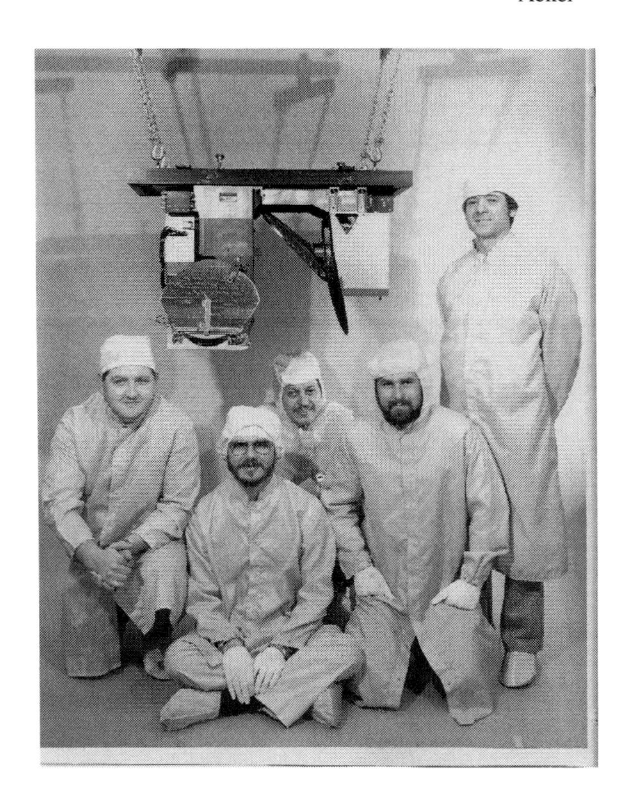

Ball Brothers CZCS engineering staff: (left to right) Gary Griffith, technician; Dave Giandinato, electronics technician; Bill (Eloy) Roybal, optics testing and design; Steve Varlese, optical alignment and testing; standing, Joe Gallegos, program test engineer. Picture provided by Randy McConaughey.

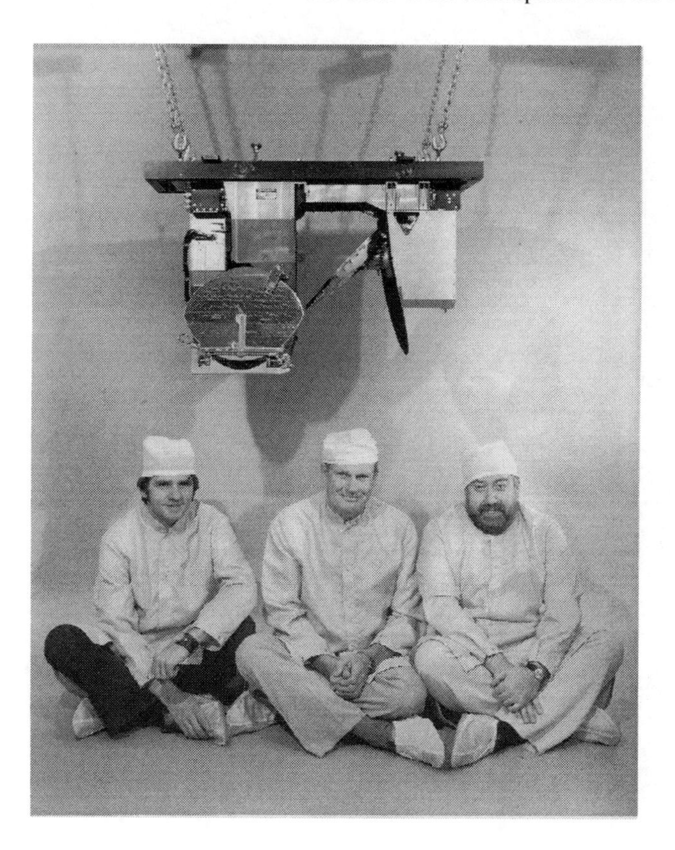

Ball Brothers CZCS engineering staff: (left to right) Joe Pem, configuration management; the middle individual could not be identified; Gavin Noble, quality assurance.

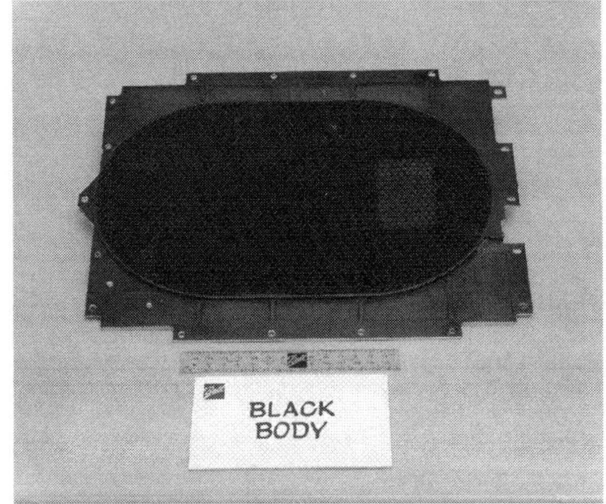
BLACK
BODY

Pictures of the CZCS: (previous page, top) Close-up of the scanning mirror assembly; (previous page, middle) black body calibration target; (previous page, bottom) power supply; (above) the assembled CZCS instrument protoflight model – the instrument which flew on Nimbus 7.

4

ADVANCING THE SCIENCE

1986 was not an easy year for NASA, primarily due to the disastrous post-launch explosion of the Space Shuttle *Challenger* on January 28, 1986.[1] The *Challenger* disaster had several ramifications for NASA space and earth science. One of the effects was on the Jupiter Galileo mission; the storage and transport of the satellite which was to be launched in May 1986 (and which finally launched in 1989) probably led to the failure of the probe's high-gain antenna to deploy properly.[2] The *Challenger* disaster also ended plans to launch Space Shuttles from Vandenberg, terminating a program that had expended a somewhat prodigious amount of money. Yet the availability of the Shuttle and its large payload capacity had already inspired NASA engineers and project planners to "think big" in terms of what could be launched into space. Ideas and plans conceived during this time would shape NASA's next ocean color missions.

At the same time, NASA was now in possession of a treasure trove of CZCS data that was highly desired by oceanographers; and they were facing the dual challenges of improving the data quality and making the data more readily available to researchers.

When Janet Campbell was working in the "good old days" of CZCS data at Bigelow Laboratory, an entire room was devoted to the storage of over a thousand CZCS data tapes that were delivered to Charlie Yentsch as a member of the CZCS NET.[3] Campbell recalls that they had boxes and boxes of tapes with just a 25-character mailing label. So in a room donated by the National Marine Fisheries Service, furnished with racks of tapes, a terminal, and a tape reader, a high school student that Campbell hired would load the tape, type a READ command, and read the data off the tape. The student loaded all of the metadata (indicating the location of the data on the tape) into a database, so that if the researchers wanted to look at data from a particular location, they could look in the database, find the number of the tape, and load it for analysis.[4] There was no way at the time to load the actual data onto a different storage medium. This recollection illustrates how the treasured tapes were primarily in the possession of the NET members, and not in formats that were conducive to easy use – and it also illustrates why it took considerable time to garner

---

[1] NASA, "Mission Archives: STS-51L", http://www.nasa.gov/mission_pages/shuttle/shuttlemissions/archives/sts-51L.html, (accessed 23 April 2009).

[2] Michael Meltzer, "The High-Gain Antenna Failure: A Disappointment and a Challenge", Chapter 7 in *Mission to Jupiter: A History of the Galileo Project, NASA SP-2007-4231*, NASA History Division, Washington, DC, 171-186, (2007).

[3] Janet Campbell, recorded interview, 14 July 2009.

[4] Janet Campbell, recorded interview, 14 July 2009.

scientific results from CZCS data.

At this stage in the history of the CZCS data, and more generally NASA Earth remote sensing data, it is important to note that much of what was becoming possible to accomplish with the data was being facilitated by the rapid advances taking place in computer technology. This aspect of the science is a continuing theme over the course of NASA's ocean color missions; each progressive step in computational technology accrued benefits that were rapidly assimilated and utilized by the scientists working with the data. Moore's Law was affecting their work and decisions continually (Moore's Law basically states that the number of transistors that can be placed on an integrated circuit, also known as a "computer chip", increases exponentially, essentially doubling every two years).[5] While it would seem that in the late stages of the 21st century's first decade the limits of nanotechnology on computer chips may have slowed the pace of Moore's Law implementation, during the 1980s, with the noteworthy of assistance of Steve Jobs and Bill Gates, Moore's Law was in full force.

During the period 1981-1982, Kendall Carder from the University of South Florida was the first manager of NASA's Ocean Productivity Program.[6] Carder recruited Wayne Esaias as his successor in the position from 1982-1984; Esaias came to NASA Headquarters from the NASA Langley Research Center in Hampton, Virginia. Esaias interacted with the CZCS and the NET while at NASA HQ, and moved to GSFC in 1984 to continue his ocean color research with the NET members and the Nimbus data processing team.[7] His move to GSFC contributed to creating a "critical mass" of personnel at GSFC dedicated to ocean color remote sensing science. In addition to the Nimbus team, Charles McClain was also at Goddard, having worked with the Ocean Color Experiment on the second Space Shuttle flight. Soon after Esaias began working at GSFC, Gene Feldman also joined the ocean color team (in 1985).[8] McClain remembers that when Esaias arrived at Goddard, "he said the first thing we had to do was reprocess the CZCS data set … that was a monumental task." [9]

OCE on OSTA-1

While the CZCS was the pioneering ocean color instrument of the Nimbus program, NASA had one other important, though very short-lived, ocean color mission during the

[5] Gordon E. Moore, "Cramming more components onto integrated circuits," *Electronics*, 38(8), 114-117, (19 April 1965).
[6] "Kendall L. Carder, College of Arts and Sciences, Curriculum Vitae", http://www.marine.usf.edu/faculty/documents/carder-kendall-cv-10-2007.pdf, acquired 23 April 2009.
[7] "Dr. Wayne Esaias", http://neptune.gsfc.nasa.gov/STAFF/brianc/iceesaias.html, (accessed 23 April 2009).
[8] "gene carl feldman", http://oceancolor.gsfc.nasa.gov/staff/gene/, (accessed 23 April 2009).
[9] Charles McClain, comments recorded at the Ocean Color Collaborative Historical Workshop, January 13-14, 2009, St. Petersburg, Florida.

1980s: the Ocean Color Experiment (OCE) on the second flight of the Space Shuttle, OSTA-1 (Office of Space and Terrestrial Applications – 1).[10] OSTA-1 was also STS-2, flown with astronauts Joseph Engle and Richard Truly. STS-2 launched at 10:10 AM EST on November 12, 1981 – the launch was originally scheduled for 7:30 AM, and this delay affected the data acquisition by the OCE.[11] The OCE was a radiometer originally built for aircraft deployment that was modified for space to fit on the Spacelab pallet. In addition to the OCE, STS-2 also carried four other Earth remote sensing experiments.[12] These instruments were:

- Measurement of Air Pollution from Satellite (MAPS)
- Shuttle Multispectral Infrared Radiometer (SMIRR)
- Shuttle Imaging Radar (SIR-A), and
- Features Identification and Location Experiment (FILE)

The OCE had eight radiometric bands, closely spaced in the range 486 – 786 nanometers. The 486 nm band, though it was not located at the peak absorption wavelength for chlorophyll (443 nm) was used for chlorophyll absorption. Subsequent algorithms for plankton chlorophyll utilize the 490 nm band when concentrations are very high, usually in the coastal zone.[13]

The OCE mission illustrates the difficulty of conducting short-term oceanic observations from space, and it also illustrated how difficult it was to utilize the Shuttle as an Earth observational platform. Visible wavelength observations of the ocean, and to a lesser extent IR wavelength observations used for sea surface temperature, are bedeviled by the presence of clouds. Gathering a full view of the oceans without the presence of clouds requires repeated observations and is aided by the use of multiple observations and statistical averaging techniques (which will be a prime subject of subsequent discussion).

The OCE collaborated with an impressive team of oceanographers for data validation. At Goddard, the OCE team consisted of McClain, Hongsuk Kim, Norden Huang, Robert Fraser, and L. Blaine, and they collaborated with Heinz van der Piepen of the DFVLR, the Deutsche Forschungs und Versuchsanstalt fur Lüft und Raumfahrt. The OCE team set up

[10] NASA Space Shuttle Launch Archive, "STS-2", http://science.ksc.nasa.gov/shuttle/missions/sts-2/mission-sts-2.html, (accessed 23 April 2009).

[11] Heinz van der Piepen, H.H. Kim, W.D. Hart, V. Amann, H. Helbig, Armando F.G. Fiuza, Michel Viollier, and Roland Doerffer, "The Ocean Color Experiment (OCE) on the Second Orbital Flight Test of the Space Shuttle (OSTA-I)," *IEEE Transactions on Geoscience and Remote Sensing*, GE-21(3), 350-357, (July 1983).

[12] NASA Space Shuttle Launch Archive, "STS-2".

[13] John E. O'Reilly, Stephane Maritorena, David A. Siegel, Margaret C. O'Brient, Dierdre Toole, B. Greg Mitchell, Mati Kahru, Francisco P. Chavez, Peter Strutton, Glenn F. Cota, Stanford B. Hooker, Charles R. McClain, Kendall L. Carder, Frank Muller-Karger, Larry Harding, Andrea Magnuson, David Phinney, Gerald F. Moore, James Aiken, Kevin R. Arrigo, Ricardo Letelier, and Mary, Culver, "Ocean Color Chlorophyll a Algorithms for SeaWiFS, OC2, and OC4: Version 4", Chapter 2 in Stanford B. Hooker and Elaine Firestone, editors, *SeaWiFS Postlaunch Technical Memorandum Series, NASA/TM-2000-206892*, Volume 11, pages 9-23, (October 2000).

data validation collaborations around the world, including observations off of Portugal, the northwest Atlantic, the South Atlantic Bight, coastal Costa Rica, and the Kuroshio Current. For the northwest Atlantic study, the lead researcher was Peter Wiebe of WHOI, who had extensively researched Gulf Stream eddy systems. In the South Atlantic Bight, the lead scientists were James Yoder and Larry Atkinson, both then located at the Skidaway Institute of Oceanography. Victor Klemas of the University of Delaware intended to examine the data for Costa Rica.[14]

The comic strip *Li'l Abner*, drawn by Al Capp, included a character named "Joe Btfsplk", known as the world's worst jinx.[15] Joe Btfsplk was always shown with a little storm cloud over his head that followed him everywhere he went, and wherever he went, things went wrong. For the OCE mission, Joe Btfsplk could have been selected as the instrument mascot.

As noted, STS-2 launched three hours late. This resulted in a significant change in the solar zenith angles (see Appendix 1) over the study areas at the time of the instrument overflight. So data from alternate orbits, not necessarily in the most fortuitous viewing location, had to be selected for the data validation effort. Furthermore, the Sun is low in the Northern Hemisphere in November, limiting the observations that could be utilized. The mission was originally scheduled to be launched a month earlier in October, but a nitrogen tetroxide spill moved the launch to November. This delay caused the mission to coincide with storm systems over both the East and West Coast of the United States, covering some of the areas that were planned for imaging. And because this was only the second flight of the Space Shuttle, malfunctions were expected, and they happened as expected, resulting in a two-day reduction in the planned four-day mission duration. The result of all these "Btfsplk" events was that the OCE only collected about 20-30 minutes of usable data out of a total of 118 minutes of observing time.[16]

Despite the litany of problems, the OCE did accomplish several noteworthy achievements. The OCE demonstrated the efficacy of ocean color imaging over deep water, low chlorophyll targets; at the time, the CZCS was still primarily observing the coast and had not yet been tasked to attempt global coverage including the open ocean basins. The OCE showed the effectiveness of conducting dedicated in-water observations at the same time as remote sensing observations. Where the ocean was visible, the OCE was able to demonstrate that the movement of high chlorophyll regions could be tracked, indicating ocean circulation. The mission detected high sediment concentrations in the plume of the Yangtze River. And the data was also used to calculate water depth on the Grand Bahamas

---

[14] Heinz van der Piepen, Hongsuk H. Kim, W.D. Hart, V. Amann, H. Helbig, Armando F.G. Fiuza, Michel Viollier, and Roland Doerffer, "The Ocean Color Experiment (OCE) on the Second Orbital Flight Test of the Space Shuttle (OSTA-I)".

[15] Lil' Abner Official Web Site, "Other Characters", http://www.lil-abner.com/other.html, (accessed 24 April 2009).

[16] Heinz van der Piepen, Hongsuk H. Kim, W.D. Hart, V. Amann, H. Helbig, Armando F.G. Fiuza, Michel Viollier, and Roland Doerffer, "The Ocean Color Experiment (OCE) on the Second Orbital Flight Test of the Space Shuttle (OSTA-I)".

Bank, an application of ocean color data that is still a subject of research for both military and geological investigations.[17]

In a 1984 proceedings paper, lead author Hongsuk Kim and his co-authors compared OCE and CZCS data using a comparison of observations acquired on November 14, 1981 over the Gibraltar Strait. This study allowed a determination of the decline in CZCS sensitivity since the beginning of the mission.[18]

Critical Decisions

Moore's Law doesn't address data storage media, but it might also apply; advances in computational technology in the 1980s were paralleled by advances in data storage media. Think back – do you remember the truly "floppy" 8-inch or 5 ¼ inch diameter floppy discs that preceded the hard case 3 ½ inch floppies? Do you remember how fragile these storage media could be? During the early 1980s, NASA was faced with critical decisions about how to transfer and reliably store the CZCS (and other Nimbus instrument data) on CRTT tapes. At the same time, critical decisions were required regarding how to process the data with the best available algorithms to facilitate effective oceanographic science. In the background to these decisions was another factor; demonstrating the pressing need for a follow-on to the CZCS mission.[19]

One of the data storage advances that became available during this crucial period was the optical platter. Very few people are unfamiliar with the Compact Disc – Read-Only Memory data storage system (also known as the CD-ROM, or just CD). Yet if you weren't intimately involved with the development of data storage technology in the 1980s, you may never have seen or heard of the CD-ROM's important predecessor, the optical disc.[20] Storage of data on optical discs occurred in conjunction with the short lifetime of the Laserdisc used for home video.[21] During the 1980s, with CD-ROM data storage still in its infancy, the optical disc was a potential data storage medium that could hold significant amounts of data. Philips and Sony brought optical disc technology into use for audio

---

[17] Heinz van der Piepen, Hongsuk H. Kim, W.D. Hart, V. Amann, H. Helbig, Armando F.G. Fiuza, Michel Viollier, and Roland Doerffer, "The Ocean Color Experiment (OCE) on the Second Orbital Flight Test of the Space Shuttle (OSTA-I)".

[18] Hongsuk Kim, Heinz van der Piepen, Michel Viollier, R. Fiedler, and S. van der Piepen, "Radiometric comparison of two ocean color scanners Nimbus-7/CZCS and OSTA-1/OCE", in *Proceedings, 18th International Symposium on Remote Sensing of Environment*, Volume 2, Paris, France, October 1-5, 1984, (A86-21101 08-43). Ann Arbor, MI, Environmental Research Institute of Michigan, 931-937, (1985).

[19] James A. Yoder, comments recorded at the Ocean Color Collaborative Historical Workshop, January 13-14, 2009, St. Petersburg, Florida.

[20] "Optical Disc", http://searchstorage.techtarget.com/sDefinition/0,,sid5_gci811276,00.html, (accessed 24 April 2009).

[21] Andy Hain, "Optical Disc", http://www.totalrewind.org/disc/disc_opt.htm, (accessed 24 April 2009).

recording and digital recording in the late 1970s and early 1980s.[22] The optical platter offered a more durable storage technology and a significantly higher data volume capability; in 1984, optical platters could store 1 Gigabyte of data.

Lacking the benefit of clear technological prescience, the ocean color data team decided to transfer the data from the CRTT tapes to optical platters. But there was a more crucial decision to be made than data storage technology – how to process the data to produce a uniform, consistent, global CZCS pigment data product. Over the course of the mission, various groups had independently developed their own systems and software (not to mention algorithms) to process the CZCS data. The primary system where the data resided was still the Nimbus data processing system, but independent systems had also been developed at Scripps and RSMAS.[23] So NASA contemplated a decision on what algorithms to use, what the algorithms had to do to produce a global data product, and what processing system to use to create the data product, which would then be placed on optical discs.

After considering the available options, Esaias and Nimbus 7 Project Scientist Al Fleig proposed to perform the global data processing in collaboration with the researchers and computers at RSMAS and the University of Miami. RSMAS had already built the DSP (digital signal processing) system and was processing CZCS data. When the proposal was accepted, GSFC and RSMAS undertook the task of creating a global algorithm for processing all the data, then processing the tapes with each individual CZCS scene (a raw number of 67,789) after screening for bad data, and then applying the global algorithm. Less than 10% of the available CZCS data were excluded from the processing.[24]

[There was, in fact, one other crucial step that had to be performed first – determining how much CZCS data was actually available. Gene Feldman described that all of the scenes had been printed on "laserfax" ribbon tapes, which was the only way to see what was available. He said "you could barely see anything just land and clouds". So each laserfax ribbon, in an archive maintained by John Sissala, was examined to find areas of open ocean, and the data was acquired for the global data processing effort, because much of it had never been requested. Feldman estimated "we probably got another 50% of the data before they shut down the Nimbus program."][25]

The creation of a global algorithm required a few hard swallows, particularly regarding the acceptance of a few assumptions that would enable a uniform processing effort. The two main assumptions were the establishment of a fixed epsilon (Appendix 1), which basically meant that a uniform standard atmosphere with a generalized aerosol concentration

---

[22] Sony History, "Chasing and Being Chased", http://www.sony.net/Fun/SH/1-20/h1.html, (accessed 24 April 2009).

[23] Stan Wilson recorded interview, 19 March 2009; University of Miami/RSMAS Remote Sensing Group, "DSP Users Manual", (13 February 1991).

[24] Gene C. Feldman, Norman Kuring, Carolyn Ng, Wayne Esaias, Chuck McClain, Jane Elrod, Nancy Maynard, Dan Endres, Robert Evans, Jim Brown, Sue Walsh, Mark Carle, and Guillermo Podesta, "Ocean color - Availability of the global data set", *Eos, Transactions of the American Geophysical Union*, 70(23), 634-635 and 640-641, (1989).

[25] Gene Feldman, comments recorded at the Ocean Color Collaborative Historical Workshop, January 13-14, 2009, St. Petersburg, Florida.

was used for all the scenes. The researchers knew that there would be places where the fixed aerosol assumption would cause errors, particularly in areas impacted by high levels of dust (like near the Sahara and Gobi deserts) and areas impacted by high concentrations of human pollution, notably the East Coast of the United States.[26] The other necessary assumption was the acceptance of a single algorithm for the CZCS pigment concentration using the normalized water-leaving radiances. The algorithm which was accepted was from Clark's 1981 paper, which had been based on data clearly lacking global coverage, as it was derived from the primary NET sampling stations.[27] By this time, however, research had established that Clark's algorithm worked pretty well in other regions.[28]

With the algorithms in place, processing commenced. The processing converted the calibrated radiances at 0.825 km spatial resolution and subsampled them, using a lyrical litany that would become familiar to subsequent missions, "every fourth pixel, every fourth scan line". This method of subsampling created the 4km Level 1A GAC product. The algorithms for pigment and K490 and normalization, to calculate the normalized water-leaving radiances, were then applied to produce the Level 2 GAC product.[29]

With the Level 2 daily scenes available, then the true fun began. The data were then converted to a daily global grid using a process called "binning". In essence, each grid point can be envisioned as a hopper accepting all the data which was acquired at that point on the Earth. The data are put into the hopper, averaged, and an average data value for that point is calculated, to produce a global product with 20 km resolution. First the data is binned at a daily temporal resolution, then at weekly, monthly, and annual temporal resolution. Using the DSP system, these composite data files were converted to a PST ("Postage Stamp") format for the archive.[30]

After the processing, the data was then put onto the optical platters. When we read about the significant storage gains that were obtained by transferring the data to the platters, and think about how much data we can now carry around in a standard laptop, we may still consider the march of computational capability with a tinge of amusement regarding what was state-of-the-art as the 20th century entered its final decade:

---

[26] Gene Feldman, comments recorded at the Ocean Color Collaborative Historical Workshop, January 13-14, 2009, St. Petersburg, Florida.

[27] Dennis K. Clark, "Phytoplankton algorithms for the NIMBUS-7 CZCS".

[28] Vittorio Barale and R.Wittenberg Fay, "Variability of the ocean surface color field in central California near-coastal waters as observed in a seasonal analysis of CZCS imagery," *Journal of Marine Research*, 44, 291-316, (1986); Gene C. Feldman, Dennis Clark, and David Halpern, "Satellite color observations of the phytoplankton distribution in the eastern Equatorial Pacific during the 1982--1983 El Niño".

[29] Gene C. Feldman, Norman Kuring, Carolyn Ng, Wayne Esaias, Chuck McClain, Jane Elrod, Nancy Maynard, Dan Endres, Robert Evans, Jim Brown, Sue Walsh, Mark Carle, and Guillermo Podesta, "Ocean color - Availability of the global data set".

[30] Gene C. Feldman, Norman Kuring, Carolyn Ng, Wayne Esaias, Chuck McClain, Jane Elrod, Nancy Maynard, Dan Endres, Robert Evans, Jim Brown, Sue Walsh, Mark Carle, and Guillermo Podesta, "Ocean color - Availability of the global data set".

"30,000 magnetic tapes in a warehouse have been reduced to three bookcases of optical discs adjacent to a computer workstation"

Nonetheless, the optical platter system made it feasible for a research institution to acquire the global CZCS dataset. Realizing that not everyone wanted to travel to Greenbelt to look at the CZCS data, NASA created a nationwide network of archive sites. These sites were the University of Rhode Island, the Bigelow Laboratory for Ocean Sciences, the University of Miami, the University of South Florida, the University of California – Santa Barbara, the Jet Propulsion Laboratory, Oregon State University, and the University of Washington.[31]

Even though the archive sites had been established, there was still one large missing piece of the data archive puzzle – a way for an oceanographer to search for CZCS scenes for a particular area and time period. Now, there has been at least one person who could scan the grooves of a phonographic disc and identify the piece of classical music that was recorded on it, but human eyes do not have this capability for laser-etched optical platters. So the next step in the process was the creation of a CZCS data browsing system. For the purpose of browsing, CZCS pigment images of each data file were created, allowing users to view them quickly for a given time and place. The system, which was built when "electronic mail" was still coming into vogue, allowed researchers to select files of interest and send a request to the data archive to get the full-resolution data for further examination and research. The files would be sent back to the researcher on magnetic tape. [A description of the browse system states that the computer "Should have ability to send Electronic mail to NASA". The PC systems were also required to have at least a 10 MB hard disk.] The browse program also allowed examination of the Level 3 images.[32]

Janet Campbell remembers that it took Bigelow Laboratory several years to get an NSF grant to buy their computing system. The price for such a system was coming down with new mini-computers, so the system was affordable for $100,000 dollars or so.[33]

The Explosive Period for CZCS Research

Over the period 1987-1989, NASA was involved with creating the global data set and archiving it more efficiently, and also creating the capability for researchers to much more readily access the data. These efforts were intended to greatly increase the use of CZCS

---

[31] Gene C. Feldman, Norman Kuring, Carolyn Ng, Wayne Esaias, Chuck McClain, Jane Elrod, Nancy Maynard, Dan Endres, Robert Evans, Jim Brown, Sue Walsh, Mark Carle, and Guillermo Podesta, "Ocean color - Availability of the global data set".

[32] Gene C. Feldman, Norman Kuring, Carolyn Ng, Wayne Esaias, Chuck McClain, Jane Elrod, Nancy Maynard, Dan Endres, Robert Evans, Jim Brown, Sue Walsh, Mark Carle, and Guillermo Podesta, "Ocean color - Availability of the global data set".

[33] Janet Campbell, recorded interview, 14 July 2009.

data for research – and they did. The availability of the CZCS data for research, and the demonstration that true global ocean color observations could be achieved, was important for two reasons.

The first reason was the clear demonstration that increased data availability made the data vital for oceanographic research, rather than just a novelty with an image or two illustrating a concept or augmenting discussion of a study region.[34] Wayne Esaias noted that data on the spatial distribution of fluorescence sent to John Walsh (at Brookhaven National Laboratory) were "kind of a breakthrough". According to Esaias, Walsh spent an entire Christmas vacation constructing contour maps and coloring them by hand with colored pencils.[35] Stan Wilson said that Walsh and the science working group "defined what needed to be done".[36] Jim Yoder indicated that oceanographers Jim McCarthy, Peter Brewer, Walsh, and John Steele were strong proponents of the importance of CZCS.[37]

Walsh himself pointed out that there was a division between physicists, representing the physical oceanography community, and biologists representing bio-oceanography. Walsh indicated that the physicists recognized the value of time-series observations to time-varying phenomena. The physicists used biological data to observe phenomena such as plume structures, color fronts and dispersion patterns.[38]

The second reason was that demonstrating the global ocean color observations could be accomplished confirmed the need for a dedicated global ocean color sensor – and Chapter 5 describes the repeated efforts to build and launch one. [39]

The focus in the following section is on the science; the scientific topics for which CZCS data enabled research. To present this progress of science, first some papers from the three "interim years" (1986 to 1989) will be highlighted. Appendix 2 further describes CZCS-enabled research on different oceanographic provinces, processes, and phenomena for which the CZCS data provided dramatic new insights. The actual significance of the CZCS mission would be minimal were it not for the remarkably broad range of oceanographic topics that were investigated with CZCS data.

The Early Years (1986-1989)

Two papers from 1986 from researchers at Scripps highlighted the waters off of California and the California Current. Vittorio Barale and R. Wittenberg Fay of the Scripps

---

[34] Stan Wilson recorded interview, 18 March 2009.

[35] Wayne Esaias, comments recorded at the Ocean Color Collaborative Historical Workshop, January 13-14, 2009, St. Petersburg, Florida.

[36] Stan Wilson recorded interview, 18 March 2009.

[37] James A. Yoder, comments recorded at the Ocean Color Collaborative Historical Workshop, January 13-14, 2009, St. Petersburg, Florida.

[38] John J. Walsh, comments recorded at the Ocean Color Collaborative Historical Workshop, January 13-14, 2009, St. Petersburg, Florida.

[39] James A. Yoder, comments recorded at the Ocean Color Collaborative Historical Workshop, January 13-14, 2009, St. Petersburg, Florida.

VisLab performed a seasonal analysis off the central California coast, and Jose Pelaez (also known as J. Pelaez-Hudlet) teamed up with zooplankton expert John McGowan to look at phytoplankton pigments in the California Current.[40] In the latter study, the researchers observed boundary fronts in the current south of San Diego, including a semi-stable feature called the Ensenada Front. The concentration of pigment north of the front was three times higher than south of the front, and the front extended offshore nearly 500 kilometers. Pelaez and McGowan also noted that a basic similarity of the appearance of sea surface temperature and ocean color remote sensing images over broad areas was much less apparent when smaller areas were examined at higher resolution.

The year 1987 included some contributions from former NET members who were still working with the data (and with NASA) to improve the data. Howard Gordon and Diego Castaño examined the issue of multiple scattering effects for the CZCS atmospheric correction algorithm; this research would be important for the more advanced atmospheric correction algorithms used for subsequent missions.[41]

Ray Smith was the lead author of a study that integrated data from surface ships, overflying aircraft, and CZCS data.[42] This study examined the dynamics of features that had become familiar to researchers using the data to study the northern Atlantic Ocean and the Gulf Stream – Gulf Stream warm-core rings. Prior to this (and also subsequently), Peter Wiebe of Woods Hole had led a large number of research expeditions to determine how the warm-core rings "worked" – with the added capabilities of remote sensing, researchers were better able to determine how often they occurred, how long they lasted, and if the biological and physical oceanographic understanding of how they evolved and aged was borne out by synoptic observations. A Gulf Stream warm-core ring forms when a meander of the Gulf Stream lengthens and narrows northward, and then gets cut off from the flow of the current, in a fashion similar to the formation of an oxbow lake by a sinuous river in flat terrain. When the ring forms, it captures in its center a volume of warm water from the low-productivity waters lying south of the Gulf Stream, and the water circulation of cooler water, generally higher productivity waters from north of the current, flows around the central low-productivity, low-chlorophyll water mass.[43] The actual observational campaign took place in April 1982. Simultaneous airborne observations utilized the Airborne Oceanographic

---

[40] Vittorio Barale and R.Wittenberg Fay, "Variability of the ocean surface color field in central California near-coastal waters as observed in a seasonal analysis of CZCS imagery"; Jose Peláez and John A. McGowan, "Phytoplankton pigment patterns in the California Current as determined by satellite," *Limnology and Oceanography*, 31, 927-950, (1986).

[41] Howard R. Gordon and Diego Castaño, "Coastal Zone Color Scanner atmospheric correction algorithm: multiple scattering effects," *Applied Optics*, 26, 2111-2122, (1987).

[42] NASA Goddard Space Flight Center Distributed Active Archive Center, "Classic CZCS Scenes: Chapter 7: Gulf Stream Rings", http://daac.gsfc.nasa.gov/oceancolor/scifocus/classic_scenes/07_classics_rings.shtml, (accessed 24 April 2009).

[43] Raymond C. Smith, Otis B. Brown, Frank E. Hoge, K.S. Baker, Robert H. Evans, Robert N. Swift, and Wayne E. Esaias, "Multiplatform sampling (ship, aircraft, and satellite) of a Gulf Stream warm core ring," *Applied Optics,* 26, 2068-2081, (1987).

Lidar (AOL before AOL meant something else in the Internet era) to perform measurements of laser-induced chlorophyll fluorescence. April 1982 provided very clear days for observation, providing excellent CZCS and SST observations which could be correlated with the ship and aircraft data (Figure 4.1).

Despite the fact that ocean color remote-sensing science using satellite data was essentially at the toddler stage, a number of research groups published papers evaluating the potential of the data to allow estimation of oceanic primary productivity. Robert Bidigare published such an initial foray as a collaborative study in the first volume of the journal *Global Biogeochemical Cycles*, which included Ray Smith as a co-author.[44] In the same year, two Canadians who were destined to become key figures in the advancement of ocean color science in the 1990s, Trevor Platt and Marlon Lewis of Dalhousie University, collaborated on a paper discussing how remote-sensing could be utilized to estimate phytoplankton productivity.[45] Other persons who would be taking a lead role published an examination of various feature sizes on the southeastern continental shelf of the United States. In this paper, James Yoder and Charles McClain collaborated with Jackson Blanton and Lie-Yauw Oey of the Skidaway Institute of Oceanography, determining that Gulf Stream eddies were the dominant size class of features in this region.[46]

The French connection now included Annick Bricaud, Cecile Dupouy, and Herve Demarcq in addition to André Morel, and they utilized CZCS data to examine upwelling off of Mauritania and Senegal[47]. In another upwelling zone, this one off the California coast, J. Pelaez-Hudlet previewed an important application of ocean color data in a description of a red-tide bloom occurring in July-August 1980.[48] It's interesting to note that a paper describing observations of a red tide was published *seven years after it occurred*, in light of the current need to have ocean color data within hours to advise affected interests of the danger of a red tide bloom in its early stages of growth.

A new milestone in the burgeoning science of ocean color occurred in 1988; a research paper utilizing CZCS data published in *Nature* merited a commentary on the paper in the same issue. The paper described observations of a short-lived phytoplankton bloom

---

[44] Robert R. Bidigare, Raymond C. Smith, K.S. Baker, and John Marra, "Oceanic primary production estimates from measurements of spectral irradiance and pigment concentrations," *Global Biogeochemical Cycles*, 1, 171-186, (1987).

[45] Trevor Platt and Marlon Lewis, "Estimation of phytoplankton production by remote sensing," *Advances in Space Research*, 7, 131-135, (1987).

[46] James A. Yoder, Charles R. McClain, Jackson O. Blanton, and Lie-Yauw Oey, "Spatial scales in CZCS-chlorophyll imagery of the southeastern U.S. continental shelf," *Limnology and Oceanography*, 929-941, (1987).

[47] Annick Bricaud, André Morel, and Jean-Michel André, "Spatial/temporal variability of algal biomass and potential productivity in the Mauritanian upwelling zone, as estimated from CZCS data," *Advances in Space Research*, 7(2), 53-62, (1987); Cecil Dupouy, and Herve Demarcq, "NIMBUS-7 CZCS as a help for understanding the mechanism of phytoplankton productivity during upwelling off Senegal," *Advances in Space Research*, 7(2), 63-71, (1987).

[48] J. Pelaez-Hudlet, "Satellite irnages of a 'red tide' episode off southern California," *Oceanologica Acta*, 10, 403-410, (1987).

in the western Mediterranean Sea.[49] The location of this study could qualify for a geography bee question, i.e. "What is the name of the miniscule island located due east of the Straits of Gibraltar between Spain and Algeria, which is home to a rare limpet?" If you said *Isla de Alboran*, you have also provided the alternate name of this marine locale: the Alboran Sea. This is an area of complex physical and biological interactions, due to the constricted exchange of water between the Mediterranean and the Atlantic Ocean, giving rise to eddies, fronts, and the occasional soliton.[50] In this region, the normally low-productivity waters (the Mediterranean diet may be famous for its seafood component and the Mediterranean Sea famous for fish, but the waters are actually productively low due to a minimal input of nutrients from the surrounding land and rivers) are stirred by mixing waters and ponderous waves beneath the surface. In the Alboran Sea, the circulatory interactions can give rise to a sudden strong upwelling, providing nutrients for a quick and hungry proliferation of plankton. That occurrence is what the research paper documented with CZCS data; as commentator Paul Falkowski noted, using "a combination of skill and luck".[51] In many of the CZCS research papers, the serendipitous convergence of observational schedule with dynamical events in the ocean certainly played a part. Luck certainly played a part in these observations, as the researchers saw the bloom between May 8 and May 11, 1986 – which was during the final period of CZCS observations when the sensor was "ON" permanently. Falkowski noted how such an observation demonstrated how short-term events such as this one significantly influenced the total primary productivity for a region – a point that is still becoming clearer to ocean color researchers today.

Cecile Dupouy and her co-authors noticed another kind of bloom – this one due to blue-green algae (cyanobacteria) in the atoll-strewn central Pacific.[52] Cyanobacteria blooms are important for two reasons: one, they can be toxic (but usually more to dogs than to people if encountered on a beach) and two, they fix nitrogen. Most phytoplankton species get their vital nutrient nitrate from what is dissolved in water, but cyanobacteria and *Trichodesmium* get it from the air – a trick shared by legumes – which also puts nitrogen in the water for other species to exploit, an important element in subsequent research. The blooms described in the Lohrenz and Dupouy research studies are illustrations of the theme explored by Trevor Platt and his collaborator/wife Shuba Sathyendranath, as they examined oceanic primary production on local and regional scales in a paper published in *Science*.[53]

---

[49] Steven E. Lohrenz, Robert A. Arnone, Denis A. Wiesenburg, and Irene P. DePalma, "Satellite detection of transient enhanced primary production in the western Mediterranean Sea," *Nature,* 335, 245-247 (1988).

[50] "Solitons: Alboran Sea", Image and caption in *Oceanography from the Space Shuttle*, Web publication of book published by the University Corporation for Atmospheric Research and the Office of Naval Research, July 1989 (published online August 1996). Image S17-34-81 (100 mm) from Flight STS-41G; Location 36N, 5W, Time 12:21:48 GMT, date 10 Oct 1984.

[51] Paul Falkowski, "Ocean productivity from space," *Nature,* 335, page 205 (1988).

[52] Cecile Dupouy, Dupouy, Michel Petit, and Yves Dandonneau, « Satellite detected cyanobacteria bloom in the southwestern tropical Pacific. Implication for oceanic nitrogen fixation," *International Journal of Remote Sensing*, 9, 389-396 (1988).

[53] Trevor Platt, and Shubha Sathyendranath, "Oceanic primary production: Estimation by remote sensing at local and regional scales," *Science*, 241, 1613-1620 (1988).

The southern seas also received coverage (despite the fact that very little CZCS observation had been accomplished in this region, due to pervasive cloud cover over the polar oceans). Cornelius "Neal" Sullivan led a group that included Chuck McClain and Goddard sea ice scientist Josefino Comiso, along with Walker Smith Jr., to examine phytoplankton along the edges of the Antarctic ice.[54] They noted that while phytoplankton growth under the ice is slow, the region immediately adjacent to the ice is vertically stable, providing optimum conditions for growth.[55] The existence and movement of the blooms at the ice edge is controlled both by biological processes (particularly the important process of getting eaten by something else) and physical processes, such as winds and currents.

Frank Muller-Karger wrote a paper with McClain and Philip Richardson to examine where the waters of the mighty Amazon go after they reach the sea. During half of the year (June to January), Amazon water appeared to flow toward Africa; during the other half of the year, the water flowed northward along the coast of South America toward the Caribbean.[56]

The NET members also continued publishing on various aspects of the data as well as topics for the future. Jim Mueller examined the important topic of cloud "ringing", which would influence design and testing of the next-generation sensors.[57] Cloud "ringing" refers to an optical effect that happens when the scan of the sensor goes over a very bright cloud and then progresses immediately to the wine-dark sea. The rapid transition between the highly contrasted areas caused the CZCS electronics to oscillate wildly for the next few pixels past the cloud, meaning that the data immediately adjacent to a cloud boundary was wildly suspect. Gordon and Castaño published an examination of the effects of the El Chichón eruption and how well the atmospheric correction algorithm handled the sulfate-laden aerosol exhalations of the volcano; they concluded that the algorithm handled the aerosol layer pretty well, with an error of only one or two digital counts.[58]

Although the CZCS was designed for the oceans, it could also view large lakes; one of the first examples of this type of research was published by C.H. Mortimer of the Center for Great Lakes Studies at the University of Wisconsin-Milwaukee.[59] He observed the changes in southern Lake Michigan occurring each spring. Earlier in his career, Mortimer also commented on a paper regarding the number of monsters in Loch Ness; unfortunately,

---

[54] Cornelius W. Sullivan, Charles R. McClain, Josefino C. Comiso, and Walker O. Smith Jr., "Phytoplankton standing crops within an Antarctic ice edge assessed by satellite remote sensing," *Journal of Geophysical Research*, 93, 12,487-12,498 (1988).
[55] George A. Knox, "Biology of the Southern Ocean," *Marine Biology Series #7*, CRC Press, 621 pages (1993).
[56] Frank Muller-Karger, Charles A. McClain, and Philip L. Richardson, "The dispersal of the Amazon water," *Nature*, 333, 56-59, doi:10.1038/333056a0, (5 May 1988).
[57] James Mueller, "Nimbus-7 CZCS: electronic overshoot due to cloud reflectance," *Applied Optics*, 27, 438-440 (1988).
[58] Howard R. Gordon, and Diego J. Castaño, D.J., "Coastal Zone Color Scanner atmospheric correction: influence of El Chichón," *Applied Optics*, 27, 3319-3322 (1988).
[59] C.H. Mortimer, "Discoveries and testable hypotheses arising from Coastal Zone Color Scanner imagery of southern Lake Michigan," *Limnology and Oceanography*, 33, 203-226 (1988).

remote sensing technology has been unable (as yet) to capture a view of this elusive legend.[60]

CZCS data would soon provide considerable insight into the phytoplankton dynamics of the Arabian Sea monsoon. Karl Banse led a study with Esaias, McClain, and Muller-Karger that examined winter phytoplankton blooms in the Arabian Sea.[61]

The more the scientific community looked at CZCS data, the more they saw; and the more they saw, the more that they realized that the data was affected by a lot of different factors, living, non-living, and even stuff somewhat in between. Emphasizing this theme was a 1989 paper by William "Barney" Balch and colleagues, which discussed the effects of coccolithophores and dinoflagellates on CZCS pigment data.[62] Coccolithophores are one of the more fascinating forms of phytoplankton in the oceans; they form microscopic plates called *coccoliths* out of calcite and cement them into microscopic balls called *coccospheres.* These structures are highly, highly reflective, so much so that ocean waters hosting a bloom of coccolithophores taken on the same transcendent turquoise of glacial mountain lakes which hold large amounts of suspended glacial flour. Unfortunately, bright turquoise waters in the middle of the ocean confound the algorithms based on water-leaving radiances, because their reflective nature produces much more water-leaving radiance than expected. Dinoflagellates, typified by the bad character *Karenia brevis* discussed earlier, also have unusual light scattering properties, and because they can move via flagellum, and this factor can also cause changes in the phytoplankton population.

Another factor that continued to receive attention was *Gelbstoff* and its components. Some of the identifiable components of *Gelbstoff* in surface waters consist of humic and fulvic acids. These organic compounds are derived from the breakdown of lignins in formerly living plant matter, and other than the fact that they have high molecular weights, they are hard to characterize. Humic acids are usually found in higher concentrations along the coast near the outflow of a river. They can be light-absorbing, affecting the estimation of remotely-sensed pigments and chlorophyll, which was discussed by Ken Carder and his associates.[63]

Other papers published in 1989 discussed new areas of investigation. Thorkild Aarup, (whose name always seems to appear first in alphabetized ocean color bibliographies) examined the frequently stormy, frequently turbid North Sea in collaboration with Steve Groom and plankton specialist Patrick Michael Holligan, who were both located at the

---

[60] C.H. Mortimer, "The Loch Ness Monster -limnology or paralimnology?" *Limnology and Oceanography*, 18, 343-345 (1973).

[61] Karl Banse, Charles R. McClain, Wayne E. Esaias, and Frank Müller-Karger, F. "Satellite-observed winter blooms of phytoplankton in the Arabian Sea," in M. F. Thompson and N. M. Tirmizi, editors, *Marine Science of the Arabian Sea. Proceedings of an International Conference,* American Institute of Biological Science, Washington, D.C., pp. 293-307 (1988).

[62] William M. Balch, Richard W. Eppley, Mark R. Abbott, and F.M.H. Reid, "Bias in satellite-derived pigment measurements due to coccolithophores and dinoflagellates," *Journal of Plankton Research*, 11, 575-581 (1989).

[63] Kendall L. Carder, Robert G. Steward, G.R. Harvey, and P.B. Ortner, "Marine humic and fulvic acids: Their effect on remote sensing of ocean chlorophyll," *Limnology and Oceanography*, 34, 68-81 (1989).

Plymouth Marine Laboratory in the United Kingdom.[64] Jim Mueller undertook the concept of bio-optical provinces – regions of the ocean defined by their biological populations and optical characteristics, a concept that required ocean color remote sensing to be applied on the basin scale – in the northeast Pacific Ocean.[65] Frank Muller-Karger, a Venezuelan by birth, examined the tranquil and inviting Caribbean Sea, and determined it to not always as pristine blue as postcards might indicate. One of the major influences on the color of the Caribbean is the outflow of Venezuela's powerful Orinoco River. During the rainy season, waters from the Orinoco carrying nutrients and probably some humic acids can move northward nearly to Puerto Rico and Cuba, generating a strong seasonal signal.[66]

Despite the fact that the name of the mission was the Coastal Zone Color Scanner, bio-optical oceanographers had determined that coastal ocean color presented significant difficulties for accurate analysis. The development of algorithms that could produce accurate results in complex coastal waters would preoccupy and then employ many different research groups during the following years. One of the first attempts to assay this model teamed up Shuba Sathyendranath with the Morel-Prieur team.[67] They presented a three-component model of coastal waters that indicated it would be possible – but not likely easy – to make estimates of chlorophyll in coastal Case 2 waters. The three components were phytoplankton, suspended particles that didn't contain chlorophyll, and the vexing, befuddling and recalcitrant component called "dissolved organic matter", which by itself could consist of *Gelbstoff*, humic acids, phaeopigments, and several other substances that had not at that time even been definitively characterized by chemical oceanographers.

A New Map of the World

1989 also marked the centennial year of the noted magazine *National Geographic.* To commemorate this event, which was marked with a December 1988 issue featuring a holographic crystal globe on the cover, the magazine produced a new map of the world. On the back of the map was another image of the world, under the title "Endangered Earth".[68] The ocean chlorophyll values depicted on this image were from the CZCS global processing effort (smaller CZCS images showed the Peru upwelling zone). Gene Feldman worked with the *National Geographic* cartographers for a month, arguing with them to keep the blank area in the southeastern Pacific Ocean that had not been observed by the CZCS – Feldman told

---

[64] Thorkild Aarup, Steve Groom, and Paul M. Holligan, "CZCS imagery of the North Sea. Remote sensing of atmosphere and oceans," *Advances in Space Research*, 9, 443-451 (1989).

[65] James L. Mueller and R. Edward Lange, "Bio-optical provinces of the Northeast Pacific Ocean: A provisional analysis," *Limnology and Oceanography*, 34, 1572-1586 (1989).

[66] Frank E. Müller-Karger, Charles R. McClain, T.R. Fisher, Wayne E. Esaias, and Ramon Varela, "Pigment distribution in the Caribbean Sea: Observations from space," *Progress in Oceanography*, 23, 23-64 (1989).

[67] Shubha Sathyendranath, Louis Prieur, and André Morel, "A three-component model of ocean colour and its application to remote sensing of phytoplankton pigments in coastal waters," *International Journal of Remote Sensing*, 10, 1373-1394 (1989).

[68] This map is available at http://www.maps.com/map.aspx?pid=16023, accessed 6 January 2010.

them "that's part of the story, that after seven-and-a-half years, our knowledge is still incomplete." Ultimately they capitulated to his persuasion and left the blank data area black.[69] It's also notable that this area is actually one of the darkest areas of ocean in the world, where chlorophyll concentrations are so low that it appears purple, and the water-leaving radiance spectrum actually peaks in the ultraviolet – the true "azure sea". [70]

Processes and Phenomena, Particles and Pollution

As the decade of the 1990s began, major events were happening with regard to NASA's remote sensing missions (Chapters 5 and 6). Attempts to establish a new ocean color mission met with *repeated* frustration. Meanwhile, the availability of the global CZCS data set and browse system fueled a marvelous expansion of knowledge regarding biological dynamics of the world's oceans, along with the occasional eye-opening surprise. Given the proliferation of published papers, it is impossible to describe more than a small fraction of them. Appendix 2 describes many papers in which researchers using CZCS data characterized and investigated various oceanic systems and phenomena as they awaited (and waited) for the next mission to launch.

The Joint Global Ocean Flux Study (JGOFS)

In the mid-1980s, an oceanographic program was in development that would affect and interact greatly with NASA's ocean color mission planning. This program was the Joint Global Ocean Flux Study (JGOFS). JGOFS blossomed as a result of a set of recommendations from a National Academy of Sciences workshop that took place in 1984.[71] The goal of JGOFS, which grew into an international program with a large component in the United States (the U.S. JGOFS program) was to vastly improve oceanographic knowledge regarding the cycling of carbon and related elements (both macro and trace nutrients) in the world's oceans. This increased knowledge was deemed to be vital to better understand the role of the oceans in climate change.[72]

---

[69] Gene Feldman, comments recorded at the Ocean Color Collaborative Historical Workshop, January 13-14, 2009, St. Petersburg, Florida.

[70] Paula Bontempi and B. Greg Mitchell, comments recorded at the Ocean Color Collaborative Historical Workshop, January 13-14, 2009, St. Petersburg, Florida.

[71] Margaret C. Bowles and Hugh D. Livingston, "U.S. Joint Global Ocean Flux Study (JGOFS)", http://www1.whoi.edu/overview.html, (accessed 29 April 2009). The text originally appeared in the *McGraw-Hill Yearbook of Science and Technology.*

[72] Margaret C. Bowles and Hugh D. Livingston, "U.S. Joint Global Ocean Flux Study (JGOFS)".

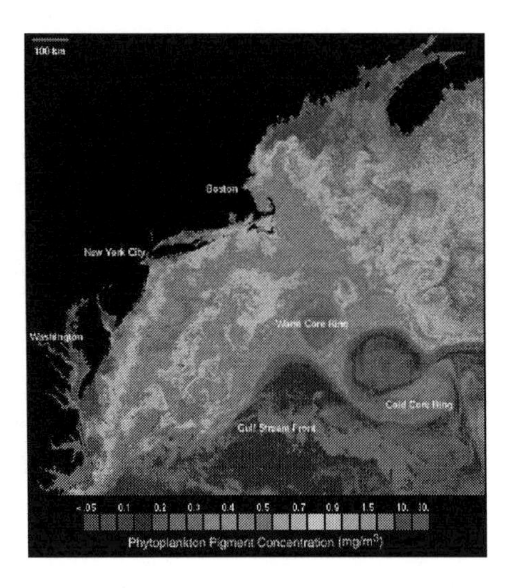

Figure 4.1. CZCS image showing Gulf Stream cold and warm core rings. This is the same image as used in *National Geographic*, but with labels. The warm core rings get the warm water from the low-productivity, low-pigment waters of the Sargasso Sea, and are strongly contrasted with higher pigment shelf waters. Cold-core rings have higher pigment waters surrounded by warmer, lower-productivity waters, and sometimes form in tandem with a warm core ring, as shown here. Cold core rings frequently "disappear" from the surface as the cold core sinks and is covered by a layer of warmer water, but the ring circulation persists for months.

An established program such as JGOFS meant that a significant amount of money would become available for oceanographic research which was closely related to NASA's goals for ocean color data. Thus, a natural parallel synergy of effort developed which would lead to both positive interactions as well as a few degrees of frustration.

GOFS, subsequently to become JGOFS, was organized in the mid-1980s following the September 1984 planning workshop. Some of the guiding minds behind the program in the United States were Neil Anderson of the National Science Foundation, Peter Brewer of Woods Hole and subsequently the Monterey Bay Aquarium Research Institute, Otis Brown and Frank Millero of RSMAS, David M. Karl of the University of Hawaii, Anthony H. Knap of the Bermuda Biological Station for Research, Hugh Ducklow of the Virginia Institute of Marine Science, and Glenn Flierl of the Massachusetts Institute of Technology.[73] The GOFS planning committee was a remarkable assemblage of oceanographic and geochemical "talent".[74] JGOFS was also an international program, under the auspices of the

---

[73] Woods Hole Oceanographic Institute, "Former and Corresponding Members of the U.S. JGOFS Scientific Steering Committee", http://usjgofs.whoi.edu/mzweb/ssc-all.html, (accessed 29 April 2009); Peter Brewer, "Why and How We Created JGOFS and the Lessons Learned", http://ijgofs.whoi.edu/Final_OSC/Brewer.pdf, (accessed 29 April 2009); "Joint Global Ocean Flux Study Research Highlights", http://ijgofs.whoi.edu/highlights.html, (accessed 29 April 2009).
[74] GOFS Planning Committee: Alice Alldredge, University of California – Santa Barbara, http://www.lifesci.ucsb.edu/eemb/faculty/alldredge/, (accessed 29 April 2009); D. James Baker,

International Geosphere-Biosphere Program (IGBP) and the Scientific Committee for Oceanic Research (SCOR). The international program was headquartered at the University of Bergen, Norway.[75]

In 1986, shortly after the cessation of observations by the CZCS in June, a crucial convergence of GOFS planning and NASA ocean color activities took place. The November 4, 1986 issue of the American Geophysical Union weekly newspaper *Eos* featured a cover image of CZCS pigment concentrations in the Atlantic Ocean, from the polar seas around Greenland to south of the equator, and the land surfaces of North America, South America, Europe, and Africa (which were colored with a Land Vegetation Index palette ranging from desert to dense vegetation). The image is colloquially referred to as the "Pac-Man image", because a zone of missing data (colored black) over central Canada has the appearance of the famous video-game character Pac-Man, which appears to be gobbling up half of Hudson Bay (Figure 4.2). This issue of *Eos*, which also contained the abstracts for presentations at the upcoming 1986 AGU Fall Meeting, featured two papers: "The Global Ocean Flux Study" and "Monthly Satellite-Derived Phytoplankton Pigment Distribution for the North Atlantic Basin".[76] According to Peter Brewer, this "coup" of a paper about GOFS and a paper about the basin-scale observations of the CZCS both appearing in the same widely-read issue of Eos "linked the NSF/NASA GOFS effort in very public ways, and cemented a long partnership".[77] This partnership was initially typified by the use of CZCS images to plan the JGOFS process studies.[78]

---

former president of Joint Oceanographic Institutions and Administrator of NOAA, http://symposia.cbc.amnh.org/archives/climate/biobaker.html, (accessed 29 April 2009); Wallace Broecker, Lamont-Doherty Geological (now Earth) Observatory, http://www.ldeo.columbia.edu/vetlesen/recipients/1987/broecker_bio.html, (accessed 29 April 2009); Richard Eppley, Scripps Institution of Oceanography, http://www.tos.org/oceanography/issues/issue_archive/issue_pdfs/3_2/3.2_weiler_et_al.pdf, (accessed 29 April 2009); Glenn Flierl, Massachusetts Institute of Technology, http://eapsweb.mit.edu/people/person.asp?position=Faculty&who=flierl, (accessed 29 April 2009); G. Ross Heath, http://www.ocean.washington.edu/people/faculty/rheath/heath.html, (accessed 29 April 2009); Susumu Honjo, Woods Hole Oceanographic Institution, http://www.whoi.edu/profile.do?id=shonjo, (accessed 29 April 2009) ; William Jenkins, http://www.whoi.edu/profile/wjenkins/, (accessed 29 April 2009); James McCarthy, http://www.oeb.harvard.edu/faculty/mccarthy/mccarthy-oeb.html, (accessed 29 April 2009) ; John Steele, former director of Woods Hole Oceanographic Institution, http://www.whoi.com/page.do?pid=20375, (accessed 29 April 2009).

[75] "International Web site of the Joint Global Ocean Flux Study", http://ijgofs.whoi.edu/, (accessed 29 April 2009).

[76] Peter G. Brewer, Kenneth W. Bruland, Richard W. Eppley, and James J. McCarthy, "The Global Ocean Flux Study (GOFS): Status of the U.S. GOFS program," *Eos, Transactions of the American Geophysical Union*, 67, 827-832, (4 November 1986); Wayne E. Esaias, Gene C. Feldman, Charles R. McClain, and Jane A. Elrod, "Monthly Satellite-Derived Phytoplankton Pigment Distribution for the North Atlantic Basin," *Eos, Transactions of the American Geophysical Union*, 67, 835-837, (4 November 1986).

[77] Peter Brewer, "Why and How We Created JGOFS and the Lessons Learned".

[78] Gene C. Feldman, James W. Murray, and Margaret S. Leinen, "Use of the Coastal Zone Color Scanner for EqPAC planning," *Oceanography*, 5, 143-145, (1992).

[Jim Yoder noted that the "Pac-Man" image was probably the single CZCS image that had the "biggest impact on biological oceanography" in three different ways. Yoder was "sure that it led Shelby Tilford [NASA's associate administrator for Earth Sciences] to put

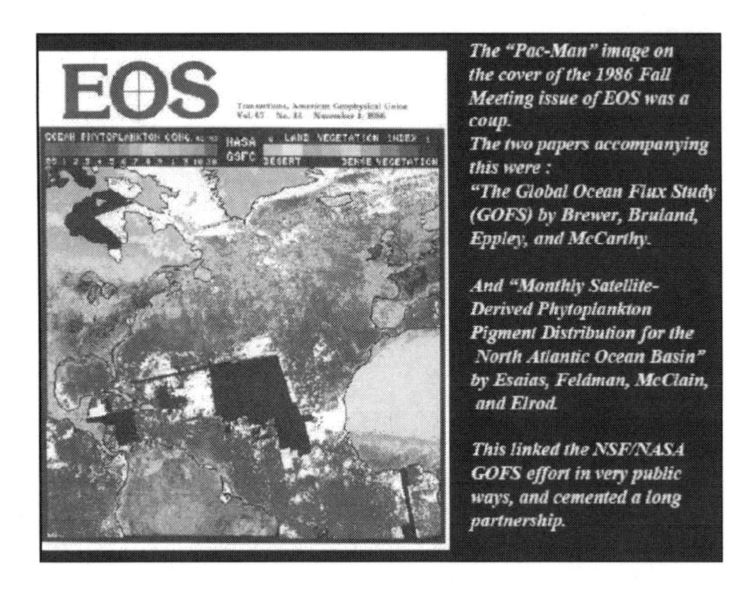

Figure 4.2. The CZCS North Atlantic basin composite image, colloquially referred to as the "Pac-Man" image. Image acquired from "Why and How We Created JGOFS and the Lessons Learned" by Peter Brewer.

up money for SeaWiFS" and gave Tilford "interest in pushing for SeaWiFS". Another reason was that it had an important impact on JGOFS planning, as Yoder indicated they used it to lay out stations along the 20° W longitude line for the North Atlantic Bloom Experiment, described below. The third reason was that it substantially influenced international opinion about both what NASA was doing and what could be done with ocean color data; Yoder recalled a meeting in Villefranche-sur-mer in 1986 where he showed the same image, and it caused a substantial impact on the audience, that NASA could produce a "whole basin view". Yoder said this image even boosted ESA's ocean color remote sensing program.[79]]

One of the first JGOFS activities was the establishment of two ocean time-series stations, HOTS (the Hawaii Ocean Time-Series) and BATS (the Bermuda Atlantic Time-Series).[80] At the location established for each station, moored arrays of data collection devices, including sediment traps at various depths, were deployed. Research cruises regularly visited the moorings, recovered the collected data and material, and returned it to the laboratory for further analyses. Data collection at the two time-series stations began in

---

[79] James A. Yoder, comments recorded at the Ocean Color Collaborative Historical Workshop, January 13-14, 2009, St. Petersburg, Florida.

[80] Margaret C. Bowles and Hugh D. Livingston, "U.S. Joint Global Ocean Flux Study (JGOFS)".

1988.

The first JGOFS process study was considered a large-scale "dress rehearsal", and was intended to characterize the North Atlantic spring bloom. The North Atlantic Bloom Experiment (NABE) took place in 1989 for the U.S. program, with international research involving four other nations continuing into 1991.[81] While NASA did not have an ocean color satellite in space at the time, NASA aircraft sensors on the NASA P-3 participated in NABE.[82] This NASA participation resulted in a memorably understated quote concerning the uncertainty of conducting aircraft oceanography at a long distance from a place to land; when NASA investigator Jim Yoder asked one of the P-3 pilots about their fuel situation during the May 2, 1989 mission, the pilot replied laconically, "If we crash on our final approach to Shannon [Ireland], there won't be much of a fire".[83]

The first full-scale process study was the Equatorial Pacific Process Study (EqPAC), which took place from 1991-1993, a period during which the equatorial Pacific Ocean experienced a peculiar "on-off" ENSO condition oscillating between normal and El Niño conditions about every six months.[84] The U.S. JGOFS program coordinators for EqPAC were Jim Murray of the University of Washington and Margaret Leinen of the University of Rhode Island.[85] As will be noted in Chapter 5, by this time researchers were fervently anticipating a new ocean color satellite mission.

NASA aircraft again took part in JGOFS by overflying the EqPac operations. The EqPAC effort included a new aircraft, the P-3B, with "twice the range of the plane used during the NABE".[86] Even so, the vastness of the Pacific required refueling operations on Christmas Island. NASA's airborne participation resulted in one of the most remarkable occurrences of the JGOFS program. While flying over the region in which the research vessel *Thomas G. Thompson* was conducting nearby oceanographic data collection, researchers on the aircraft collecting chlorophyll fluorescence data noted that only a few miles away, there was an extraordinarily sharp demarcation of the color of the ocean.[87] The airborne research team radioed the ship, and the ship made an unplanned excursion to the location of

---

[81] Margaret C. Bowles and Hugh D. Livingston, "U.S. Joint Global Ocean Flux Study (JGOFS)".
[82] Woods Hole Oceanographic Institution, "5. Process Studies", http://www1.whoi.edu/strategy5.html, (accessed 39 April 2009).
[83] James McCarthy, "Why and How We Created JGOFS and The Lessons Learned", http://ijgofs.whoi.edu/Final_OSC/McCarthy.pdf, (accessed 19 March 2009).
[84] Margaret S. Leinen, "U.S. EqPAC Encounters El Niño", *U.S. JGOFS News (Newsletter),* 3(3), pages 1 and 10, (March 1992). [This issue also included an article by Stanford Hooker and Wayne Esaias, "Measuring Ocean Color from Space: An Overview of the SeaWiFS Project", pages 7-9.]; Richard T. Barber, "Fall Survey Cruise Finds Cooling Conditions in Equatorial Pacific", *U.S. JGOFS News (Newsletter),* 4(1), pages 1 and 5 (September 1992).
[85] U.S. JGOFS Program, "Equatorial Pacific Process Study", http://www1.whoi.edu/eqpac.html, (accessed April 30, 2009); James W. Murray, Margaret W. Leinen, Richard A. Feely, J. R. Toggweiler and Rik Wanninkhof. "EqPac: A Process Study in the Central Equatorial Pacific," *Oceanography*, 5(3), 134-142 (1992).
[86] Jim Yoder, "EqPAC at 500 Feet: The JGOFS Air Corps Returns to Action", *U.S. JGOFS News (Newsletter)*, 4(1), pages 5-6 (September 1992).
[87] Jim Yoder, "EqPAC at 500 Feet: The JGOFS Air Corps Returns to Action"; University of Rhode Island Graduate School of Oceanography (URI-GSO), "A Line in the Sea", http://www.po.gso.uri.edu/color/yoder2.html, (accessed 30 April 2009).

the remarkable boundary. The ship researchers discovered that the boundary was a sharp convergence zone, with much higher plankton populations on one side of the boundary than the other. The surface appearance of the boundary was caused by a bloom of the planktonic species *Rhizoselenia*, which is particularly buoyant.[88]

A few days before these observations were taking place at sea, the Space Shuttle was in orbit overhead. Astronauts took photographs of the boundary, which appeared from space as a sharp, clear "Line in the Sea" (Figure 4.3).[89] The boundary line was so strong that waves actually broke where the density of ocean waters changed sharply, and the front was detectable by radar.[90] The photographs, confirming that the length of this line extended hundreds of kilometers, confirmed the need for an ocean color satellite mission. SST measurements also showed the location of this remarkable oceanic front.[91] David Archer of the University of Chicago was the first author of a paper which fully characterized this remarkable oceanic feature.[92]

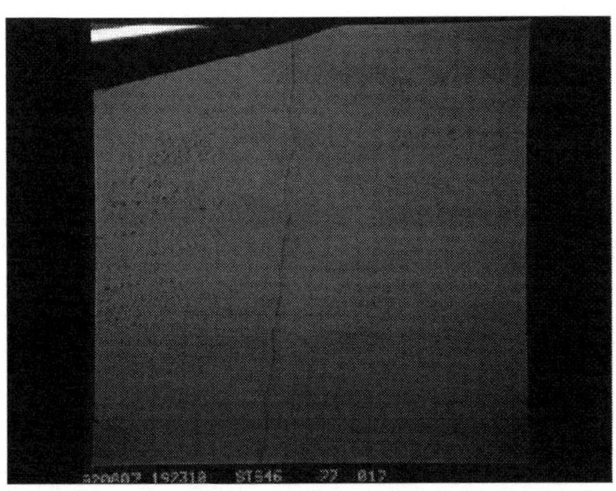

Figure 4.3. The Line in the Sea, photographed from the Space Shuttle.

While NASA's ocean color mission was delayed (as described in the next chapter), JGOFS proceeded with its third process study, this one designed to study the remarkable Arabian Sea monsoon. The program coordinator for the Arabian Sea Process Study was Sharon Smith of RSMAS.[93] The Arabian Sea expeditions took place from 1994-1996. The

[88] James A. Yoder, Steven G. Ackleson, Richard T. Barber, Pierre F. Flament, and William M. Balch., "A *line in the sea,*" *Nature*, 371, 689-692, (1994).
[89] Lyndon B. Johnson Space Center, astronaut photograph STS046-77-017.
[90] URI-GSO, "A Line in the Sea".
[91] Richard T. Barber, "Fall Survey Cruise Finds Cooling Conditions in Equatorial Pacific".
[92] David Archer, James Aiken, William Balch, Richard Barber, John Dunne, Wilford D. Gardner, Chris Garside, Catherine Goyet, Eric Johnson, David Kirchman, Michael McPhaden, Jan Newton, Edward Peltzer, Leigh Welling, Jacques White, and James Yoder, "A meeting place of great ocean currents: shipboard observations of a convergent front at 2° N in the Pacific," *Deep-Sea Research* II 44, 1827-1849 (1994).
[93] U.S. JGOFS Program, "Arabian Sea Process Study", http://www1.whoi.edu/arabian.html, (accessed 30 April 2009).

1991 planning report for the Arabian Sea Process Study indicated that it was expected that the next NASA ocean color mission would be in orbit for this phase of JGOFS.[94] There was expressed disappointment in the JGOFS community and the oceanographic community with the continuing mission delays (only briefly alluded to in a few documents and meeting minutes) because the Arabian Sea Process Study would have been an ideal test bed for ship-based research combined with satellite observations. [95] NASA still participated in the Arabian Sea Process Study, led by aircraft researchers Frank Hoge and Charles Yungel from the Wallops Flight Facility.[96]

At the end of 1996, with JGOFS deeply immersed in the planning stages for its next process study in the Southern Ocean, JGOFS scientists waited for a hopeful word from NASA regarding the long-anticipated and much-delayed CZCS follow-on mission, which they had read about in the September 1992 issue of the U.S. JGOFS Newsletter; the Sea-viewing Wide Field-of-view Sensor, SeaWiFS.[97]

Summary: The data from the CZCS were truly unprecedented, and for a large part of the oceanographic community, unexpected. The oceanographic community had been somewhat removed during the 1970s from the concept of viewing environmental variables from space, and even NASA was still largely in the "gee-whiz" phase of instrument development; the CZCS was truly a proof-of-concept instrument. Because it proved the concept, and then provided eight years of remarkable oceanic observations, the CZCS provided a great deal of oceanographic insight. Yet this insight was garnered only after the perceptive campaign to produce a usable global CZCS data set, combined with the serendipitous increases in computational capabilities that enabled researchers to utilize the CZCS data set. This combination of circumstances resulted in the CZCS transforming from an oceanographic novelty act into a powerful tool for oceanographic science, a tool which ultimately was perceived as a necessity for observing the oceans and improving our understanding of the global carbon and climate systems of which the oceans are a fundamental component.

---

[94] U.S. JGOFS Program, "Arabian Sea Process Study", U.S. JGOFS Planning Report Number 13, (November 1991); Wayne E. Esaias, "SeaWiFS Status Report", http://modis.gsfc.nasa.gov/sci_team/meetings/199310/presentations/x8_seawifs_status.pdf, (accessed 30 April 2009).

[95] Hugh D. Livingston, "National Science Foundation Final Project Report: Planning and Implementation for the U.S. Joint Global Ocean Flux Study", http://www.osti.gov/bridge/servlets/purl/820629-ukeQPV/native/820629.pdf, (accessed 30 April 2009); Ocean Carbon Transport, Exchanges and Transformations Workshop Report, "Modern Observing Systems Working Group Summary", http://www.msrc.sunysb.edu/octet/mod_obs_sys.html, (accessed 30 April 2009); U.S. JGOFS Program, "U.S. JGOFS Scientific Steering Committee meeting: October 25-27, 1995", http://usjgofs.whoi.edu/sc10-95minutes.html, (accessed 30 April 2009).

[96] U.S. JGOFS Program, "Arabian Sea Process Study", http://www1.whoi.edu/arabian.html, (accessed 30 April 2009).

[97] Stanford Hooker and Wayne Esaias, "Measuring Ocean Color from Space: An Overview of the SeaWiFS Project".

5

HIGH STAKES AND BRIGHT SEAS

3… 2… 1… drop…

1… 2… 3… ignition…

The date: June 27, 1994. The event: the first satellite launch of the Pegasus XL launch vehicle, the belly-mounted launch vehicle built by Orbital Sciences Corporation of Dulles, Virginia.[1] The innovative Pegasus, and its larger version, the Pegasus XL, eliminated the need for a traditional launch vehicle first stage.[2] Instead, to gain the first 30,000 feet or so of altitude, the Pegasus or Pegasus XL was mounted underneath a large aircraft, first a B-52 Stratofortress and subsequently a modified L-1011. The Pegasus XL was required to carry larger payloads, including the much-awaited NASA SeaWiFS ocean color satellite mission. When the aircraft had reached the launch altitude, the rocket went through its preflight checks, and then it was released and dropped before ignition and flight. Several successful Pegasus launches had preceded this first Pegasus XL launch.[3]

The first moments of the Pegasus XL launch appeared perfect. About 5 heart-stopping seconds after the rocket was released from underneath the aircraft, the engine ignited explosively, and the white arrow of the rocket shot forward in front of a brilliant white contrail, arcing upward toward the deep violet blue of space. Accelerating at amazing speed, the rocket soon became a white dot ahead of the smoothly lengthening contrail.

Until 35 seconds had elapsed. Suddenly, the contrail was not smooth – it appeared to be spiraling, oscillating, wavering, certainly not the hallmark of a launch vehicle rising smoothly into space. Something appeared very wrong.

Something was indeed wrong. The Pegasus XL launch vehicle was out of control. Little more than three minutes into its inaugural mission, the Pacific Launch Range controller at Vandenberg Air Force Base issued the fateful command to self-destruct the vehicle and its satellite payload – in that moment dashing the high expectations of the ocean

---

[1] Orbital Sciences Corporation, "Pegasus Mission History", http://www.orbital.com/SpaceLaunch/Pegasus/pegasus_history.shtml, (accessed 2 March 2010).
[2] "Pegasus Fact Sheet", http://www.orbital.com/NewsInfo/Publications/Pegasus_fact.pdf, (accessed 9 May 2009).
[3] Orbital Sciences Corporation, "Pegasus Mission History".

color science community in the United States and abroad.[4]

That's how it could have happened. Fortunately for the ocean science community, the payload on the inaugural Pegasus XL launch was *not* the Sea-viewing Wide Field-of-view Sensor (SeaWiFS) – which it was originally scheduled to be (rather, it was the Air Force STEP-1 satellite). So instead of a heart-wrenching loss – and NASA was soon to experience more than a few of those, as well as the space science community in other nations – the failure of the first Pegasus XL meant that the SeaWiFS mission was going to be delayed yet again. And surprisingly, delay in this case, while not necessarily desirable, would turn out to be not entirely a bad thing, even beyond the fortunate reality that SeaWiFS was not the payload of the first Pegasus XL.

Our history now flashes back to 1984, with the CZCS still in operation. Waiting in the wings was another CZCS, the flight model. NASA scientists and associated oceanographers conceived that the CZCS flight model could be used for the follow-on CZCS mission – provided funding and a satellite platform to carry it could be procured. At the Planning Workshop for the Global Ocean Flux Study (GOFS), the participants considered the idea of the second CZCS, referred to as the Ocean Color Imager (OCI). They indicated that the OCI was identified as an *essential component* of GOFS.[5]

Plans to build and launch the OCI were already underway. Prior to 1983, NASA had attempted to place a CZCS II (the flight version sitting on the shelf) on the National Ocean Satellite System (NOSS) and the Navy Remote Ocean Sensing System (NROSS) satellites, but this effort did not get far. [6] The next three attempts involved putting the CZCS II on one of the series of NOAA polar orbiters, starting with NOAA H (NOAA 11 in orbit), NOAA I (13), and NOAA J (14).[7]

NASA initially engaged with the National Ocean Satellite Service (NOSS), an office of NOAA, to build an Advanced TIROS-N satellite (ATN), which would include the OCI. Specifications for OCI had actually already been drawn up in 1983 and the specifications were updated in 1985.[8] At this point in time, it was generally realized that the CZCS flight model, despite possessing advanced optics, was saddled with 1970s era electronics, and

[4] "Panel to Investigate Pegasus Failure – launch delays expected," *Defense Daily*, 29 July1994, http://findarticles.com/p/articles/mi_6712/is_n62_v183/ai_n28640320/ , (accessed 9 May 2009).
[5] GOFS Executive Committee, *Ocean Color Imager*, brochure prepared for the GOFS Executive Committee in cooperation with Joint Oceanographic Institutions, produced by InterNetwork, Inc., Del Mar, CA, June 1985; "NASA Oceanic Processes Program: Annual Report, Fiscal Year 1985", http://czic.csc.noaa.gov/czic/TL521.3.U53_FY_85_1986/89FF63.pdf, (accessed 9 May 2009).
[6] Goddard Space Flight Center, "Specification for the Advanced Tiros-N (ATN) Ocean Color Instrument (OCI)", Report, 102 pages. (Note about NOSS is attached to front.) Also, Dennis Clark, comment recorded at the Ocean Color Collaborative Historical Workshop, January 13-14, 2009, St. Petersburg, FL; Space Studies Board, National Academy of Sciences, *Mission to Planet Earth: Space Science in the Twenty-First Century -- Imperatives for the Decades 1995 to 2015*, http://books.nap.edu/openbook.php?record_id=753&page=73 (1988).
[7] Wayne Esaias, unpublished notes on CZCS follow-on attempts; Ocean Color Science Working Group, John J. Walsh, Chairman, "The Marine Resources Experiment Program (MAREX)", Report, NASA Goddard Space Flight Center, Sections A-G, (December 1982).
[8] Goddard Space Flight Center, "Specification for the Advanced Tiros-N (ATN) Ocean Color Instrument (OCI)".

updating the electronics boards would prove impossible. So the CZCS flight model was shelved permanently, and currently resides on the second floor of GSFC Building 28.[9]

Even though the second CZCS instrument was abandoned, in 1985, eminent oceanographer Roger Revelle referred to NASA's attempts to launch a new ocean color satellite in a short editorial in *Science*.[10] Revelle stated: "NASA is currently investigating flight opportunities for ocean color measurements that allow determination of chlorophyll content of the surface layers and provide flow visualization. This mission has been recommended by the ocean science community for launch in 1990." Revelle's mention of the plans for a new ocean color mission indicates the awareness of the broader oceanographic community of the success of CZCS and the changes it had made in studying the ocean, and the need to expand such study. Revelle's 1985 editorial was echoed in 1988 by another *Science* editorial, this one penned by Philip Abelson, deputy editor for engineering and applied sciences, in which he politely lamented the loss of CZCS and the lack of a follow-on mission.[11] Abelson wrote: "The outlook for some of the most important observations is chancy. We are highly dependent on the Nimbus 7 satellite that has long outlived its expected usefulness. A gap in ocean color observations has already occurred, owing to termination of the coastal zone color scanner measurements on Nimbus 7."

With the OCI in possession of a preliminary design, the next attempts to garner a CZCS follow-on mission involved this sensor. There were some preliminary discussions with the Japanese space service, but these did not get any traction.[12] The next attempt to find a berth to space involved the French SPOT series, a dual mission on SPOT-3. This effort also met with a quick termination, even though it did generate a study report which included a draft Memorandum of Understanding with the Centre Nationale d'Etudes Spatiales (CNES).[13] So NASA and NOAA again went into negotiations to put OCI on either NOAA K (15), NOAA L (16), or NOAA M (17).[14] Even though NOAA proposed OCI as a new start in 1987, in each of these cases, NOAA was ultimately unwilling to undertake the additional programmatic costs associated with adding OCI to the satellite manifest.[15] Also in this period, there was some discussion of a NOAA-D mission; NOAA-D had been planned as a longer satellite that could accommodate more instruments. This idea also didn't pan out, and NOAA-D was launched and became NOAA 12.[16]

---

[9] Gene Carl Feldman, "Nimbus 7 Coastal Zone Color Scanner", http://seawifs.gsfc.nasa.gov/ops/czcs.pl, (accessed 13 May 2009).
[10] Roger Revelle, "Oceanography from Space", *Science*, 228(4696), 133, (12 April 1985).
[11] Philip Abelson, "Space science: past and future," *Science*, 241(4864), 397, (22 July 1988).
[12] Stan Wilson recorded interview, 19 March 2009.
[13] NASA/CNES, "Ocean Color Instrument on SPOT-3: Phase A Study Report, Technical Summary", 7 chapters plus appendices, (February 1985).
[14] GOFS Executive Committee, "Ocean Color Imager".
[15] GOFS Executive Committee, "Ocean Color Imager"; Dennis Clark, comments recorded at the Ocean Color Collaborative Historical Workshop, January 13-14, 2009, St. Petersburg, Florida.
[16] National Environmental Satellite, Data, and Information Service, "NOAA 12 Spacecraft Status Summary", http://www.oso.noaa.gov/poesstatus/spacecraftStatusSummary.asp?spacecraft=12, (accessed 13 May 2009).

As Robert Evans stated it: "We came so close so many times."[17]

With the count standing at eleven possibilities (some better than others) that hadn't resulted in a CZCS follow-on launch, and with considerable expectancy emanating from other agencies and international programs (such as JGOFS), NASA and the Reagan Administration and the Joint Oceanographic Institutes (JOI) finally decided it was time to get serious. At the time, the presidential administration of Ronald Reagan was advocating public-private partnerships as part of an overall effort to reduce government size and costs (Reagan "took an ideological cue from Britain's Prime Minister [Margaret] Thatcher"), and so the focus of an ocean color mission turned to one of the current public-private satellite remote sensing collaborations, the Earth Observation Satellite Company (EOSAT).[18] EOSAT's main corporate activity at the time was marketing and selling remote sensing imagery from the Landsat missions.[19] Landsat 4 had launched on July 16, 1982, and Landsat 5 launched on March 1, 1984.

At that time, EOSAT was in the early planning stages of the Landsat 6 mission. A serious effort was launched to place an ocean color sensor on the same platform with Landsat 6. A two-day meeting in February 1987 initiated the production of a report released in August 1987 entitled "System Concept for Wide Field-of-view Observations of Ocean Phenomena from Space".[20] The report was created by the Joint EOSAT/NASA SeaWiFS Working Group, with Dr. James Baker of the Joint Oceanographic Institutions and Dr. Kenneth Ruggles of Systems West, Inc. as co-chairmen.[21] One of the key players in the meeting was Robert "Bob" Kirk, who had been hired at GSFC to head the effort to start a new ocean color mission. Kirk had originally come to Goddard to work on the NOSS and NROSS efforts, then worked on a scatterometer mission that eventually went to JPL, and then the OCI.[22]

The EOSAT report highlighted the new concept of a public-private partnership, which would allow the company responsible for the mission to sell real-time data commercially. The value of the real-time data would be preserved by encryption; commercial users were to be given keys for purchased data that would be used to decrypt the data broadcast directly from the satellite. Target markets included the most obvious, the fishing industry and fisheries research, but also included the oil and gas industry (with respect to offshore drilling platforms) and the shipping industry. Another potential market

---

[17] Robert Evans, comments recorded at the Ocean Color Collaborative Historical Workshop, January 13-14, 2009, St. Petersburg, Florida.

[18] David Seader, "The United States' experience with outsourcing, privatization and public-private partnerships", http://www.ncppp.org/resources/papers/seader_usexperience.pdf, accessed 23 January 2011.

[19] Jet Propulsion Laboratory Mission and Spacecraft Library, "Landsat 4,5 Quicklook", http://msl.jpl.nasa.gov/QuickLooks/landsat4QL.html, accessed May 13, 2009; NASA Goddard Spaceflight Center, "The Landsat Program: Landsat 5", http://landsat.gsfc.nasa.gov/about/landsat5.html, accessed 13 May 2009.

[20] Joint EOSAT/NASA SeaWiFS Working Group, "System Concept for Wide Field-of-view Observations of Ocean Phenomena from Space," Report, NASA/EOSAT, August 1987, 92 pages.

[21] Joint EOSAT/NASA SeaWiFS Working Group, "System Concept for Wide Field-of-view Observations of Ocean Phenomena from Space."

[22] Robert Kirk, email message received 14 May 2009.

that was identified was the military industry, as the Office of Naval Research was an active partner in ocean optics research (see Chapter 7). The scientific users of the data would be required to wait for a certain period of time (initially estimated as 7-10 days) before the unencrypted data would be available for research use.[23]

The initial SeaWiFS specifications called for an 8-band sensor, with 6 visible and 2 infrared bands. Two of the main new visible-range bands were to be used for atmospheric correction, at 765 and 865 nm. The sensor was definitely going to be a true wide field-of-view sensor; the total scan angle would be ±58°, with scientific quality data expected to be collected in the middle ±45° of the full swath (i.e., the full-swath width was 116°, and the scientific-quality data would be collected in the central 90°). The width of the scan, projected on the Earth's surface, would thus be a remarkable *2800 km.* The EOSAT report indicated that even though it would be difficult to collect science-quality data at the edges of the wide swath, due to several optical effects that were likely to degrade the data quality, commercial interests could still utilize imagery, and the wider swath provided much more daily ocean coverage than a narrower swath. Placing the sensor on the Landsat platform would mean that it would orbit at an altitude of 705 km, resulting in a 1.13 km surface resolution, which was deemed acceptable. The satellite would directly broadcast the high resolution data to ground stations (High Resolution Picture Transmission stations, called HRPT stations), many of which were already operating around the world and receiving AVHRR data from NOAA satellites. Onboard the satellite, the data would be "aggregated" to produce a global 4.5 km resolution data product, and this data would be broadcast to the EOSAT ground receiving facility.[24]

Experience from the CZCS mission indicated the clear necessity of regular instrument calibration. Calibration would be accomplished by regular viewing of a solar diffuser target while the scanner was rotating, when the scan view was not aimed at the Earth. The possibility of using the Moon as a calibration target was also suggested, but in the EOSAT report it was noted that viewing the Moon required a satellite maneuver, so the Moon was to be a "target of opportunity".[25]

One of the key factors in improving SeaWiFS over CZCS involved increasing the signal-to-noise ratio. The initial SeaWiFS design used a rotating scanner which acquired the data, which was followed in the optical path by a telescope and an image "derotator". The immediate problem recognized with this design was increased polarization sensitivity up to 5%, quickly assessed as unacceptable. So the adopted baseline design was a remarkable innovation; instead of a mirror-telescope combination, the SeaWiFS scan mechanism would actually become a rotating telescope. According to an optical analysis, this design reduced the polarization sensitivity to well below the 2% specification. The instrument was required to tilt rapidly, to reduce the amount of scan time lost while slewing from different

[23] Joint EOSAT/NASA SeaWiFS Working Group, "System Concept for Wide Field-of-view Observations of Ocean Phenomena from Space," pp. 42-44.
[24] Joint EOSAT/NASA SeaWiFS Working Group, "System Concept for Wide Field-of-view Observations of Ocean Phenomena from Space."
[25] Joint EOSAT/NASA SeaWiFS Working Group, "System Concept for Wide Field-of-view Observations of Ocean Phenomena from Space," page 55.

tilt positions. Tilt capability was recognized as necessary to reduce the impact of sun glint.[26]

The EOSAT report also discussed the berthing considerations required for Landsat 6 and SeaWiFS to share the same satellite. Two of the main considerations were data storage and power budget. It was expected that SeaWiFS would collect data for about 40 minutes per orbit, with 10 minutes allocated to high-resolution data. The power aspect is found in a footnote in the EOSAT report, and suggested that there might be necessary "compromises" between the Enhanced Thermal Mapper and SeaWiFS operations.[27] Expansion of power resources, i.e., bigger solar panels, was considered as an option.

With the EOSAT report out on the streets, NASA set about negotiating the actual contract for the SeaWiFS mission. One of the principals in these negotiations was Marlon Lewis, who began serving as program manager in the NASA Ocean Biology program office in 1988, succeeding Jim Yoder.[28] According to Lewis, attempting to finalize this mission contract required an "enormous amount of work", partly because NASA's primary focus is space exploration.[29] Also occurring at this time was planning and initial sensor development work for the Earth Observation System (EOS), which will be described more extensively in Chapter 6. Optimism was high that the EOSAT collaboration would work; Jim Yoder, in the first issue of the magazine *Oceanography* published by The Oceanography Society, describes the mission.[30] In the same issue, Stan Wilson states that a contract with the Department of Commerce and EOSAT was signed, but that NASA "is currently seeking support for this initiative". [31] Marlon Lewis indicated that Bob Winokur, at the time the technical director to the Oceanographer of the Navy, was instrumental in selling the SeaWiFS Project.[32]

NASA sought support for the initiative over most of the next two years, but was stymied again and again. EOSAT was under criticism for raising the prices of Landsat 4 and 5 data, so part of the difficulty was coming up with an acceptable cost structure for the government.[33] As conceived, NASA would buy the data from EOSAT; EOSAT was undertaking the additional costs of berthing the satellite on the Landsat 6 satellite.

As Marlon Lewis approached the end of his tenure at NASA HQ, he became increasingly discouraged and convinced that the SeaWiFS mission was not going to work, despite its promising beginning, as negotiations with EOSAT stalled – they were concerned

---

[26] Joint EOSAT/NASA SeaWiFS Working Group, "System Concept for Wide Field-of-view Observations of Ocean Phenomena from Space," pages 49-55.

[27] Joint EOSAT/NASA SeaWiFS Working Group, "System Concept for Wide Field-of-view Observations of Ocean Phenomena from Space," page 59.

[28] Marla Cranston, "Understanding ocean colour – satellite gives scientists a view of ocean changes", Dal News, http://dalnews.dal.ca/2007/09/24/nasa-ocean.html, 24 September 2007 (accessed 13 May 2009).

[29] Marla Cranston, "Understanding ocean colour – satellite gives scientists a view of ocean changes."

[30] James A. Yoder, Wayne E. Esaias, Gene C. Feldman, and Charles R. McClain, "Satellite Ocean Color – Status Report," *Oceanography*, 1(1), 18-20 & 35, (1988).

[31] W. Stanley Wilson, "Status Report of NASA's Oceanographic Activities," *Oceanography*, 1(1), 34, (1988).

[32] Marlon Lewis, comments recorded at the Ocean Color Collaborative Historical Workshop, January 13-14, 2009, St. Petersburg, Florida.

[33] NASA Goddard Space Flight Center, "The Landsat Program: Landsat 5".

that SeaWiFS would delay the Landsat 6 schedule.[34] As his successor B. Greg Mitchell was preparing to move from California to Washington DC in 1990, Lewis indicated to Mitchell that the SeaWiFS mission was lost. But another player had entered the conversation; an aggressive, young, and forward-looking company named Orbital Sciences Corporation (OSC). OSC was formed in 1982 by three Harvard Business School classmates – Dave Thompson, Bruce Ferguson, and Scott Webster – and the corporation's avowed goal was to make space technology more available and more affordable.[35] To accomplish this, Orbital came up with some innovative strategies, including a less expensive launch vehicle technology: the aircraft belly-mounted, altitude-launched Pegasus rocket, conceived in 1986. The Pegasus was actually conceived in the wake of the Challenger disaster, as NASA decided subsequent to that event that there would no longer be commercial payloads in the Shuttle manifest.[36] This impacted Orbital's first space vehicle, the Transfer Orbit Stage, which had been designed to launch satellites (mainly commercial communications satellites) deployed in near-earth orbit into geostationary orbit, or Earth escape trajectories, as in the case of the Mars Observer.[37] OSC signed significant contracts for military Pegasus launches in 1988. In 1989, they completed more contracts, primarily military including development of a ground-based Taurus launch vehicle. In 1990, Orbital was in a very successful growth phase, marked by an initial public stock offering.[38]

Thus, in 1990, NASA abandoned the EOSAT collaboration, which was part of the reason for Marlon Lewis' despair. The circumstances surrounding the next steps of the SeaWiFS procurement are somewhat remarkable. At NASA HQ, Greg Mitchell was still unpacking boxes in his office. Marlon Lewis had been aware that Hughes Aerospace and the Hughes Santa Barbara Research Center were working together on a new payload, and OSC was expected to be the subcontractor for the launch and space segment. Warren Nichols of SBRC (described as "the brick wall who hates you" for his unsmiling and disciplined business demeanor, though otherwise "a very nice fellow") continued to push for a Hughes-led mission, and even got Shelby Tilford to say "Done! to a proposed mission cost of only $25 million dollars.[39] However, the Hughes board of directors voted not to pursue it because a budget that minimal for the project was considered too risky.[40] This decision was

---

[34] Marlon Lewis, comments recorded at the Ocean Color Collaborative Historical Workshop, January 13-14, 2009, St. Petersburg, Florida.

[35] Funding Universe, "Orbital Sciences Corporation", http://www.fundinguniverse.com/company-histories/Orbital-Sciences-Corporation-Company-History.html, (accessed 13 May 2009); Howard Runge, comments recorded at the Ocean Color Collaborative Historical Workshop, January 13-14, 2009, St. Petersburg, Florida.

[36] Howard Runge, comments recorded at the Ocean Color Collaborative Historical Workshop, January 13-14, 2009, St. Petersburg, Florida.

[37] Howard Runge, comments recorded at the Ocean Color Collaborative Historical Workshop, January 13-14, 2009, St. Petersburg, Florida.

[38] Funding Universe, "Orbital Sciences Corporation."

[39] Marlon Lewis and Carl Schueler, comments recorded at the Ocean Color Collaborative Historical Workshop, January 13-14, 2009, St. Petersburg, Florida. (Warren Nichols was Schueler's supervising manager at SBRC.)

[40] Marlon Lewis, comments recorded at the Ocean Color Collaborative Historical Workshop, January 13-14, 2009, St. Petersburg, Florida.

"crushing", according to Lewis. On the other hand, OSC, led on the effort by John Mehoves in a continuing collaboration with Lewis, was still considering it, partly because they thought they had a better plan; this involved using the Pegasus, buying the sensor from Hughes SBRC, and selling the data commercially. In essence, Howard Runge said that the Pegasus program "begat" SeaWiFS, because OSC wanted a partner for the mission.[41] When Hughes ultimately declined, Mehoves asked Dave Thompson if they would "jump into the driver's seat and be the prime contractor". Thompson agreed to the plan, even though it was going to cost more.[42]

At NASA Headquarters, there were plenty of issues for Greg Mitchell to juggle. NASA was in favor of OSC's bid for SeaWiFS, but for legal reasons they had to compete the bid, though only OSC was capable of doing it. Mitchell became involved with Congress to determine ways to fund the mission, and Congress was apparently "angry" with NASA for various decisions at the time.[43] As Mitchell was discovering room numbers and was talking with both Congress and his NASA supervisors (Stan Wilson, Shelby Tilford, Bill Townsend, and even NASA Associate Administrator for Space Science and Applications Lennard Fisk), Congress wanted some more information.[44] Gene Feldman suddenly received a request from D. James Baker for a lunch meeting at the glitzy Italian restaurant *I Ricci* on Capitol Hill with Maryland Senator Barbara Mikulski and high-ranking science committee staffers. Feldman recalls arriving in his standard working oceanographer garb of jeans and a T-shirt, confronting lunch with fine linens, three-piece suits and ties, and excellent Italian food – which he didn't get to eat, as he spent the entire lunch showing CZCS images to the staffers.[45]

Feldman apparently provided plenty of food for thought, even if he missed the pasta; Congress proceeded to appropriate $40 million for SeaWiFS, within the funding for the JGOFS program, as a line item (the budget earmarks also included funding for the State of Oregon Aquarium). Now Mitchell had to run an RFP (Request for Proposal) by his bosses, Diane Wickland and Bob Murphy; then he had to get approval from higher-ups, including Shelby Tilford, Wilson, and Fisk. The short RFP did not even provide for a data system, as they expected to use existing hardware, despite a data volume expected to be ten times greater than CZCS. In the language of appropriations, Mitchell said that they stated, "billions for EOS (the Earth Observing System), millions for SeaWiFS".[46]

All of these events took place during the appropriations process, likely early in the latter

[41] Howard Runge, comments recorded at the Ocean Color Collaborative Historical Workshop, January 13-14, 2009, St. Petersburg, Florida.
[42] Marlon Lewis, comments recorded at the Ocean Color Collaborative Historical Workshop, January 13-14, 2009, St. Petersburg, Florida.
[43] B. Greg Mitchell, comments recorded at the Ocean Color Collaborative Historical Workshop, January 13-14, 2009, St. Petersburg, Florida.
[44] B. Greg Mitchell, comments recorded at the Ocean Color Collaborative Historical Workshop, January 13-14, 2009, St. Petersburg, Florida.
[45] Gene Feldman, comments recorded at the Ocean Color Collaborative Historical Workshop, January 13-14, 2009, St. Petersburg, Florida.
[46] B. Greg Mitchell, comments recorded at the Ocean Color Collaborative Historical Workshop, January 13-14, 2009, St. Petersburg, Florida.

half of 1990. With money in hand, Bob Kirk took the lead in releasing an Announcement of Opportunity and RFP with an amazingly quick turnaround.[47] Kirk recalls that Goddard Director John Klineberg would not sign a contract that he characterized as saying "give us the money, meet you at the launch pad".[48] In fact, the Goddard Flight Project Directorate wouldn't take on the program, so the Earth Sciences Directorate took it instead.[49] The AO came out on the day after Thanksgiving, and the RFP was released in the middle of December 1990. Howard Runge said that OSC thus had to work on the proposal over Christmas vacation in order to respond to the RFP in a month.[50] Carl Schueler, manager of Advanced Development Programs at Hughes SBRC at the time, said one reason they were able to turn around the proposal in a month was that Hughes had already been spending internal resources on SeaWiFS in parallel with work on Landsat and other instruments, including the EOS-bound Moderate Resolution Imaging Spectrometer (MODIS). The actual idea for the rotating telescope had been conceived in 1987 or 1988. They had included thermal bands and four detectors (instead of one), to allow "Time Delay and Integration" (TDI), which increased the signal-to-noise ratio by accumulating the signal from four detectors before passing it on to the instrument electronics. The thermal bands were no longer needed when the Navy abandoned participation in the SeaWiFS program, which helped keep the cost under $25 million.[51]

The fast-and-furious events in late 1990 and early 1991 ultimately paid off for oceanographers. NASA awarded OSC the "data-buy" contract in March 1991 for the SeaWiFS mission[52]. Now SeaWiFS would be a free-flyer, a single sensing system on its own satellite, which was given the optimistic appellation *SeaStar*.[53]

The contract with Orbital Sciences retained most of the specifications that had been stated in the EOSAT report in 1987. However, there were significant changes. The thermal infrared bands were removed from the sensor, and replaced by two additional visible bands, including one in the problematic blue zone of the visible spectrum at 412 nm. The band was problematic due to the increased difficulty of atmospheric correction in the blue; the band was desired because it allowed increased discrimination of *Gelbstoffe* and chlorophyll. And rather than have one 500 nm band, two bands were placed in close spectral proximity at 490 and 510 nm. So SeaWiFS would now only be a visible-range

---

[47] Robert Kirk, telephone comments recorded at the Ocean Color Collaborative Historical Workshop, January 13-14, 2009, St. Petersburg, Florida.
[48] Robert Kirk, telephone comments recorded at the Ocean Color Collaborative Historical Workshop, January 13-14, 2009, St. Petersburg, Florida.
[49] Vincent V. Salomonson recorded interview, December 2, 2008
[50] Howard Runge, comments recorded at the Ocean Color Collaborative Historical Workshop, January 13-14, 2009, St. Petersburg, Florida.
[51] Carl Schueler, comments recorded at the Ocean Color Collaborative Historical Workshop, January 13-14, 2009, St. Petersburg, Florida.
[52] NASA History Office, *Exploring the Unknown: Volume VI: Space and Earth Science*, John M. Logsdon, Stephen J. Garber, Roger D. Launius, and Ray A. Williamson, editors, U.S. Government Printing Office, Washington, DC, (2004), page 469; Tim Furniss, "Orbital Sciences wins ocean-monitor deal," *Flight International*, 139 (4258), 20(1), (13 March 1991).
[53] Forecast International, "SeaStar – Archived 4/2003", http://www.forecastinternational.com/archive/sp/sp11502.htm, (accessed 14 May 2009).

sensor. Another significant change was that the "target of opportunity" of a lunar view for calibration was now stated as a requirement; a requirement requiring a regular satellite maneuver, essentially flipping the satellite completely upside down.[54]

The finalization of the contract meant that NASA now had a SeaWiFS Project. Robert Kirk was designated as the Project Manager. Former astronaut Mary Cleave, who had orbited the Earth twice on the Space Shuttle and had deployed the Magellan Venus orbiter on one mission, was hired as the Deputy Program Manager.[55]

PC-Seapak

While NASA HQ had been working on the SeaWiFS contract, the ocean color scientists at the University of Miami and at Goddard were still very busy. One of the issues that they were addressing was one of the greatest handicaps to the utilization of the CZCS data – the lack of processing and analysis software for the data. As a result, various groups developed their own data processing software. In the U.S., the DSP (mentioned in Chapter 5) and SEAPAK systems were developed independently.[56] Both software packages used DEC VAX computers with separate image processing hardware. SEAPAK was developed to be user-friendly using NASA's Transportable Applications Executive (TAE) as the user interface.[57] Judy Chen and Mike Darzi were the initial SEAPAK developers at Goddard. In SEAPAK, all of the applications from Level-1 ingest through data analysis used the same user menu and program parameter input interface. This made the system very easy to use, allowing visitors to learn the system quickly.

SEAPAK was eventually expanded to include AVHRR SST processing and GEMPAK, the GSFC Severe Storms Branch software which used the same computer and image/graphics display systems (International Imaging Systems, IIS), and TAE interface. Jim Firestone, who had worked on the GEMPAK software, handled the linkages between the two software packages. This expansion allowed for the integration of satellite imagery and meteorological data products.[58]

---

[54] Charles R. McClain, Wayne E. Esaias, William Barnes, Bruce Guenther, Daniel Endres, Stanford B. Hooker, B.Greg Mitchell, and Robert Barnes, "Volume 3: SeaWiFS Calibration and Validation Plan", in S.B. Hooker and E.R. Firestone, editors, *NASA Technical Memorandum 104566*, NASA Goddard Space Flight Center, Greenbelt, 42 pages, (September 1992); William Barnes, "Acceptance Tests Baseline", presentation at the 1st SeaWiFS Science Team Meeting January 19-22, 1993 (in participant notebook entitled "Volume 2: SeaWiFS Project Presentations").

[55] NASA Johnson Space Center Oral History Project, *Mary Cleave*, Biographical Data Sheet, (accessed 14 May 2009).

[56] University of Miami/RSMAS Remote Sensing Group, "DSP Users Manual".

[57] Charles R. McClain, Michael Darzi, James Firestone, Eueng-nah. Yeh, Gary Fu and Daniel Endres, "SEAPAK Users Guide, Version 2.0, NASA Technical Memornadum 100728, *Volume I - System Description*, NASA Goddard Space Flight Center, Greenbelt, MD, 158 pages, (1991); Charles R. McClain, Michael Darzi, James Firestone, Eueng-nah Yeh, Gary Fu and Daniel Endres, SEAPAK Users Guide, Version 2.0, NASA Technical Memorandum 100728, *Volume II--Descriptions of Programs*, NASA Goddard Space Flight Center, Greenbelt, MD, 586 pp., (1991).

[58] Charles R. McClain, email message received 7 July 2009.

During the 1980s, McClain hosted a number of visiting scientists, graduate students, and postdoctoral students, including Gene Feldman (State University of New York/Stony Brook), Karl Banse (University of Washington), Jim Yoder (Skidaway Institute of Oceanography), Vittorio Barale (Scripps Institute of Oceanography), Frank Muller-Karger (University of Maryland), Joji Ishizaka (Texas A&M University), and John Brock (University of Colorado/Boulder). Initially, all of these researchers were using the Severe Storm Branch's computer system and had to schedule time. This led to some very interesting negotiations one hot summer when Feldman, Banse, Barale, and McClain were all politely vying for time on the system. The group eventually moved to another facility in Building 28, which was noteworthy because the available computers were abysmally slow. One day, McClain found Judy Chen with tears in her eyes because she would type a command and have to wait minutes for a response. Muller-Karger finally broke the proverbial camel's back when he racked up a bill for computer time ranging somewhere between $50,000 and $100,000. At that point, the ocean color team decided it was advisable to buy their own system![59]

The primary disadvantage of SEAPAK was the overall expense of the VAX, IIS, and disk drive storage systems, which together was of the order of $100,000 or more. Eventually, SEAPAK was ported to a UNIX-based Silicon Graphics workstation (Brian Schieber primarily handled this with guidance from Gary Fu) without the graphics hardware requirement. Several years later, the CZCS processing functions were incorporated into the SeaWiFS Data Analysis System (SeaDAS). At that point in the early 1990s, SEAPAK development and support ended. The success of SEAPAK was noted more widely than just in the oceanographic community; an article about the use of the software to observe "marine life" with the software package appeared in Government Computer News in 1991.[60]

During Jim Yoder's first tenure at NASA HQ in the late 1980's, having become familiar with the advantages of SEAPAK, he approached Chuck McClain about the feasibility of porting SEAPAK to a PC-based system and making the software available to the community. Jim Firestone and Gary Fu, two of the SEAPAK software developers, researched the existing PC computer and graphics display hardware, peripherals (disk drives, 9-track tape drives, color printers, monitors, Ethernet cards, etc) and window multi-tasking software products (remember, this was when DOS was the standard). They developed a system concept using Intel 386 systems with a Matrox graphics card and DESQview window multitasking environment, which could be augmented with a Weitek chip (a special socket on most 386 motherboards). Running under the 386 protected mode circumvented the DOS 640KB memory limit.[61]

Yoder agreed with the design concept and provided additional SEAPAK support for the development while he was at NASA Headquarters (his first stint as the head of the Ocean Biology and Biogeochemistry program). It took about a year to port the SEAPAK

---

[59] Charles R. McClain, email message received 7 July 2009.

[60] Judith Silver, "Seapak analyzes satellite data to chart marine life," *Government Computer News*, 10(21), 78, (14 October 1991).

[61] Judith Silver, "Seapak analyzes satellite data to chart marine life."

programs with the same graphics overlay capabilities and user interface analog (PC-SEAPAK did not include GEMPAK routines) to personal computers. McClain recalled that "Processing time comparisons showed that the PC with the Weitek chip blew the socks off a MicroVAX-II!" The software was made available using a Space Act agreement through COSMIC (NASA's software distribution center). Gary Fu was the lead developer and handled user support. Many ocean color researchers cut their teeth using PC-SEAPAK.[62] It also laid the groundwork for SeaDAS, discussed later in this chapter.

The rapid advances in computing occurring in the late 1980s, which now included the adolescent Internet, impacted the availability of ocean color data and its associated software – because PC-SEAPAK could actually be obtained from a *Web site*! The PC-SEAPAK User's Guide goes into considerable detail about how a PC system could be set up to run the software; back in these dark ages of PC computing, an unfortunate aspect of advanced software was that if the system wasn't precisely set up according to fairly detailed specifications, the software probably wouldn't run. And back in those ancient times, PC-Seapak required a demanding *20 Mb* of hard disk storage space.[63]

Despite these considerations, and the daunting aspects of a 332-page users' manual, PC-SEAPAK was another noteworthy expansion of the availability of ocean color data to the scientific community. And in a couple of years, it would become a critical aspect of the SeaWiFS mission.

Fabrication phase

OSC contracted with the Hughes Aircraft Santa Barbara Research Center (SBRC) to build the SeaWiFS instrument. Dick Roberts was the program manager, with Dave Rogers the lead electronics engineer and Alan Holmes the systems engineer. The team reported to Aram Mika, Vice President/General Manager of the Systems Division, Hughes SBRC.[64] According to John Mehoves, one of the critical elements of the program was that Mika agreed to a firm fixed-price contract with OSC for the sensor; had he not done this, "the risk would have been too great for Orbital to take on alone".[65] The SeaWiFS Project designated instrument calibration specialist Robert Barnes to interact with SBRC on the optical design, performance, and calibration issues; Barnes set up a standing hotel reservation in Goleta and made numerous trips to the West Coast.

The preliminary design review (PDR) for SeaWiFS was held on July 6, 1991, and the Critical Design Review was held on December 16, 1991. Alan Holmes recalled that the PDR was scheduled to start at 8:30 AM, and that when it was time to begin, Project Manager

---

[62] Charles R. McClain, Gary Fu, Michael Darzi, and James K. Firestone, "PC-Seapak User's Guide Version 4.0," *NASA Technical Memorandum 104557*, NASA Goddard Space Flight Center, 332 pages, (January 1992).
[63] Charles R. McClain, Gary Fu, Michael Darzi, and James K. Firestone, "PC-Seapak User's Guide Version 4.0."
[64] Carl Schueler, email message received 13 January 2010.
[65] John Mehoves, email message received 13 January 2010.

Kirk was not present. They decided not to wait, and Kirk entered the meeting room in the middle of the first presentation. Holmes said this showed that the fast-track SeaWiFS program was "leaving the station".[66] For the next 10 months, SBRC painstakingly built two SeaWiFS sensors "in parallel", and the engineering unit was completed in September of 1992. The flight unit was completed shortly thereafter.

The fabrication phase for SeaWiFS provided several puzzles that required significant pondering by skilled and experienced instrument designers. One of the first concerns was the motor that would rotate the SeaWiFS telescope – this motor would be required to turn for more than a billion rotations. The supplier delivered a motor for SeaWiFS that had never run, requiring significant checks on its performance. SBRC discovered that the motor drag was way too high by "a lot" – which meant too much power consumption. The problem was identified as the drag in the bearings. With the design group at El Segundo stumped, an expert on ball bearings from Hughes suggested using 1/4 - inch bearings rather than the 1/8-inch bearings currently in the motor. The expert had been present at three design meetings and had never said anything (which was amazing to Alan Holmes), but when he was finally asked about what should be done, he immediately confirmed that the 1/4 – inch bearings would very likely fix the problem. Holmes said that this created a standard rule in his experience for future reference, the grammatically incorrect "4B" rule, regarding the size and efficacy of bigger ball bearings being better in instruments.[67]

Holmes also noted that the original motor failed 30 days into testing, with the rest of SeaWiFS almost finished. This was not good for a motor that needed to run for at least five years. The problem turned out to be the lubricating oil. Holmes said that many designers of space instruments are concerned about oil vapor pressure – which is rarely the problem. Rather, oil has a tendency to "wick" through any passageway into an adjoining small space, in a physics-based effort to reduce surface tension in any way possible. So when the failed motor was opened up, oil was everywhere. This required a complete redesign of the bearings; Holmes noted that the bearings can only be touched with a "knife edge" to prevent wicking.

One of the other events that Holmes noted was a near-disaster prevented by prior experience. SeaWiFS was built in a Class 10,000 clean room, a room similar to where the MODIS instrument was built, though MODIS was in a larger room. Holmes took pains to cover SeaWiFS with static-proof plastic every night after work was done, and the MODIS engineers questioned the need for that. Holmes told them that when he had been at Ball Aerospace, he had been working in a two-story clean room in a four-story building. A toilet on an upper floor flooded, cascading water into the clean room. According to Holmes, the MODIS engineers laughed at the need for such precautions at the SBRC facility – until an unexpected massive rainstorm in southern California caused the roof to leak, casting water onto the protective plastic covering SeaWiFS – and creating a puddle on the floor of the clean room where MODIS was being built. By the end of that week, it was standard

[66] Alan Holmes, telephone interview recorded on 14 April 2009.

[67] Alan Holmes, telephone interview recorded on 14 April 2009.

operating procedure to cover MODIS with static-proof plastic at the end of the work day.[68]

This completion commenced a strenuous program of testing. One of the first things that had to be done was to compare the radiometric "integrating spheres" used at Goddard and at SBRC.[69] An integrating sphere is an apparatus that has an optically-uniform coating on the interior of the sphere that provides a diffuse reflectance target for optical testing. Integrating spheres come in various sizes, from tabletop models to spheres 4 meters in diameter. Figure 5.1 shows an integrating sphere used at the Jet Propulsion Laboratory.[70] Once the comparison was done, SBRC could use their sphere for ongoing testing of SeaWiFS.

Figure 5.1 Integrating sphere for the Multi-angle Imaging Spectroradiometer (MISR) at the Jet Propulsion Laboratory.

SeaWiFS was rolled out to view the outside world on December 8, 1992, and performed measurements of the light from the Sun and the Moon. Thermal vacuum and vibration testing took place in January, a second field test was undertaken in March, and on April 27, 1993, SBRC performed a "pre-ship" review.[71]

This review revealed two serious concerns. The most significant concern was the impact of "stray light", which is essentially reflected light bouncing around the interior of the instrument before landing on the instrument's focal planes. [This discovery of stray light effects also had implications for MODIS, which SBRC was also fabricating.] This was a particular problem when the instrument was viewing particularly bright parts of the Earth surface, or even brighter atmospheric clouds (so bright that they were nearly akin to direct sunlight).[72] Stray light from bright targets created optical cross talk between the "blue" side detectors and the "red side" detectors, which were situated on a focal plane and separated with a precise saw cut. To decrease the cross talk for bright targets, Holmes added a blue filter and a red filter; reflections from the filter surfaces created more stray light. (Holmes said that he reminded the reviewers that stray light had been identified at the PDR as a

---

[68]Alan Holmes, telephone interview recorded on 14 April 2009.

[69] Robert A. Barnes, William L. Barnes, Wayne E. Esaias, and Charles R. McClain, "Volume 22, Prelaunch Acceptance Report for the SeaWiFS Radiometer", in S.B. Hooker, E.R. Firestone, and James G. Acker, editors, *NASA Technical Memorandum 104566*, NASA Goddard Space Flight Center, Greenbelt, 32 pages, (September 1994).

[70] NASA Jet Propulsion Laboratory, http://www-misr.jpl.nasa.gov/mission/images/sph00.jpg.

[71] Robert A. Barnes, William L. Barnes, Wayne E. Esaias, and Charles R. McClain, "Volume 22, Prelaunch Acceptance Report for the SeaWiFS Radiometer."

[72] Alan Holmes, telephone interview recorded on April 14, 2009.

potential problem, and there had been no comment at that time. When he brought out the slide from the PDR to show when they were reworking the instrument, "it got a lot more attention".[73])

The other major concern, which had been on the minds of the ocean remote-sensing scientists for years since the CZCS, was the problem of sensor "ringing" – the electro-optical reverberations that happened when the instrument scan transitioned from the bright target (particularly a land surface or cloud) to the darkness of the oceanic blue. Both of these problems were related to pre-ship review's finding that SeaWiFS did not meet the bright-target recovery (BTR) specification.[74] The problem had arisen because the rapid process of building SeaWiFS meant that the engineering test model did not exactly duplicate all the characteristics of the flight unit. It was also significant; if unaddressed, the problems would mean a loss of data for 10 pixels after the bright target was seen, which translated to a distance of more than 10 km on the ocean surface. Unacceptable amounts of data in crucial coastal zones and near clouds would be lost.

Because the fabrication of both the satellite that would carry SeaWiFS and the launch vehicle were delayed, there was time to address these two issues. Thus, a month after the pre-ship review, a stray light review was held at GSFC.[75] Detailed analysis of the light paths within the instrument was presented, and a variety of alternatives to address the problem were presented. Each of the options was reviewed in terms of the impact on the sensor and the potential difficulty of making the required changes. After reviewing the alternatives, NASA and OSC modified the contract to allow the changes to be made. This was one of the few times that NASA found it necessary to allow a SeaWiFS contract modification.[76] Mark Abbott noted that it was very important that the SBRC team, under the guidance of Aram Mika, was willing to add the stray light requirement to the SeaWiFS contract.[77] Mika was the Vice President heading the Systems Division of Hughes SBRC.[78]

The course of action required to address the problem required both modifications to the instrument and changes to the data processing scheme for the data. In the electronics of the instrument, a *bilinear gain* was added (Figure 5.2)[79] The bilinear gain utilized the four detectors that detected the light for each of the SeaWiFS bands and the TDI electronic circuitry. As the telescope scanned the Earth, the detectors acquired light from four adjacent pixels. The signal from the detector for each pixel was temporarily stored. A pixel was thus viewed consecutively, in a very short time period, by each of the detectors, and the total signal from each pixel was summed electronically. Three of the four detectors were set to saturate at a maximum radiance just a bit higher than the maximum radiance predicted for

---

[73] Alan Holmes, telephone interview recorded on April 14, 2009.
[74] Alan Holmes, telephone interview recorded on April 14, 2009.
[75] Alan Holmes, telephone interview recorded on April 14, 2009.
[76] Mary Cleave and Gene Feldman recorded interview, 21 November 2008.
[77] Mark Abbott, recorded interview, November 28, 2008.
[78] Carl Schueler, email message received January 13, 2010.
[79] Robert A. Barnes, Robert E. Eplee, Jr., G. Michael Schmidt, Frederick S. Patt, and Charles R. McClain, "Calibration of SeaWiFS. I. Direct techniques," *Applied Optics*, 40(36), 6682-6700, (20 December 2001).

open ocean waters, but the fourth detector was set with a much higher saturation radiance – approximately the radiance predicted for bright clouds. So over anything brighter than the ocean surface, the three detectors would saturate and provide the value of their saturation setting to the sum, while the fourth detector would continue to detect higher radiance values, and contribute them to the final sum.[80] At the point where the three detectors saturate, there is an abrupt change in the sensor response. This point was called the *knee radiance* because the plot of instrument response (Figure 5.2) bends like a knee at that point. The bilinear gain would allow SeaWiFS to view bright targets, though with less optical sensitivity than that required for the ocean surface. (The bilinear gain also had a serendipitous consequence that would be fully realized only after the sensor commenced its remote-sensing mission.) In order to fix the ringing effect, also called the *electronic recovery tail*, a specific capacitor in the instrument was removed from the electronics path, which reduced the problem by a factor of 100. [81]

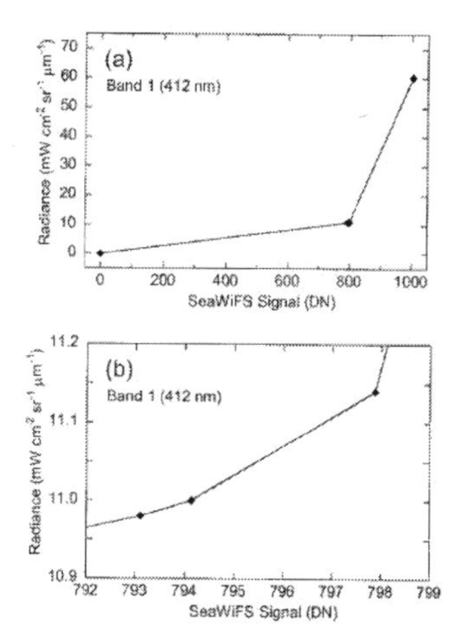

Figure 5.2. Diagram of the SeaWiFS bilinear gain response, from Barnes et al. (2001). Accompanying caption: "The bilinear calibration curve for SeaWiFS band 1. The signals are the net outputs from the band _after the zero offset has been removed. (a) Measurements over the entire dynamic range of the band. The spectral radiances for ocean measurements occur below the knee and for land occur above the knee. The knee is actually three knees in one. (b) Measurements in the region of the three knees. Each knee occurs when one of the high sensitivity channels goes into digital saturation (1023 DN)."

The more difficult aspect of the BTR fix took place inside the instrument itself. One of the options considered was to place small mechanical separations (*septums*) between the

[80] Robert A. Barnes, Alan W. Holmes, and Wayne E. Esaias, "Volume 31, Stray Light in the SeaWiFS Radiometer," in S.B. Hooker and E.R. Firestone, editors, *NASA Technical Memorandum 104566*, NASA Goddard Space Flight Center, Greenbelt, MD, (July 1995), 76 pages.
[81] Robert A. Barnes, Alan W. Holmes, and Wayne E. Esaias, "Volume 31, Stray Light in the SeaWiFS Radiometer."

focal planes. But because these were already held fast in place with epoxy, moving them to insert septums was impossible. In fact, during the fabrication of the instrument, one of the focal planes failed and had to be replaced, and when the overlying filter was removed, it was chipped and had to be replaced.[82] Had septums been part of the initial design, the optical cross talk problem would not have occurred.

So the instrument team came up with a clever and relatively simple solution. The flat surface of the polarization scrambler (hearkening back to the polarization scrambler of the CZCS) produced ghost reflections. By tilting the scrambler just a little – giving it a wedge surface for reflection rather than a flat surface – the ghost reflections were summarily exorcised.[83] Similarly, the filter for the blue side of the detector array was tilted slightly, which reduced the stray light reflections that leaked over to the red side.

Following the review, OSC undertook the modifications through the summer and fall, and performed one more field test on November 1, 1993. During this field test, the optical team performed the SeaWiFS Transfer-to-Orbit Experiment (an experiment suggested by calibration guru Philip Slater of the University of Arizona) where a preflight solar calibration was done, and the results were, as the name of the experiment suggests, transferred to the instrument when it was in orbit.[84] The experiment was performed and all other tests looked good, so on December 2, 1993, a second pre-ship review took place. SeaWiFS passed, and was ready for delivery in time for Christmas (Figure 5.3)[85]

The satellite and the launch vehicle, were not ready, however – and recent events had taken place that suddenly made the stakes appear quite a bit higher.

---

[82] Robert A. Barnes, Alan W. Holmes, and Wayne E. Esaias, "Volume 31, Stray Light in the SeaWiFS Radiometer."

[83] Robert A. Barnes, Alan W. Holmes, and Wayne E. Esaias, "Volume 31, Stray Light in the SeaWiFS Radiometer."

[84] Robert A. Barnes, Robert E. Eplee, Jr., Stuart F. Biggar, Kurtis J. Thome, Edward F. Zalewski, Philip N. Slater, and Alan W. Holmes, "Volume 5, The SeaWiFS Solar Radiation-Based Calibration and the Transfer-to-Orbit Experiment," in S.B. Hooker and E.R. Firestone, editors, *NASA Technical Memorandum 1999-20689, SeaWiFS Postlaunch Technical Report Series*, 28 pages, (June 1999); Robert A. Barnes, Robert E. Eplee, Jr., Frederick S. Patt, and Charles R. McClain, "Changes in the radiometric sensitivity of SeaWiFS," *Applied Optics*, 38(21), 4649-4664, (1999).

[85] Robert A. Barnes, William L. Barnes, Wayne E. Esaias, and Charles R. McClain, "Volume 22, Prelaunch Acceptance Report for the SeaWiFS Radiometer"; Ocean Color Biology Processing Group, http://oceancolor.gsfc.nasa.gov/SeaWiFS/SEASTAR/seawifs_bench.gif.

Figure 5.3 SeaWiFS.

NOAA-13, Mars Observer, and Landsat 6

In a relatively short period of time in 1993, three confounding events took place – cogent reminders that launching and operating satellites are not simple and straightforward operations. The first of these events involved the ongoing NOAA-NASA collaboration to build and launch a continuing series of polar-orbiting environmental observation satellites. This highly successful collaboration began in the late 1970s, with NASA building the satellites and NOAA operating them once they had launched and reached orbit. The concept of the series was that each satellite and instrument on the satellite would be duplicated as closely as possible, so that the data would be familiar and (hopefully) very similar, even though different sensors were collecting it. The satellites were designated with letters (i.e. NOAA-K) before launch, and with numbers (NOAA-11) after launch. (The numbers and letters don't correspond like an alphabet; NOAA-A was redesignated NOAA-6, NOAA-B didn't reach orbit, so NOAA-C was NOAA-7.)[86]

The first numbered NOAA satellites were the TIROS series. Starting with number 8, the satellites were designated as the Advanced TIROS-N satellites (ATN).[87] (If the CZCS-II or OCI had reached orbit, they would have been on this series of satellites.) The ATN series carried the AVHRR and TOVS instruments, and a solar proton monitor. NOAA-13 also carried ERBE and the SBUV radiometer, and the Search and Rescue Satellite-aided Tracking system (SARSAT) system.[88]

---

[86] National Climate Data Center, "Satellite Metadata", http://www.ncdc.noaa.gov/oa/documentlibrary/satellite-doc.html, (accessed 15 May 2009).

[87] "Advanced TIROS-N: NOAA-G", *NASA Technical Memorandum 102977*, NASA GSFC, 21 pages, http://ntrs.nasa.gov/archive/nasa/casi.ntrs.nasa.gov/19900066687_1990066687.pdf, (accessed 15 May 2009); National Oceanic and Atmospheric Administration, "NOAA's Geostationary and Polar-Orbiting Weather Satellites", http://noaasis.noaa.gov/NOAASIS/ml/genlsatl.html, (accessed 15 May 2009).

[88] National Space Science Data Center, "Experiment Search Results – NOAA 13", http://nssdc.gsfc.nasa.gov/nmc/experimentSearch.do?spacecraft=NOAA%2013, (accessed 15 May 2009).

So NOAA-I on the ground became NOAA-13 in space on August 9, 1993 after a launch from Vandenberg. The satellite operated nominally for the first 12 days. On August 21, 1993, the satellite controllers suddenly noticed low battery voltages and high battery temperatures on all three of the satellites battery systems. An orbit later, no communications were received from the satellite.[89]

On that very same day, NASA also lost contact with the Mars Observer spacecraft, which was on approach to go into orbit around Mars.[90] Mars Observer had launched 11 months earlier on the Space Shuttle and was propelled to Mars on the Orbital Sciences Transfer Orbit Stage. August 21 happened to be a Saturday; a lot of people probably went into their NASA and NOAA offices in a bit of a shell-shocked state that next Monday.

Each satellite failure was extensively investigated. In the case of NOAA-13, the telemetry received before the satellite ran out of power indicated that the cause of the failure was the common electrical malady known as a "short circuit". In the case of NOAA-13, a screw had penetrated the aluminum insulation that was used to dissipate heat. The screw came in contact with a radiator plate, causing the short circuit, and dooming NOAA-13 to silence.[91]

The Mars Observer failure was somewhat more exciting – if that can actually be said of a satellite failure. The investigation board determined that the most likely cause was the rupture of a fuel line. In the course of pressurizing the fuel tanks to prepare for the propulsion burn that would slow down the satellite – and enable it to be captured by the gravity of Mars – the fuel line ruptured, perhaps due to the leakage of a small amount of the oxidant, nitrogen tetroxide, in the cold of space during the cruise from Earth to Mars. When the nitrogen tetroxide mixed with the monomethyl hydrazine fuel, the result was a highly energetic reaction, commonly referred to as an "explosion". The resulting fuel line leak induced the satellite to spin at a much higher rate than expected by the satellite's onboard computer, causing it to revert to a "safe" or "contingency" mode. At the same time, the high spin rate meant that the solar arrays couldn't orient and supply power to the batteries, but this was also likely a moot point because the leaking fuel would have likely ravaged the satellite's electronic systems.[92]

With the investigational boards for these two failures still getting organized in September, the much-anticipated launch of Landsat 6 was set for October 1993. The new Enhanced Thermal Mapper (ETM) had been augmented with a 15-meter resolution panchromatic band to augment the standard 30-meter resolution bands.[93] The long-lived Landsat 5 had launched in 1984 and was still operating effectively (in fact, as this was written in 2008, Landsat 5 was still working, despite showing signs of age with a failed transmitter

[89] Brian Dunbar, "Investigation panel releases report on NOAA-13 failure", http://www.nasa.gov/home/hqnews/1994/94-157.txt, (accessed 15 May 2009).
[90] Malin Space Science Systems, "The Loss of Mars Observer", http://www.msss.com/mars/observer/project/mo_loss/moloss.html, (accessed 15 May 2009).
[91] Brian Dunbar, "Investigation panel releases report on NOAA-13 failure".
[92] National Space Science Data Center, "Reports on the loss of Mars Observer", http://nssdc.gsfc.nasa.gov/planetary/text/marsob.txt, (accessed 15 May 2009).
[93] NASA, "Landsat 6", http://landsat.gsfc.nasa.gov/about/landsat6.html, (accessed 15 May 2009).

and a broken solar array drive.[94]) Landsat 6 was designated as the next satellite to carry on the successful Landsat earth imaging program, a program that is required by Congressional statute.[95]

[Chuck McClain noted that he and Bob Kirk actually worked with Landsat project manager Rick Obenschain and project scientist Darrel Williams on getting a wide-field sensor called EOS Color onto Landsat 6, who were "enthusiastic" about this plan. McClain was the EOS Color project scientist. The EOS Color team showed that they could provide a sensor for less than $10 million dollars that would fit on Landsat 6, with no risk to the Landsat 6 schedule, but NASA and EOSAT were very concerned about the schedule, and turned it down on that basis. This decision ended up, in hindsight, being a stroke of luck for the SeaWiFS Project. EOS Color and its eventual transformation into a collaborative international research project are described in context in Chapters 6 and 7.][96]

Landsat 6 was not destined to be the heir apparent of the Landsat series. The satellite launched on October 5, 1993 on a reliable Titan-III and soared out of sight from Vandenberg. Out of sight, and nearly out of the atmosphere, the "kick motor" required to push the satellite into Earth orbit suffered a rupture of a hydrazine fuel line, and the satellite failed to achieve the necessary velocity to stay in space.[97] Somewhere over and then into the Pacific Ocean, Landsat 6 returned to Earth.

Thus, in this period of sudden and shocking launch and satellite failures, the ambitious engineers of Orbital Sciences Corporation were building both an *untested satellite* to carry SeaWiFS, and a *new launch vehicle* to convey it to space! And watching their preparations from the perspective of NASA experience was a former astronaut, Mary Cleave, who knew from personal experience the importance of checking and testing, and testing and checking repeatedly, every component of a space vehicle.

The First SeaWiFS Science Team Meeting

The ocean color community had been considerably heartened by the commencement of the SeaWiFS Project in 1991, and scientific excitement was heightened when NASA issued an RFP (Request for Proposals) in 1992 that would result in the selection of the SeaWiFS Science Team. The proposals were submitted in the summer of 1992, and the team was selected late in the year. At the time, the schedule was very tight; NASA and OSC were still pushing for a SeaWiFS launch in 1993. So in the first month of 1993, the SeaWiFS Science Team convened for the very first time on January 19-22, at the Governor Calvert House historic inn located just off State Circle in Annapolis, MD.[98]

---

[94] NASA, "Landsat 5", http://landsat.gsfc.nasa.gov/about/landsat5.html, (accessed 13 May 2009).
[95] United States Code Title 15, Chapter 82, Land Remote Sensing Policy, (accessed 15 May 2009).
[96] Charles R. McClain, email message received 7 July 2009.
[97] NASA, "Landsat 6".
[98] Stanford B. Hooker, Wayne E. Esaias, and Lisa A. Rexrode, "Volume 8, Proceedings of the First SeaWiFS Science Team Meeting", in S.B. Hooker and E.R. Firestone, editors, *NASA Technical Memorandum 104566*, NASA Goddard Space Flight Center, Greenbelt, MD, 61 pages, (March 1993).

The Science Team attendees at the meeting consisted of 27 researchers from the United States and 18 members from foreign countries. In addition to the SeaWiFS science team, 11 members of the MODIS-Oceans team – a collaboration that would be increasingly important – also attended the meeting. The SeaWiFS Project staff from GSFC attended *en masse*, along with a large number of interested invitees.[99]

The agenda was packed, covering the basic goals of the mission, the data processing system, calibration and validation of the data products, the all-important issue of data encryption and public data availability (both short-term and long-term). In addition to the presentations by the SeaWiFS Project staff, each team member presented a short description of the research they were planning to conduct utilizing SeaWiFS data.[100]

Yet the issue on everyone's minds was – when would it actually launch? It wasn't just on everyone's minds; cruise schedules and dedicated observational activities were dependent on the timing of the launch. At the time of the meeting, Project Manager Kirk reported, happily, that the sensor fabrication was ahead of schedule, and not so happily that the spacecraft fabrication was delayed. Notes indicate that one of the key aspects was "personnel conflicts" within OSC. Mary Cleave noted that OSC essentially brought in a new management team for the SeaWiFS Project, headed by John McCarthy, to improve the project management structure.[101] (More discussion of the satellite issues follows in the next section.) The result of the delays was the euphemistic "schedule erosion", also known as being behind schedule. So now the launch timeframe was August to October of 1993. And SeaWiFS was, at that time, scheduled to be the first satellite on the Pegasus XL launch vehicle (at the time called the *Stretch* Pegasus), and it would also be the first time a Pegasus was dropped from Orbital's modified L-1011, rather than from the B-52 Stratofortress.[102]

The science team presentations covered an extraordinary gamut of planned effort around the world. The fact that SeaWiFS would provide actual global coverage was highly anticipated, so campaigns in the Southern Ocean were described. The calculation of primary productivity in a multitude of environments, especially in the problematic coastal zone, was a theme in several science team proposals. Involvement of the SeaWiFS team and scientists with the next JGOFS process study in the Arabian Sea was included in several proposals. Algorithm development, calibration and validation of the products, improvement

---

Supplementary material for the 1st SeaWiFS Science Team Meeting January 19-22, 1993 (in participant notebook entitled "Volume 2: SeaWiFS Project Presentations") indicates the meeting was at the Governor Calvert House. Official documents indicate only the Historic Inns of Annapolis.

[99] Supplementary material for the 1st SeaWiFS Science Team Meeting January 19-22, 1993 (in participant notebook entitled "Volume 2: SeaWiFS Project Presentations"). With so many attendees, name tags were color-coded. Science team members from the U.S. wore "Mars Magenta". Foreign science team members were given "Lunar Blue", and MODIS Science Team members had "Orbit Orange" tags. Invitees wore "Neptune Blue", a deeper-space shade of blue, and the SeaWiFS Project staff was notable with their shiny "Gold" name tags.

[100] 1st SeaWiFS Science Team Meeting January 19-22, 1993, participant notebook entitled "Volume 2: SeaWiFS Project Presentations".

[101] Mary Cleave and Gene Feldman recorded interview, 21 November 2008.

[102] Robert Kirk, "SeaWiFS Project Baselines, Status, and Introductions", 1st SeaWiFS Science Team Meeting January 19-22, 1993, presentation in participant notebook entitled "Volume 2: SeaWiFS Project Presentations".

of the all-important atmospheric correction methods utilizing new bands (and faster computers) were all aspects of the Science Team activities.[103]

The SeaWiFS Project covered virtually every aspect of the satellite's systems and data processing plans. One of the critical aspects was how the data was going to be delivered, and how it was going to be distributed. Because of the partnership with OSC, the stated plan was that any researcher would be able to get data – all they had to do was to submit a signed agreement to become a "SeaWiFS Authorized Research User", with stipulations that the data could only be used for research, and that it couldn't be given to other parties who were not Authorized Research Users. OSC controlled the rights to the commercial use of the data, both for commercial activities like fishing and ship guidance, and also for the use of images in publications sold for profit. The SeaWiFS Project was working with the Goddard Distributed Active Archive Center (DAAC) regarding how the data would be delivered and distributed.[104] The DAAC system had been created in the early 1990s in anticipation of the data stream from the Earth Observing System (EOS) missions, which were also in the planning and development stages, exemplified by the participation of the MODIS Oceans team members at the SeaWiFS Science Team meeting.[105] The DAAC was setting up a system that would utilize an innovative (and at the time, somewhat controversial) self-describing data format called HDF, for Hierarchical Data Format.[106] "Self-describing" meant that all of the associated data, or metadata, required to process the actual remote-sensing data in a data file was contained in the same data file. Previous remote-sensing data processing systems had required additional data in other files and formats to allow processing, including data as seemingly simple, yet fundamentally vital, as the solar zenith angle and the satellite viewing angle. HDF provided a capability for subsetting, which was one of the main reasons it was adopted as the SeaWiFS and EOS data format. With regard to SeaWiFS data, the DAAC would get the data two weeks after acquisition. The data embargo period required by OSC was actually seen as a good thing; this period would allow the data to be fully processed before delivery to the DAAC. The DAAC would provide the data to researchers on magnetic tapes.[107]

Another issue complicated by the OSC partnership was the ground, or HRPT, stations. SeaWiFS would broadcast its full-resolution (1.13 km) Local Area Coverage (LAC) data directly to any properly-equipped HRPT station, but would only save a small

---

[103] Stanford B. Hooker, Wayne E. Esaias, and Lisa A. Rexrode, "Volume 8, Proceedings of the First SeaWiFS Science Team Meeting".
[104] Stanford B. Hooker, Wayne E. Esaias, and Lisa A. Rexrode, "Volume 8, Proceedings of the First SeaWiFS Science Team Meeting".
[105] Supplementary material for the 1st SeaWiFS Science Team Meeting January 19-22, 1993 (in participant notebook entitled "Volume 2: SeaWiFS Project Presentations").
[106] Stanford B. Hooker, Wayne E. Esaias, and Lisa A. Rexrode, "Volume 8, Proceedings of the First SeaWiFS Science Team Meeting"; During this time, the author was present at a meeting at NASA HQ with Dixon Butler in which it was firmly decided that HDF would be the data format for SeaWiFS as a precursor to EOS, despite arguments in favor of other data formats, such as CDF or netCDF.
[107] Dorothy A. Zukor, "Data Archive and Delivery Baseline", 1st SeaWiFS Science Team Meeting January 19-22, 1993, presentation in participant notebook entitled "Volume 2: SeaWiFS Project Presentations".

amount of the LAC on-board for specific requests, which were expected to consist of open ocean areas with operating research cruises (or perhaps an unusual event). There were to be two classes of HRPT station licenses: standard and real-time.[108] There would be 13 rotating real-time licenses to support ongoing research operations close enough to the station, in order to provide directions to oceanic features of interest spied by the sensor.[109]

It was also noted that PC-Seapak would be modified to do two things: support the data processing requirements of the SeaWiFS Project, and then it would be distributed to HRPT stations for their own data processing. (Gene Feldman noted that this important connection between the stations and the SeaWiFS Project evolved significantly in the favor of science and SeaWiFS; the stations agreed to send their data to the SeaWiFS Project for processing, and the SeaWiFS Project provided the software to process the Level 0 raw telemetry to Level 1, the calibrated radiances, which was the product sent to the SeaWiFS Project. Originally the data was to be sent on magnetic tapes.[110]) Programmers at the SeaWiFS Project, including Gary Fu, Karen Settle, and Brian Schieber undertook the task of converting SeaPAK for ground station use.

One other topic of interest was a look beyond SeaWiFS – the EOS Color mission. The EOS Color mission, essentially an expanded and augmented SeaWiFS, was intended to bridge a possible gap between SeaWiFS, if it only lasted for 3 years or so, and the EOS ocean color sensors. A white paper on EOS Color was already in preparation. One of the perceived problems was if everything worked as planned, including other international missions from Japan and Europe, there could possibly be five ocean color satellite missions operating at the same time in 1998. Therefore, justifying EOS Color against the overseas competition was deemed to be a potential problem.[111]

The final plenary session of the meeting covered a variety of recommendations. Atmospheric correction was a critical topic; many investigators were concerned about dust and clouds, absorption by oxygen, and interference from stratospheric aerosols – no one at the meeting had forgotten El Chichón and the CZCS, and the aerosols from the June 1991 massive eruption of Mount Pinatubo in the Phillippines were still detectable two and a half

[108] Gene C. Feldman, "Data Processing Baseline", 1st SeaWiFS Science Team Meeting January 19-22, 1993, presentation in participant notebook entitled "Volume 2: SeaWiFS Project Presentations";
[109] When the first SeaWiFS Science Team meeting took place, there was no requirement on any of the stations to archive the data. The only LAC data that was expected to be collected routinely was the data that would be received by the GSFC HRPT station, covering the U.S. East Coast. Stan Wilson asked if it would be possible to archive data from all of the United States stations: the answer was that the Project was not funded to archive data from more stations. This resulted in a recommendation to formulate a deal for LAC from other stations. It was also recommended that the DAAC have a catalog of information from the stations, and to browse the data users could contact the individual stations – providing the station operators decided to archive their data! – Gene Feldman, comments recorded at the Ocean Color Collaborative Historical Workshop, January 13-14, 2009, St. Petersburg, Florida.
[110] Gene Feldman, comments recorded at the Ocean Color Collaborative Historical Workshop, January 13-14, 2009, St. Petersburg, Florida.
[111] Notes taken during the final plenary meeting of the 1st SeaWiFS Science Team meeting, January 19-22, 1993, by the author.

years later.[112]

Data distribution was also a major priority. There were recommendations that regional data subsetting be enabled, because it was recognized that not all of the scientific users wanted all of the global data, or even a full-size data file. It was also recommended that a plan be formulated to archive as much LAC data as possible, even if it just meant "dry storage" until funds could be fund to unpack the tapes.[113]

The team formed several subgroups at the end of the meeting: atmospheric correction headed by Howard Gordon (not a surprise to anyone); calibration protocols, headed by Jim Mueller; primary productivity algorithms, headed by Paul Falkowski; bio-optical algorithms, possibly headed by Chuck McClain or Dennis Clark, and a sensor calibration team.[114]

Anchors aweigh; the SeaWiFS Science Team was putting out to sea.

SeaStar

The satellite that was built to carry SeaWiFS was originally named SeaStar, taking the name from the class of satellites that OSC built to be carried by the Pegasus and Pegasus XL, the *PegaStar.* When OSC spun off its affiliate Orbimage in 1997 as an affiliate devoted to remote sensing, SeaStar was renamed Orbview-2. Orbimage subsequently became GeoEye with the acquisition of Ikonos. (Orbview-2 was not renamed GeoEye-2.) The PegaStar was a modified satellite platform designed for an optimal 3-5 year lifetime, and was built out of a honeycomb aluminum framework. The satellite operating section, with the power, control systems, and solar power array, is connected to the payload platform. The spacecraft utilized the now-familiar 3-axis stabilization pioneered by the Nimbus series, maintaining attitude with momentum wheels and torque rods. The satellite's attitude determination system featured horizon and sun sensors, as well as magnetometers. Orbit raising (to be discussed subsequently) and orbit control were accomplished with a hydrazine propulsion system, and orbit telemetry included GPS data from onboard GPS receivers. The satellite had multiple communications capability in L-band and S-Band. The solar power array provided about 170 watts of power.[115] In fact, placing sufficient hydrazine on the spacecraft to reach remote-sensing altitude was the main design change from the original

[112] Stephen Self, Jing-Xia Zhao, Rick E. Holasek, Ronnie C. Torres, and Alan J. King, "The atmospheric impact of the 1991 Mount Pinatubo eruption", in C.G. Newhall and R.S. Punongbayan, editors, *Fire and mud: Eruptions and lahars of Mt. Pinatubo, Philippines,* Philippine Institute of Volcanology and Seismology, Quezon City and University of Washington Press, Seattle, 1126 p, (1996).

[113] Stanford B. Hooker, Wayne E. Esaias, and Lisa A. Rexrode, "Volume 8, Proceedings of the First SeaWiFS Science Team Meeting".

[114] Notes taken during the final plenary meeting of the 1st SeaWiFS Science Team meeting, January 19-22, 1993, by the author.

[115] "Pegastar", http://www.spaceandtech.com/spacedata/satellites/pegastar_sum.shtml, (accessed 15 May 2009); NASA, "Access to Space: Orbital Pegastar", http://accesstospace.nasa.gov/ats3/accessmodes/spacecraft_buses/Orbital/PegaStar.asp?proxyid=guest&xsection=3&subxsection=5&subsubxsection=6, (accessed 15 May 2009).

satellite bus; the electronics were moved to pallets on the side of the spacecraft.

During the preparations for the SeaWiFS launch, OSC had time and motivation to make considerable refinement to the Pegastar space vehicle. Howard Runge said that Orbital had the luxury under the contract of designing as they saw fit, and said "if NASA didn't like it, they would at least know about it".[116] However, Runge noted specifically, the low budget for the spacecraft was always a dominant consideration. Runge said that for such items as data recorders, error detection and correction schemes (EDACs), and solid-state microswitches, NASA thought that OSC was overconfident in the reliability of new technologies and that the real issues would be with the space vehicle, and not the launch vehicle. The Pegastar had to be sufficiently maneuverable to accomplish the requirements for solar and lunar calibration, and the angular momentum of the spinning telescope also had to be managed. Runge stated that the Pegastar was a *redundant* satellite; it had a backup system for every avionics function. One of the innovations was a "tape measure" hinge, alternatively called a *strain energy hinge)*, that locked the solar panels in place after they were deployed. (Imagine a standard metal tape measure, which is flexible but curved. With sufficient effort, the tape measure can be bent. But as it straightens out, the curve of the metal band makes it fairly rigid. The same principle was used to maintain the extension of the solar arrays on Pegastar after they were deployed. In fact, the key element was actually a tape measure.)[117]

There were also doubts about the solid-state recorders and new batteries; some of the NASA battery people disagreed strongly with OSC's battery life testing and projections, and they reportedly thought the satellite "wouldn't last a year".[118] Radiation hardness was a particular concern; despite some compromises, the satellite in orbit only had a few radiation problems. One aspect of the detection of "bit flips" (or single event upsets, SEUs) by the solid-state recorders was that it enabled a map of the main place where they occurred, the problematic South Atlantic Anomaly (SAA).[119] Runge also said that the plan emphasized on-orbit calibration; they discovered that the solar calibration was too strong, so the sensor actually viewed the diffuser plate with a periscope covered by a plate with holes in it, so that only a fraction of the light intensity came through! Because the solar diffuser changed due to paint aging in the harsh sunlight hitting it in the vacuum of space, the space vehicle had a backup diffuser plate, but during the entire mission, they never popped the cover on it. Pegastar also had a spin-up momentum wheel as a momentum compensator, and "very simple" horizon sensors were used for attitude control.

SeaStar was one of the first in-house satellite projects at OSC. The PegaStar was

---

[116] Howard Runge comments recorded at the Ocean Color Collaborative Historical Workshop, January 13-14, 2009, St. Petersburg, Florida.
[117] Howard Runge comments recorded at the Ocean Color Collaborative Historical Workshop, January 13-14, 2009, St. Petersburg, Florida.
[118] Howard Runge and Charles McClain, comments recorded at the Ocean Color Collaborative Historical Workshop, January 13-14, 2009, St. Petersburg, Florida
[119] Christian Poivey, George Gee, Janet Barth, Ken Lebel, and Harvey Safren, "Lessons Learned from Radiation Effects on Solid State Recorders (SSRs) and Memories", (2002), http://radhome.gsfc.nasa.gov/radhome/papers/2002_SSR.pdf, (accessed 1 June 2009).

first used for the APEX mission, which launched in August 1994. SeaStar was fabricated in collaboration with the SeaWiFS Project. Initially, this collaboration was not particularly well-received by OSC, and NASA was unimpressed with OSC's design review process, characterizing the first Preliminary Design Review as akin to a "graduate thesis defense".[120] It was clear fairly soon that OSC had very little or no experience with some issues. NASA's relationship with OSC on the Project was one of insight, not oversight; NASA could provide on very short notice skilled engineers with decades of spacecraft fabrication experience. OSC's attitude appeared to change in 1993, with a new management structure, placing John McCarthy as Project Manager. NASA advised and critiqued OSC on multiple aspects of the spacecraft, including the radiation hardness of the electronics, the placement and operation of horizon sensors, and the power system.

With the sensor finished, the delays to the Project involved the spacecraft and the Pegasus XL launch vehicle. While Kirk had indicated to the SeaWiFS Science Team in January 1993 that SeaWiFS would be the first payload on the Pegasus XL, the Project nixed that idea in short order; Mary Cleave said that she wanted SeaWiFS "to be on the first launch after the first successful launch".[121] So with SeaWiFS finished and delivered in December 1993, the flight manifest was adjusted, and the military STEP-1 was designated for the first ride on the Pegasus XL, which took place on June 27, 1994.

And which – to put it mildly – did not go as smoothly as hoped and intended (as described at the beginning of the chapter).

Because the Pegasus and Pegasus XL were new options for the launch of small and mid-sized satellites by the United States, getting them to a flightworthy status was important to the many parties with vested interests, not just the SeaWiFS Project. A failure investigation board was convened quickly to analyze the reasons for the failure. The determination of the cause was fairly straightforward; OSC had not performed any wind-tunnel testing of the Pegasus XL aerodynamics, relying instead on computer simulations extrapolated from the flight characteristics of the Pegasus.[122] With inaccurate aerodynamics used in the flight control software, the software could not adequately control the rocket, and thus, as clearly seen by the launch observers, the rocket went out of control. The solution was fairly straightforward, too: perform wind tunnel aerodynamics tests. OSC undertook four corrective actions, and readied the next Pegasus XL for launch.

Only five days before the anniversary of the first attempt to launch the Pegasus XL, the next one, with another military satellite, STEP-3, on-board, was dropped and launched from the OSC L-1011 on June 22, 1995. This time, everything looked nominal (i.e., good) as the rocket soared into the deep blue of the sky toward Earth orbit and out of sight.

Unfortunately, soon after it was out of sight, another problem developed. This one occurred at first and second stage separation. The Pegasus XL was a three-stage rocket, not requiring a heavy first stage to reach 30,000 feet altitude or so. When the first stage

---

[120] Mary Cleave and Gene Feldman recorded interview, 21 November 2008.

[121] Mary Cleave and Gene Feldman recorded interview, 21 November 2008.

[122] I-Shih Chang, "Investigation of Space Launch Vehicle Catastrophic Failures", American Institute of Aeronautics and Astronautics, AIAA-1995-3128, presented at 31st ASME, SAE, and ASEE, Joint Propulsion Conference and Exhibition, San Diego, CA, July 10-12, 1995.

separated, the interstage ring which connected the two stages and which protected the second-stage engine, designed to fall off after explosive bolts fired to release it, didn't release. The thrust of the steerable second-stage engine was now directed against the sides of the ring, rather than directly from the engine, which severely restricted its ability to steer. So the second stage tumbled out of control, and again had to be destroyed by the range safety officer.[123]

This launch failure had more ramifications than just delaying the launch of SeaWiFS again. NASA had committed several small satellites to the Pegasus XL, and now faced the potential need to investigate other launch vehicle options.[124] Insurance companies, having watched two satellites worth an aggregate total of $100 million dollars or so disappear in less than a year, were getting reluctant about insuring another launch.[125] So Orbital was faced with an urgent need to get the launch vehicle reliable, and to have some successful launches.

Furthermore, Chuck McClain noted that as the SeaWiFS schedule slipped, and slipped again, Goddard director Klineberg came close a few times to canceling the program altogether. But he desisted.[126]

Just to help remind everyone of the perils of spaceflight, the movie *Apollo 13* premiered in June 1995, too.[127] Mary Cleave, who had taken over as SeaWiFS Project Manager from Robert Kirk, pulled in a spacecraft engineer to advise OSC on how to achieve a clean stage separation. As Cleave noted, it is very hard in space to achieve a "clean sep", and there are guys "who do that for a living".[128] One of these "guys" briefly became an OSC consultant.

Waiting for Launch: SeaWiFS Project and Science Team Activities

After the initial 1991 meeting, the SeaWiFS science team convened a number of subcommittees and dedicated themselves to assessing and improving a large variety of scientific topics relate to the acquisition of remotely-sensed ocean radiances. The SeaWiFS Project had decided to attempt something truly unprecedented in the history of satellite remote sensing; to extensively document just about everything related to the mission in a

---

[123] Federal Aviation Administration, "Special Report: U.S. Small Launch Vehicles: 1st Quarter 1996", http://www.faa.gov/about/office_org/headquarters_offices/ast/media/sr_96_1q.pdf, (accessed 18 May 2009); Futron Corporation, "Design Reliability Comparison for SpaceX Falcon Launch Vehicles", http://www.spacex.com/FutronDesignReliability.pdf, September 2004, (accessed 18 May 2009).

[124] "NASA eyes alternate launchers; Pegasus XL remains grounded". Satellite News, August 14, 1995, http://www.accessmylibrary.com/coms2/summary_0286-6342837_ITM, (accessed 18 May 2009).

[125] "Orbital Sciences Suffers Another Pegasus XL Launch Failure", Satellite News, June 26, 1995, http://www.accessmylibrary.com/coms2/summary_0286-6341683_ITM, (accessed 18 May 2009).

[126] Charles McClain, comments recorded at the Ocean Color Collaborative Historical Workshop, January 13-14, 2009, St. Petersburg, FL.

[127] David Brandt, "NSS Promotes, Hosts Premiere of Apollo 13," *SpaceViews*, June 1995, http://seds.org/archive/spaceviews/9507.html, (accessed 18 May 2009).

[128] Mary Cleave and Gene Feldman recorded interview, 21 November 2008.

series of detailed Technical Memoranda, which was called the *SeaWiFS Technical Report Series*. The Technical Report Series was produced by the Project under the editorial leadership of Stanford Hooker and Elaine Firestone.[129]

The Technical Report series covered both topics directly related to the spacecraft and sensor (such as Volume 22, the Prelaunch Acceptance Report for the SeaWiFS Radiometer) and a broad variety of other topics[130]. One of the critical first volumes was Volume 5, Ocean Optics Protocols for SeaWiFS Validation, authored by Jim Mueller and Roswell Austin.[131] This volume provided exact specifications for how to take ocean optical measurements at sea, including solar radiance instruments on the deck of the ship; in-water instrumentation at the surface; profiles of the ambient light environment from the surface down to much deeper depths; how to collect and analyze data from instrumented moorings and drifting buoys; how to calibrate the instruments used to collect the data; how to measure the plankton pigments and chlorophyll, including how many samples to collect. The memorandum even described how to avoid the shadow of the ship from which the in-water instrumentation was deployed! Volume 5 immediately became an essential handbook for ocean optical scientists around the world, because if the data wasn't collected according to the protocols, it probably wasn't going to be useful to, or used by, the SeaWiFS Project.

Volume 13 was entitled Case Studies for SeaWiFS Calibration and Validation, Part 1, covering a comparison of existing bio-optical algorithms; corrections for ozone in the atmosphere; CZCS pigments in the Southern Ocean; and studies of sensor ringing and sensitivity to sun glint.[132] Volume 14 covers the first SIRREX (SeaWiFS Intercalibration Round-Robin Experiment), a continuing effort to examine every aspect of radiometric instrument calibration. SIRREX, which eventually involved scientists and engineers from the National Institute for Standards and Technology (NIST, formerly the National Bureau of Standards) was a painstaking program, with Hooker managing the effort for the SeaWiFS Project. SIRREX analyzed integrating spheres, lamps used to calibrate the radiometers – even power sources used to deliver a supposedly constant current to the calibration lamps.

The TM series also included meeting minutes; Volumes 18 and 24, the third and fourth, cumulative indices respectively, include the proceedings of the third, fourth, and fifth BAOPWs, the Bio-Optical Algorithm Optical Protocols Workshops. Volume 18 also

---

[129] Stanford B. Hooker and Elaine R. Firestone, editors, "The SeaWiFS Technical Report Series", *NASA Technical Memorandum 104566*, (other editors assisted on specific volumes of the Technical Report Series). Available online: SeaWiFS Prelaunch Technical Report Series, http://oceancolor.gsfc.nasa.gov/cgi/tech_memo.pl ; SeaWiFS Postlaunch Technical Report Series, http://oceancolor.gsfc.nasa.gov/cgi/postlaunch_tech_memo.pl.

[130] Robert A. Barnes, William L. Barnes, Wayne E. Esaias, and Charles R. McClain, "Volume 22, Prelaunch Acceptance Report for the SeaWiFS Radiometer".

[131] James L. Mueller and Roswell W. Austin, "Volume 5: Ocean Optics Protocols," in S.B. Hooker and E.R. Firestone, editors, NASA Technical Memorandum 104566, NASA Goddard Space Flight Center, Greenbelt, Maryland, 43 pages, (July 1992).

[132] Charles R. McClain, Josefino Comiso, Robert Fraser, James K. Firestone, Brian Schieber, Eueng-nan Yeh, Kevin R. Arrigo, and Cornelius W. Sullivan, "Volume 13: Case Studies for SeaWiFS Calibration and Validation, Part 1" . in S.B. Hooker and E.R. Firestone, editors, *NASA Technical Memorandum 104566*, NASA Goddard Space Flight Center, Greenbelt, Maryland, 52 pages plus color plates, (January 1994).

includes the proceedings of a particularly important program review, the MOBY software review. MOBY stands for the Marine Optical Buoy; and MOBY was a key element to the hoped-for success of the SeaWiFS mission.[133]

At the same time, many of the oceanographers on the SeaWiFS Science Team were also involved in other oceanographic research efforts, most notably the JGOFS Arabian Sea Process Study, which commenced in October 1994 and ended in January 1996.[134]

MOBY and the MODIS Science Team

As noted earlier, there were several different groups represented at the first SeaWiFS Science Team, sporting different-colored nametags. One of the groups was the members of the MODIS science team (a few members of the SeaWiFS science team were also members of the MODIS science team – the notes from the first SeaWiFS Science Team meeting don't indicate if these individuals were allowed to wear two different colors).

Because the SeaWiFS Project was a data-buy mission with a programmatic goal of keeping costs down, NASA HQ directed the MODIS Science Team, which had a considerably higher budget, to work with the SeaWiFS science team on matters of mutual benefit to both teams.[135] One of the areas of mutual benefit was atmospheric correction; Howard Gordon was designated to lead the atmospheric correction algorithm development for both teams, but he could apply some of this work on the MODIS algorithms to SeaWiFS where possible. In a likewise fashion, the bio-optical algorithm teams could also work together. Mission operations shared computational facilities and examined different scenarios for viewing geometry and coverage, particularly regarding the differences between sensors capable of tilt and sensors that did not have that capability. The evolution of MODIS for ocean observations will discuss this in more detail.

One of the clearest areas of synergy was the MODIS Optical Buoy, aka MOBY.[136] Dennis Clark headed the program to develop a moored in-water radiometer with the specific objective of collecting radiometric data at and below the sea surface at the same time that satellite sensor was overhead, scanning the same patch of ocean where the buoy was moored. Clark had actually begun collecting and creating components of the buoy in 1986,

---

[133] Elaine R. Firestone and Stanford B. Hooker, "Volume 18: SeaWiFS Technical Report Series Cumulative Index: Volumes 1-17," in S.B. Hooker and E.R. Firestone, *NASA Technical Memorandum 104566*, NASA Goddard Space Flight Center, Greenbelt, Maryland, 47 pages, (February 1995); Elaine R. Firestone and Stanford B. Hooker, "Volume 24: SeaWiFS Technical Report Series Cumulative Index: Volumes 1-23," in S.B. Hooker and E.R. Firestone, editors, NASA Technical Memorandum 104566, NASA Goddard Space Flight Center, Greenbelt, Maryland, 36 pages, (June 1995).

[134] "Arabian Sea Process Study", http://www1.whoi.edu/arabian.html, (accessed 18 May 2009); Abstracts of Deep Sea Research special issue, http://www1.whoi.edu/mzweb/arabdsr.htm, (accessed 18 May 2009).

[135] Stanford B. Hooker, Wayne E. Esaias, and Lisa A. Rexrode, "Volume 8, Proceedings of the First SeaWiFS Science Team Meeting".

[136] David Herring, "MODIS/SeaWiFS team deploys marine optical buoy, continues Marine Optical Characterization Experiment," *The Earth Observer*, 6(1), 17-22, (January/February 1994).

developing and refining the buoy each year – on a very limited budget, many times utilizing "end of year money", which means money left over at the end of the fiscal year that needs to be spent – if there is any.[137] One improvement was the shape of the float that supported the buoy on the surface; after watching a presentation by a company (Gilman Corporation) that specialized in such floats, a design with a blunt conical shape at the top was selected. This shape worked well because as the buoy bobbed downward, pressure from the increased immersion depth built up rapidly, bringing the buoy back to the surface and maintaining good stability. The MOBY figures (Figure 5.4) show the old and new float design.[138]

Comparison of the satellite measurements to the buoy measurements of the same radiometric variables, particularly the upwelling light field that creates the water-leaving radiances measured by the satellite sensor (when those few plucky surviving quanta finally made it out of the atmosphere into space) would allow a continuous validation target to maintain the calibration of the sensor. This would reduce the reliance on measurements taken from research cruises (though these were still important) and would provide data from an area of ocean with well-defined optical characteristics.

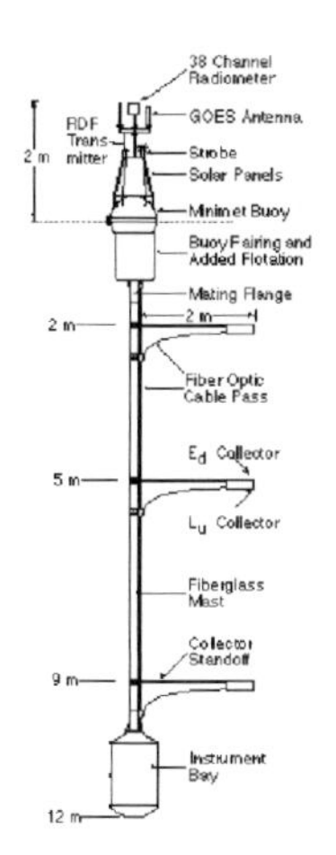

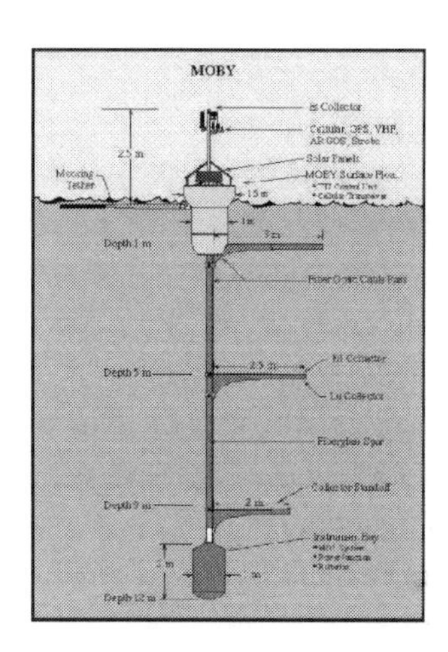

Figure 5.4. Early MOBY (left), refined operational MOBY (right).

Choosing where to deploy MOBY was an initial predicament. The desired characteristics of the area included low wave action, so a location in the "wind shadow" of an island was indicated. A region with clear skies as often as possible was also desired; it was undesirable to have a cloud over the area where the buoy was located when the sensor had a chance to view that area. Because the main focus of the measurements was on the radiometric accuracy of the sensor, an area of open ocean, with low chlorophyll

[137] Dennis Clark, interview transcript notes.

[138] Black and white image from David Herring, "MODIS/SeaWiFS team deploys marine optical buoy, continues Marine Optical Characterization Experiment". Color image from David Harring (actually Herring), "Marine Optical Buoy (MOBY) Evolves, While Marine Optical Characterization Experiment (MOCE) continues in Support of SeaWiFS, MODIS, and OCTS", http://eospso.gsfc.nasa.gov/eos_observ/9_10_97/p15.html, (accessed 8 January 2010).

concentrations, deep water, and little or no terrestrial turbidity was also desired. Finally, a nearby island with the amenities of civilization – specifically, communications capable of monitoring the state of the buoy, and with a location that would allow regular servicing of the buoy, was very important. Given that set of desired attributes, the choices very quickly narrowed down to two locations: Hawaii and Bermuda, both of which were also near the locations of the JGOFS time-series sites. After weighing the merits of both locations, the choice was Hawaii, partly due to the lower amount of cloud cover. A location in the lee of the island of Lanai was chosen for the mooring site.[139]

Regular servicing was a necessary requirement for MOBY because seawater is both physically and biologically hostile to anything that sits in it for a long time. The salty sea will corrode anything metal, so the main boom and fairings of MOBY were fabricated out of fiberglass. Three arms extending from the boom are equipped with radiance sensors, and the above-water part also has radiance sensors, radio antennas, solar panels, and a GPS receiver. At the end of the boom, providing a weighted stability as well as a place for the electronics, was the actual radiometer and computer compiling the measurements from all of the sensors.

The problem with having sensitive optical sensors deployed in ocean waters is also a problem familiar to boaters with hulls in the water for an extended period of time; ocean organisms are great opportunists, and they will grow on just about anything sitting in the water for an extended period of time! This gunk (euphemistically referred to as "bio-fouling") can quickly degrade, to the point of uselessness, any optical instrumentation. There are ways around this, such as painting surfaces with materials that contain chemicals toxic to the biotic denizens trying to live on said surfaces, but these materials can also interfere with optics, and can also be detrimental to the general marine environment. So rather than one MOBY, there were three; each replacement cruise would pull one buoy out of the water and put a fresh, clean and polished MOBY in, while the third was being serviced and recalibrated.[140] Divers inspected and serviced the buoys periodically between the buoy swaps.

MOBY was initially deployed in 1992 and 1993 off of Monterey Bay for testing, and went into the waters off of Lanai in 1994, waiting for the launches of SeaWiFS and MODIS.[141] Dennis Clark and the MODIS Ocean Characterization Experiment (MOCE) team members kept in touch with MOBY by cell phone; the buoy was in range of the cell phone tower on Lanai. Clark kept regular tabs on MOBY and made sure that it kept phoning home while collection prelaunch data for the site and learning about MOBY's quirks – and also learning about the multifarious hazards that a moored buoy could encounter in supposedly idyllic tropical seas. The not-exactly-idyllic adventures of MOBY are described in greater detail in Chapter 6.

---

[139] David Herring, "Marine Optical Buoy (MOBY) Evolves, While Marine Optical Characterization Experiment (MOCE) continues in Support of SeaWiFS, MODIS, and OCTS."
[140] David Herring, "Marine Optical Buoy (MOBY) Evolves, While Marine Optical Characterization Experiment (MOCE) continues in Support of SeaWiFS, MODIS, and OCTS."
[141] David Herring, "Marine Optical Buoy (MOBY) Evolves, While Marine Optical Characterization Experiment (MOCE) continues in Support of SeaWiFS, MODIS, and OCTS."

ADEOS and OCTS

While SeaWiFS was on the fast-track that led to launch vehicle failures, review boards, and inevitable and undesirable delays, there was increased awareness – and even a hint of envious hopefulness – regarding a mission that was becoming increasingly prominent in the minds of many ocean color scientists: the Japanese National Space Development Agency (NASDA) flagship mission ADEOS – Advanced Earth Observation Satellite.[142] ADEOS was moving forward at a faster rate than NASA's EOS program, and ADEOS was an ambitious potential preview of what EOS might expect. EOS (described in more detail in Chapter 6) had planned large satellites with a large number of instruments; ADEOS was slated to carry eight instruments into orbit. NASA had a direct involvement in building two: one was NSCAT, the NASA Scatterometer from JPL, which would measure wind speeds over the ocean by bouncing radio waves off the little wavelets that form on the ocean surface when winds blow over it.[143] Also onboard ADEOS was the Advanced Visible and Near Infrared Radiometer (AVNIR); a Total Ozone Mapping Spectrometer (TOMS), also provided by NASA; the French-built Polarization and Directionality of the Earth's Reflectance (POLDER), which was also capable of low-spatial-resolution ocean color measurements; the Improved Limb Atmospheric Spectrometer (ILAS); the Retroflector in Space (RIS), which was a cubic mirror similar to those on the Apollo moon-deployed reflectors, and which would bounce a laser back to Earth to allow measurement of light absorption in the laser beam; and the Interferometric Monitor for Greenhouse Gases (IMG).[144]

That's seven instruments: it was the eighth, OCTS (which stands for Ocean Color and Temperature Scanner) that was on the minds of the ocean color community.[145] OCTS was going to make the observations that the ocean color community wanted; the radiances that could be translated into chlorophyll concentrations, and in addition, it would also collect simultaneous thermal data to indicate sea surface temperatures. OCTS had a top-notch science team that was also collaborating with SeaWiFS and MODIS, particularly Drs. Motoaki Kishino and Hajime Fukushima. Kishino was the Japanese parallel to Dennis Clark; his specialty was bio-optical algorithms and he was also the head of a team designing and building an in-water radiometric buoy for the purpose of OCTS data validation. Kishino's prototype buoy was tested from January to March 1995 in Sugura Bay off of

---

[142] Japan Aerospace Exploration Agency, "Advanced Earth Observing Satellite "Midori" (ADEOS)", http://www.jaxa.jp/projects/sat/adeos/index_e.html, (accessed 19 May 2009).
[143] Jet Propulsion Laboratory, "Missions-NSCAT", http://winds.jpl.nasa.gov/missions/nscat/index.cfm, (accessed 19 May 2009).
[144] Japan Aerospace Exploration Agency, "Advanced Earth Observing Satellite "Midori" (ADEOS)".
[145] Hiroshi Kawamura and the OCTS Team, "OCTS Mission Overview," *Journal of Oceanography*, 54(5), 383 -399 (1998).

Shizuoka, Japan.[146] Fukushima was akin to OCTS what Howard Gordon was to CZCS, SeaWiFS, and MODIS; the leader of the atmospheric correction team.[147]

Even before the SeaWiFS Project had commenced, ocean color researchers were looking at OCTS as a data source, as indicated in the conclusions of Feldman et al. 1989 (the global CZCS data set). In that publication, a 1995 ADEOS launch was expected.[148] ADEOS fell behind schedule as well, but not by a lot; on August 17, 1996, ADEOS was carried into space on an H-II launch vehicle from the Tanegashima Space Center and began instrument start-up and check-out activities. Following the successful launch, the satellite was renamed "Midori", which means "green".[149] The Japanese earth observation mission was described in *Science* a week later and received a brief mention in *Aviation Week & Space Technology* in late September.[150] The *Science* article succinctly captured the mood of the ocean color community: "Hiroshi Kawamura, a physical oceanographer at Tohoku University, says he and his colleagues around the world have been waiting for an instrument like the Ocean Color and Temperature Scanner (OCTS) ever since the last major satellite-bome ocean color sensor, part of a U.S. mission, gave out 10 years ago."

NASA did not have a direct connection to OCTS data – but NOAA did, to acquire OCTS data for United States waters. However, NOAA was still working on getting their data archive operational, and until that was done, they planned to ignore the data that was being broadcast from Midori. The SeaWiFS Project cajoled NOAA to at least get all the data that they could from OCTS, even if they weren't ready to process it, and NOAA reluctantly agreed[151]. OCTS began its imaging mission in October 1996, at a time when strange winds were beginning to blow across the Pacific Ocean.[152]

---

[146] Motoaki Kishino, Takashi Ishimaru, Ken Furuya, Tomohiko Oishi and Kiyoshi Kawasaki, "In-Water Algorithms for ADEOS/OCTS," *Journal of Oceanography*, 54 (5), 431-436 (1998); Motoaki Kishino, Joji Ishizaka, Sei-ichi Saitoh, Yasuhiro Senga, Masayoshi Utashima, "Verification plan of ocean color and temperature scanner atmospheric correction and phytoplankton pigment by moored optical buoy system," *Journal of Geophysical Research,* 102(D14), 17,197–17,207, (1996).
[147] Hajime Fukushima, Akiko Higurashi, Yasushi Mitomi, Teruyuki Nakajima, Toshimitsu Noguchi, Toshio Tanaka and Mitsuhiro Toratani, "Correction of Atmospheric Effect on ADEOS/OCTS Ocean Color Data: Algorithm Description and Evaluation of Its Performance," *Journal of Oceanography*, 54 (5), 417-430, (1998).
[148] Gene C. Feldman, Norman Kuring, Carolyn Ng, Wayne Esaias, Chuck McClain, Jane Elrod, Nancy Maynard, Dan Endres, Robert Evans, Jim Brown, Sue Walsh, Mark Carle, and Guillermo Podesta, "Ocean color - Availability of the global data set".
[149] Jane's Information Group, "ADEOS/MIDORI series (Japan), Spacecraft – Earth Observation", http://www.janes.com/articles/Janes-Space-Systems-and-Industry/ADEOS-MIDORI-series-Japan.html, (accessed 19 May 2009).
[150] Dennis Normile, "New satellite puts Japan in top tier of remote sensing," *Nature,* 273 (5278), 1038 (23 August 1996); Michael Mecham, "Instruments coming alive on Adeos-1 Earth-scanner," *Aviation Week & Space Technology*, 145(13), 26, (23 September 1996).
[151] Mary Cleave and Gene Feldman recorded interview, 21 November 2008.
[152] Hiroshi Kawamura and the OCTS Team, "OCTS Mission Overview".

The Simulated SeaWiFS Data Set

Another activity of the SeaWiFS Science Team was the creation of a simulated SeaWiFS data set from CZCS data. Watson Gregg, Fred Patt, and Bob Woodward collaborated on this particular task. The data set included simulated global GAC data, simulated on-board recorded LAC data, and simulated high-resolution HRPT station data. All of the aspects of ocean color data as well as data from other sources – which usually meant land, clouds, and ice – were included.[153] The data set was needed to run the data processing system through its full gamut of operating capability. The data set was given to the receiving ground stations, particularly the Wallops Island ground station that would receive the twice-daily downlink of the GAC data from the satellite. The data was transmitted from Wallops to GSFC, where the data processing system swung into action. The first operation was to format the data for "ingest", which is a technical term for submitting, accepting, recording, and recognizing each data file in the system. Once ingested, the algorithms cooked artfully on each data pixel, first providing the bland but nourishing Level 1 calibrated radiance data, and then producing the gourmet items, the Level 2 geophysical data products, which were the menu items that the hungry researchers were slavering for. After the Level 2 data products were created, they were then mapped into the global Level 3 data products at daily, 8-day, monthly, and annual intervals. Then they sat, at times awaiting the arrival of necessary ancillary data, particularly atmospheric ozone concentrations from either TOMS or TOVS, and also meteorological data that provided atmospheric pressure. All of these elements provided necessary corrections for band absorption interference and atmospheric correction. It was anticipated that due to the fact these data were from other sources, they might not be immediately available to the data system – which was not a large problem, because there was a two-week embargo on the release of the data to the scientific community anyway. When everything had been received, the data was processed with the "best available" data, and then subject to a variety of quality analysis and quality control (QA/QC) tests.[154] Finally, when everything was done and the two-week embargo had expired, the data then was transferred to the archive and distribution facility: the Goddard Distributed Active Archive Center (DAAC), part of the Earth Observing System Data and Information System (EOSDIS).

---

[153] Watson W. Gregg, Frank C. Chen, Ahmed L. Mezaache, Judy D. Chen, and Jeffrey A. Whiting, "Volume 9: The Simulated SeaWiFS Data Set, Version 1," in S.B. Hooker and E.R. Firestone and A.W. Indest, editors, *NASA Technical Memorandum 104566*, NASA Goddard Space Flight Center, Greenbelt, Maryland, 17 pages, (May 1993); Watson W. Gregg, Frederick S. Patt, and Robert H. Woodward, "Volume 15: The Simulated SeaWiFS Data Set, Version 2," in S.B. Hooker and E.R. Firestone, editors, *NASA Technical Memorandum 104566*, edited by S.B. Hooker, E.R. Firestone, NASA Goddard Space Flight Center, Greenbelt, Maryland, 51 pages, (January 1994).
[154] Watson W. Gregg, Frederick S. Patt, and Robert H. Woodward, "Volume 15: The Simulated SeaWiFS Data Set, Version 2".

The Goddard DAAC: CZCS Browser, Preparations for SeaWiFS

The algorithm is defined above, but the question is repeatedly asked: what exactly is a DAAC? The reasons for the existence of DAACs go back to the evolution of computer and data archive and storage technology which was occurring at breakneck pace during the late 1980s and early 1990s. This was a period when the terms "electronic mail", "Internet", and "World Wide Web" were just entering into the public parlance; computer power was both escalating and miniaturizing well beyond the wildest extrapolations of science-fiction writers; and a time when old ways of archiving, storing, and transmitting data were being replaced more rapidly than planners could plan for.

The DAAC system was created by NASA in preparation for the vast volumes of Earth remote sensing data that would be acquired and delivered by the EOS constellation of satellites.[155] In the early 1990s, it was truly unimaginable that this volume of data could be delivered by a network of interconnected computers: in fact, data storage on portable CD-ROMs was still fairly novel, and even CD data volumes were not capable of carrying the expected gigabytes of EOS data. So the DAACs were created to receive the data from the NASA missions, to store it in silos of tapes, and to send data tapes to researchers around the world, as recommended by the EOSDIS Science Advisory Panel.[156] The Goddard DAAC got started well in advance of EOS by archiving data from other missions, such as the Upper Atmosphere Research Satellite (UARS), which provided atmospheric context for the processes contributing to ozone depletion and the Antarctic ozone hole; ozone data from TOMS and TOVS; the AVHRR Normalized Differential Vegetation Index (NDVI); and also the ocean color pioneer, CZCS.[157] As noted earlier, one of the initial functions of the DAAC with regard to CZCS data was to receive the data requests generated by the optical platter CZCS browser, put the requested data files on tape cassettes, and send the tapes to the requesting scientist.

SeaWiFS was slated to be an important precursor test of the operational DAAC system; a system to receive data from a mission and send it to researchers around the world, more researchers than had ever been provided data before. So the DAAC created an ocean color data team headed by Rebecca Farr, and including Angela Li, Ravi Kartan, Robert Simmon (responsible for Web page design) and several "ingestors" – system engineers who would monitor the incoming data stream and make sure the data was going where it was supposed to go and its location was still known.[158]

---

[155] John H. McElroy and Ray A. Williamson, "The Evolution of Earth Science Research from Space: NASA's Earth Observing System," Chapter 4 in John M. Logsdon, Stephen A. Garber, Roger D. Launius, and Ray A. Williamson, editors, *Exploring the Unknown, Volume 6: Space and Earth Science, NASA-SP-2004-4407*, subchapter "EOS Data and Information System", 464-470, (2004).
[156] Science Advisory Panel for EOSDIS, "Initial Scientific Assessment of the EOS Data and Information System (EOSDIS)", Washington, DC: NASA, (1989).
[157] Center for Health Applications of Aerospace Related Technologies, "Data Sources on the Internet: The Goddard DAAC", http://geo.arc.nasa.gov/sge/health/links/links.html, (accessed 19 May 2009).
[158] Rebecca A. Farr, Ravi Kartan, Angela W. Li, and Robert Simmon, "SeaWiFS ocean color data products and services at the NASA Goddard Distributed Active Archive Center (DAAC)," in Steven G. Ackleson, editor, *Proceedings SPIE, "Ocean Optics XIII, Volume 2963,"* 670-673, (1997).

As the amazingness of the World Wide Web became much better known, first through the Mosaic browser and then the advent of Netscape, it was realized that giving researchers the option to browse and select data files online ("online" still being a relatively new term) was important. So the ocean color DAAC staff undertook the conversion of the CZCS DSP data files into HDF, and created a CZCS online data browser, as a prototype for the online SeaWiFS data browser.[159] This effort was taken to completion by "alpha geek" Angela Li, assisted by Ravi Kartan.

The DAAC continued to work with the SeaWiFS Project on the transfer and ingest process. The embargo and encryption of SeaWiFS data presented several unique aspects: the DAAC would have to insure that users were Authorized Research Users before sending them data, so access to the data would have to be password-protected. Furthermore, the users were to be provided with data subscription options; some users wanted Level 1 data, others wanted Level 2, and many of them indicated that a geographical subset was also useful. A scientist studying the waters around Australia, for example, didn't want to receive all of the global Level 2 data and have to throw away nine-tenths of it; he only wanted to receive data files with data for Australian waters. And it was also necessary to provide ancillary data subscription capability to the users who wanted to process the data themselves.[160]

In 1995 and 1996, the SeaWiFS Project began routinely processing the simulated SeaWiFS data set, over and over, daily, to rigorously test the system. The DAAC added a scientific user support scientist, James Acker, in mid-1996, and added a scientific programmer, Suhung Shen, and a data archive specialist, Gregory Leptoukh, soon after. George Serafino, the DAAC data manager, after consultations with the SeaWiFS Project management, decided to place Leptoukh at the head of the ocean color data support team to push the SeaWiFS data readiness effort into an accelerated mode.[161] As the SeaWiFS Project watched the Pegasus launch schedule with keen interest, and as ocean color researchers watched in amazement as OCTS ocean color data suddenly renewed the exploration of the biological dynamics of the oceans, Leptoukh readied the DAAC system and the DAAC ocean color staff for the anticipated arrival of SeaWiFS data. One of the main activities was receiving Authorized User registrations, creating and sending passwords, and recording the data preferences of increasing numbers of Authorized Research Users awaiting the launch of SeaWiFS.

---

[159] Rebecca A. Farr, Ravi Kartan, Angela W. Li, and Robert Simmon, "SeaWiFS ocean color data products and services at the NASA Goddard Distributed Active Archive Center (DAAC)."
[160] Rebecca A. Farr, Ravi Kartan, Angela W. Li, and Robert Simmon, "SeaWiFS ocean color data products and services at the NASA Goddard Distributed Active Archive Center (DAAC)"; James G. Acker, "SeaWiFS Data Available at the GSFC DAAC", http://www.ioccg.org/reports/gsfcdaac/gsfcdaac.html, 1997, (accessed 19 May 2009).
[161] James G. Acker, Suhung Shen, Gregory Leptoukh, George Serafino, Gene Feldman, and Charles McClain, "SeaWiFS Ocean Color Data Archive and Distribution System: Assessment of System Performance," *IEEE Transactions on Geoscience and Remote Sensing*, 40(1), 90-103 (January 2002).

The SeaWiFS Data Analysis System: SeaDAS

Another aspect of the SeaWiFS Project's prelaunch efforts was the development of a community data processing system. While other aspects of the Project had been planned for, with appropriate budgets, there was no money designated for the creation of a software package that could be used by the ocean color community for SeaWiFS data processing – and the success of SEAPAK (despite its technological limitations) indicated that this was a good model to follow. McClain suggested creating a SEAPAK-analogue to Frank Muller-Karger, who had taken the rotating Ocean Biology program position at NASA HQ, and who had acquired a sense of the cost of centralized computing during his earlier collaborations at GSFC. Muller-Karger prompted McClain to submit a proposal, which involved programmers Gary Fu and Brian Schieber. The proposal was accepted with some modifications; despite the Project staff's familiarity with TAE, Muller-Karger required the Project to use IDL (Interactive Data Language), which among other things was compatible with Fortran libraries that were being developed for SeaWiFS data processing.[162] This decision turned out to be presciently fortuitous, as TAE was eventually dropped. Thus, the Project undertook the creation of the SeaWiFS Data Analysis System (SeaDAS) for the user community. The first SeaDAS beta version was released in July 1994, and the 1.0 version was released in June 1995. The 2.0 version was released in May 1996. SeaDAS ended up so successfully emulating the data processing system of the SeaWiFS Project that most of the SeaWiFS-receiving HRPT ground stations popping up like daffodils around the world installed SeaDAS to process the data that they would acquire. All of these SeaDAS versions ran on workstations which had a UNIX operating system; there was also a drawback in that an IDL license also had to be purchased, but this wasn't a significant deterrent. Still, scientists wishing to use SeaDAS would have to get similar UNIX workstations, which were considerably more expensive than personal computers.[163]

The remarkable usefulness of SeaDAS to duplicate the data processing system of the mission caused researchers to jump through hoops to get a compatible workstation; this was particularly difficult for users in countries that did not have the computational infrastructure of the United States and Europe. Researchers at the Second Institute of Oceanography in Hangzhou, China proudly showed off their UNIX workstation in December 1999.[164] McClain observed, "Literally, thousands of users, some in locations so remote it's hard to imagine how they ever heard of SeaWiFS or SeaDAS, rely on SeaDAS for

---

[162] ITT, "IDL – Data Visualization Solutions", http://www.ittvis.com/ProductServices/IDL.aspx, (accessed 24 September 2009).
[163] Ocean Biology Processing Group, "SeaDAS History of Events", http://oceancolor.gsfc.nasa.gov/seadas/history.html, (accessed 19 May 2009).
[164] In 1999, the author journeyed to China to complete an adoption. In serendipitous fashion, the two cities in China which were visited on this trip, Hangzhou and Guangzhou, had institutes of oceanography with researchers who had received SeaWiFS data through the auspices of the author. The author visited both institutes, and was shown the workstation and SeaWiFS HRPT antenna at the Second Institute of Oceanography in Hangzhou. His host in Hangzhou was Dr. Mao Zhihua, and in Guangzhou was Dr. Chuqun Chen of the South China Sea Institute of Oceanography.

access to SeaWiFS data."[165]

Adapting SeaDAS to the Windows operating system was pretty much impossible, for a lot of reasons, but the availability of the Linux operating system for PCs allowed SeaDAS to release a PC version (beta) in May 1999, and SeaDAS 4.0 was the first official SeaDAS version to work in the Linux environment in June 2000, about the same time that McClain turned over management of SeaDAS to Gene Feldman.[166] The SeaDAS programmers have continued to add new functions and new data types in parallel with development of the SeaWiFS Project (and eventually the OBPG) data processing system. [Advances in computer software packages actually enabled a Windows-compatible release of SeaDAS in 2009.][167]

The availability of SeaDAS was a truly unprecedented milestone for the SeaWiFS Project. Rather than rely on analyzing processed data products from a mission, scientists could actually do it themselves, and make adjustments and tests and experiments on the data to facilitate their own research and work in their own regional environment. After a couple of nominations, SeaDAS was recognized in 2003 by receiving NASA's Software Invention of the Year Award.[168]

A Horse, a Horse, My Kingdom for a Launch Vehicle

Richard III needed a ride, any ride at all, in the midst of the battle of Bosworth Field; SeaWiFS was waiting for a ride into space on the Pegasus XL. On March 6, 1996, it appeared that the extra-large version of the rocket named for the winged steed of Greek mythology had surmounted its birthing pains when it successfully launched the U.S. Air Force REX-II (Radiation Experiment II) into space, to the relief of Orbital executives, NASA managers, spacecraft insurers, Orbital stockholders, and quite a few members of the U.S. Air Force.[169] Success begat success: on May 17, Pegasus XL proved itself again, this time with the MSTI-3 (Miniature Sensor and Technology Integration-3) satellite, also an Air Force mission.[170]

NASA crossed its fingers and hoped for a Pegasus XL hat trick on July 2, 1996; the XL delivered again, this time with the critical Earth Probe TOMS, to continue the vital atmospheric measurements begun by the Nimbus 7 TOMS.[171] EP-TOMS was similar to SeaWiFS as a single sensor free-flyer; the data was sent to the Goddard DAAC, and this data was also slated to be crucial atmospheric ozone ancillary data for SeaWiFS atmospheric

---

[165] Charles R. McClain, email message received 7 July 2009.
[166] Ocean Biology Processing Group, "SeaDAS History of Events".
[167] Ocean Biology Processing Group, "The SeaDAS Virtual Appliance – SeaDAS VA 5.4", http://oceancolor.gsfc.nasa.gov/seadas/seadasva.html, (accessed 24 September 2009).
[168] NASA GSFC, "NASA Technology Award Winners", http://techtransfer.gsfc.nasa.gov/awards-won-NASA.html, (accessed 19 May 2009).
[169] "Pegasus," http://www.astronautix.com/lvs/pegasus.htm.
[170] "Pegasus," http://www.astronautix.com/lvs/pegasus.htm.
[171] "Pegasus," http://www.astronautix.com/lvs/pegasus.htm.

correction algorithms. If a lot of people on the ground started breathing again after the Apollo 11 touchdown on the Moon, there was quite a bit of similar relief to spread around after EP-TOMS reached orbit.

With increased confidence, NASA employed the Pegasus XL again on August 21, when it launched FAST (the Fast Auroral Snapshot Explorer).[172] This mission launch was four days after the launch of ADEOS/Midori; now Mary Cleave had not just one, but four successful Pegasus XL launches to lay the path for SeaWiFS. But various factors involving mission and fiscal priorities had to be satisfied, so the next mission payload for the XL was the High Energy Transient Experiment (HETE), a mission with international participation based out of the Massachusetts Institute of Technology.[173] HETE was designed to detect and find the source of the enigmatic gamma-ray bursts which had been detected, but not assigned to a particular source, by the massive Compton gamma-ray observatory released from the Space Shuttle. In space, when HETE detected a burst, it was capable of turning rapidly and acquiring high-resolution multiwavelength data from the exact detected location of the burst.

HETE never got a chance to demonstrate this remarkable capability. The entire Pegasus XL launch vehicle functioned flawlessly on November 4, 1996, and HETE reached orbit. But the explosive bolts that had doomed the second launch came into play again. This time, the fairing protecting the satellite failed to release, and though HETE was transmitting on battery power, it was trapped in its protective shell, and could not unfold its solar power wings. The plaintive telemetry from HETE faded after a short lifetime in space.[174]

The HETE failure required more failure investigations, but these moved quickly, and Pegasus XL was ready to try again a few months into 1997, on April 21.[175] This launch was slated to carry the Minisat-1 of Spain's Instituto Nacional de Tecnica Aerospacial (INTA), with two radiation detector experiments and a micro-gravity experiment. This launch also carried something else; a few grams of cremated human ashes from the Celestis company, a company which offered the chance for an individuals and families to deliver the cremated ashes of deceased loved ones into space. This launch, the "Founders Flight" of Celestis, carried the ashes of twenty-three individuals. Two Celestis participants in particular caught the attention of the media: hippy guru and advocate of psychodelic drug use Timothy Leary, and *Star Trek* creator Gene Roddenberry.[176]

The Pegasus XL launch of Minisat-1 and the Celestis founders was successful. Gene Roddenberry might have been pleased to know that his journey into space set the

---

[172] "Pegasus," http://www.astronautix.com/lvs/pegasus.htm.
[173] Massachusetts Insitute of Technology, "The History of the HETE-2 Mission," http://space.mit.edu/HETE/history.html, (accessed 19 May 2009).
[174] Massachusetts Insitute of Technology, "The History of the HETE-2 Mission."
[175] David Steitz, "Pegasus Launch Anomaly Under Investigation," http://www.nasa.gov/home/hqnews/1996/96-227.txt, (accessed 23 June 2009).
[176] "Pegasus", http://www.astronautix.com/lvs/pegasus.htm; Celestis, Inc., "The Founders Flight", http://www.memorialspaceflights.com/memorial/founders/default.asp, (accessed 19 May 2009).

stage for SeaWiFS; Timothy Leary may have just wished SeaWiFS a nice long trip.

Shock

As SeaWiFS and SeaStar were in the Orbital launch preparation facility being mated to the launch vehicle and tested in advance of an August launch date, fateful events on June 21 provided a sudden and unsettling shock to the ocean color community and the Japanese remote sensing community. With little warning, a power cable connecting the large solar array of Midori to the instrument payload suddenly degraded and caused a power failure for the whole satellite. The cause of the failure was ultimately decided to have been structural deficiencies in the solar array, and a maneuver soon after launch had apparently triggered a slowly-developing problem. Demonstrating the rigorous demands of satellite designs, one of the related problems was that the solar array thermal blanket had higher thermal expansivity than expected.[177]

August 1, 1997: High Stakes

With OCTS gone, the stakes riding on the SeaWiFS launch increased markedly. Mary Cleave said that she felt a little more pressure to make SeaWiFS work from the oceanographic community.[178]

Orbital had initially not wanted a chase plane to observe the SeaWiFS launch; Cleave insisted on one, in order to have tape documentation if anything went wrong, and arranged for Space Shuttle pilot (STS-3) and commander (STS-51F) Gordon Fullerton to fly the chase plane. Cleave said she had excitedly told the OSC management that "they got a chase plane from Eddy!" (Edwards Air Force Base), and OSC did not share her excitement. Fullerton was already familiar with Pegasus; he had flown on six of the B-52 Pegasus launches.[179] On August 1, 1997, the L-1011 rolled out on the Vandenberg runway and took off, with Fullerton following in an F-18.

As the L-1011 reached cruising altitude at 38,000 feet and lined up along the release path, Fullerton flew on both sides of the aircraft, allowing the camera to examine every meter of the launch vehicle.[180] Gene Feldman said that as he was watching the tape, first the chase plane viewed one side, then the other side – "you could see everything". Everything

---

[177] "Adios, ADEOS: Japanese Satellite Lost", *Science*, (1 July 1997), http://sciencenow.sciencemag.org/cgi/content/full/1997/701/2, (accessed 19 May 2009); David Michael Harland and Ralph Lorenz, "The Loss of ADEOS-1", in *Space Systems Failures: disasters and rescues of satellites, rockets and space probes*, Springer, New York, 368 pages, (2005).
[178] Mary Cleave and Gene Feldman recorded interview, 21 November 2008.
[179] Mary Cleave and Gene Feldman recorded interview, 21 November 2008; Lyndon B. Johnson Space Center, "C. Gordon Fullerton: NASA Astronaut (Former)", http://www.jsc.nasa.gov/Bios/htmlbios/fullerton-cg.html, (accessed 19 May 2009).
[180] Mary Cleave and Gene Feldman recorded interview, 21 November 2008.

appeared nominal. The rocket and satellite checked out, and the range officer gave the GO for launch. The white rocket released, and fell for five endless seconds, until the engine ignited and the rocket blasted forward under the body of the L-1011, and began arcing upward, with Fullerton in pursuit, viewing the Pegasus XL over the Pacific clouds (Figure 5.5).[181] The rocket ascended smoothly, and the camera continued to observe it as it visibly dwindled to a white point ahead of a lengthening contrail.[182]

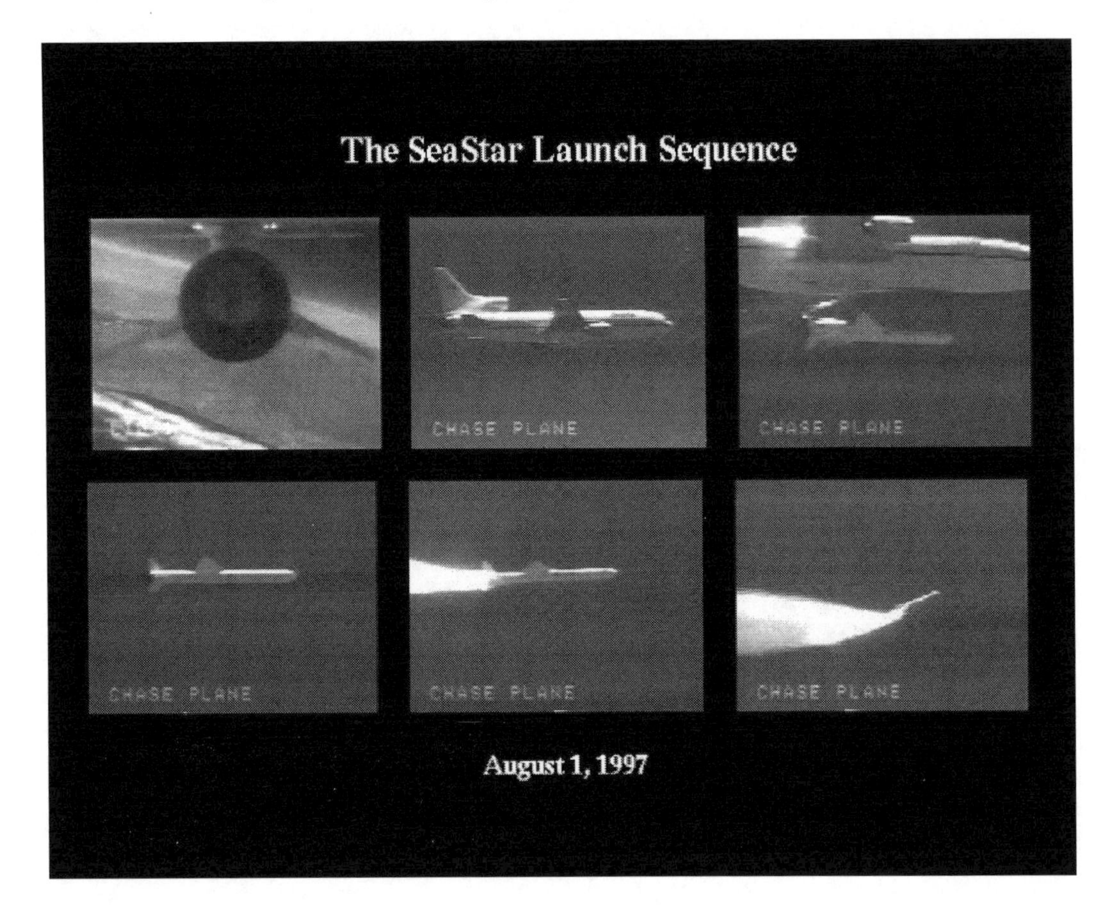

Figure 5.5. Launch of SeaWiFS on the Pegasus XL.

It appeared to be a successful launch. Telemetry was received that indicated the stage 1 separation, the stage 2 ignition, burnout, and separation, and the successful operation of the third stage. SeaWiFS radioed back that it was released into space (not suffering the ignominious demise of HETE), into orbit, initially at an altitude of 312 km. In Gene Feldman's office at the SeaWiFS Project, in the midst of the entire staff watching the launch

[181]Ocean Biology Processing Group, http://oceancolor.gsfc.nasa.gov/SeaWiFS/BACKGROUND/Gallery/launch.jpg

[182] James Yoder, "On a Winged Horse and a Prayer, SeaWiFS Launches," *Backscatter*, 8 (4), November 1997; SeaWiFS Launch Status Page, http://oceancolor.gsfc.nasa.gov/SeaWiFS/ANNOUNCEMENTS/current_launch_status.html, including digital video of the SeaWiFS launch, (accessed 19 May 2009).

on television, a triumphant and relieved Chuck McClain opened a bottle of champagne, and the plastic cork launched into the ceiling tiles and stuck. (It hasn't moved since.)[183]

An apparently healthy SeaWiFS then established contact with the ground station at McMurdo Sound, Antarctica.

And, in the blithe ignorance of computers that only do what they are supposed to do, SeaWiFS sent telemetry to McMurdo that indicated it was still running on battery power, because…

… it was UPSIDE DOWN. The sensor designed to image the oceans was pointed toward space, and the solar arrays were pointed down toward the Earth, which despite its brightness in the dark abyss of space, did not provide nearly enough light to keep SeaWiFS running. Sensing that this was a problem, the satellite went into "safe hold" to conserve power and communicated its status to Poker Flats, Alaska.[184]

The SeaWiFS Project had the full global ("expensive") capabilities of NASA at its immediate command, and now they were urgently needed. Cleave, sitting at Orbital's mission operations center, had planned to be at the operations center and not the launch, because if something went wrong with the launch, there was nothing she could do, but if something went wrong in space, she would be in the right place to respond. In the right place at the right time, Cleave called on the NASA receiving station in Greenland, and Orbital transmitted the necessary commands to Greenland to transmit to the spacecraft.[185]

In the language of spacecraft software, the basic command was "turn OVER – right NOW!"

And it did. At Vandenberg, Science Officer Jim Yoder, who had watched the SeaWiFS rollout, takeoff, and launch and then monitored the telemetry drama, was relieved enough that he joined the beer-drinking celebration fully clothed in the local motel pool.[186]

Rising To Its Mission

Though in orbit, SeaWiFS was not in the imaging orbit, which was about 400 kilometers higher. The next month involved a series of burns to raise the orbit to the altitude of 705 km.[187] Even these basic maneuvers are not without peril; in 2002, when the Comet Nucleus Tour (CONTOUR) satellite lighted its solid-fuel engine to leave Earth orbit and travel to Comet Encke and Comet Schwassmann Wachmann-3, it disappeared, probably

---

[183] Mary Cleave and Gene Feldman recorded interview, 21 November 2008.

[184] James Yoder, "On a Winged Horse and a Prayer, SeaWiFS Launches"; Mary Cleave and Gene Feldman recorded interview, 21 November 2008.

[185] Mary Cleave and Gene Feldman recorded interview, 21 November 2008.

[186] James Yoder, "On a Winged Horse and a Prayer, SeaWiFS Launches".

[187] Stanford B. Hooker, Wayne E. Esaias, Gene C. Feldman, Watson W. Gregg, and Charles R. McClain, "Volume 1, An Overview of SeaWiFS and Ocean Color", in S.B. Hooker and E.R. Firestone, editors, *NASA Technical Memorandum 104566*, NASA Goddard Space Flight Center, Greenbelt, Maryland, 25 pages, (1992).

due to another energetic propulsion system anomaly.[188] But SeaWiFS used its reliable and proven hydrazine thrusters to increase its altitude, and they worked as planned. In this orbit, the sensor was exposed to the heat and vacuum of space to release volatile compounds that could coat the polished and perfect sensor optics, and several days later, the scan system was opened to view the Earth.

First Light

SeaWiFS began collecting data. The first test was designated for a single view of the U.S. East Coast. On September 4, 1997, SeaWiFS turned on every system, and light from the Earth hit the focal planes for the first time, becoming an electronic signal, which was recorded and beamed to the Wallops ground station.

On the ground, the SeaWiFS Project turned off the data processing system that had faithfully processed the week-long simulated SeaWiFS data set over and over again for more than a year, and waited for the actual data from Wallops to arrive. Computers hummed as they had before. Data monitors showed the data progressing through the system. And then the processing was done. Someone reminded Gene Feldman that the system would produce a low-resolution browse file when it was finished. Feldman checked the data directory, and there it was.[189]

Figure 5.6 SeaWiFS "First Light" image.

And it was very cloudy. Yet there were glimpses of the open ocean, with bright circulation features, and even a glimpse of greener coastal waters in the Gulf of Maine, southwest of Nova Scotia. When the image was released to the press, Feldman superimposed a faint image of the NASA logo to celebrate the NASA's resumption of ocean

---

[188] NASA,"Discovery Missions: Contour", http://discovery.nasa.gov/contour.html, (accessed 19 May 2009).

[189] Mary Cleave and Gene Feldman recorded interview, 21 November 2008.

color imaging, 11 years after the end of the CZCS mission (Figure 5.6).[190]

Over the next 13 days, SeaWiFS was on for longer and longer periods, up to full orbits. The data processing system continued to work efficiently with the larger volumes of data. No hiccups or anomalies were apparent. The SeaWiFS Project stayed in close touch with the schedulers at OSC, and indicated that they were ready for a full day of data.

That day was September 17-18, 1997 (because the day commenced at the International Date Line). SeaWiFS collected data over the entire Earth, transmitted it to Wallops, which transmitted it to the SeaWiFS Project, which processed it, waited for all the ancillary data files, and subsequently transmitted it to the Goddard DAAC a week later, on September 24. The DAAC received it, ingested it, divided it into the preferences of the Authorized Users – online data transmission for some, magnetic tapes for others – and less than 14 hours later, the first full day of successfully received, processed, and archived NASA ocean color data was distributed to the global ocean color community via the World Wide Web.[191]

September and October 1997: Bright Seas

SeaWiFS continued to view the globe day after day, with the repeated coverage that had been long hoped for by remote-sensing scientists and biological oceanographers. As more and more data were collected, the clouds over the ocean began to part. Not literally – a close look at the Blue Planet on a given day looks much more like a whitish planet with patches of blue, green, and brown – but the process of binning and averaging the daily orbital swaths into global compilations gradually removed the clouds, replacing them with more and more pixels of the open ocean. The first 8-day global product was created about eight days after September 17, according to the calendar. And the first "monthly" product was actually a short month, September 17-September 30.

And with virtual immediacy, just about everyone looking at this first still-blurry SeaWiFS image of the world – not just ocean color scientists and skilled, experienced oceanographers – could tell that the sensor was viewing something unusual occurring in the world's oceans. The brilliant harbinger of these conditions glowed blue-white in the Bering Sea off Alaska, a color indicating an unprecedented and gigantic bloom of coccolithophores in a region where they are not normally found in large blooming concentrations (Figure 5.7).[192] As Barney Balch of the Bigelow Laboratory said, "The Bering Sea blooms happened at the wrong place at the wrong time."[193] The bloom in the Bering Sea was a bright blue indicator of strange things to come. Oceanographers had been ready for what was happening for years with buoys, weather satellites, tide gauges, and other satellite sensors –

---

[190] Ocean Biology Processing Group, http://oceancolor.gsfc.nasa.gov/SeaWiFS/IMAGES/S1997247162630.555_490_443.jpg.

[191] James G. Acker, "SeaWiFS Data Available at the GSFC DAAC".

[192] NASA, http://kids.earth.nasa.gov/seawifs/images/beringsea_lg.gif.

[193] John Weier, "Changing Currents Color the Bering Sea a New Shade of Blue", http://earthobservatory.nasa.gov/Features/Coccoliths/, and references therein, (accessed 20 May 2009).

and they knew something remarkable was in full swing – but now a view of the biology of the oceans was also possible as the Pacific Ocean showed its full range of colors.

It didn't take long for the scientific press to discover the promising early success of SeaWiFS; *Science News, Science Daily,* and *Nature* touted results from the new satellite in late September and early in October, and *Laser Focus World* covered SeaWiFS in November.[194]

Figure 5.7 Excerpt from first SeaWiFS true color global image, September 1997, showing the dramatic coccolithophorid bloom in the Bering Sea.

Flip: The First Look at the Moon

SeaWiFS did have one more operational maneuver to demonstrate; the lunar calibration maneuver. The SeaWiFS Project had insisted on the necessity of a look at the moon just about every month to monitor the inevitable changes to the sensor's electronics and optical bands; the omnipresent Howard Gordon had first suggested this approach in 1987.[195] The Moon had to be viewed a few days before or after the full moon; when the Moon is truly full, an unusual optical property of the lunar regolith occurs, called the "Heiligenschein", which markedly increases the brightness of the lunar orb when sunlight is directly reflected back toward its source.[196] (If you think a full Moon looks considerably brighter than a nearly-full waxing or waning gibbous moon, you're right!) The SeaWiFS Project had determined that even though it only occupied a few pixels in a SeaWiFS scan, the nearly-full moon was a constant radiance, diffuse reflector, calibration target.[197]

---

[171]Anonymous, "First global ocean-color images from new sensor show promise for climate, biological studies," *Science Daily*, (25 September 1997), http://www.sciencedaily.com/releases/1997/09/970925042208.htm, (accessed 30 October 2009); Richard Monastersky, "Satellite views Earth's living plumage," *Science News*, 152(14), 212, (4 October 1997); Karen Southwell, "Shades of the Sea," *Nature*, 389 (6650), 444, (2 October 1997); Laurie Ann Peach, "SeaWiFS colors the world's oceans," *Laser Focus World*, 33(11), 30-31, (November 1997).

[195] Howard R. Gordon, "Calibration requirements and methodology for remote sensors viewing the ocean in the visible," *Remote Sensing of the Environment*, 22, 103-126 (1987).

[196] Robert Wildey, "The Moon in Heiligenschein," *Science*, 200(4347), 1265 – 1267, (16 June 1978).

[197] Robert H. Woodward, Robert A. Barnes, Charles R. McClain, Wayne E. Esaias, William L. Barnes, and Ann T. Mecherikunnel, "Volume 10, Modeling of the SeaWiFS Solar and Lunar Observations", in S.B. Hooker and E.R. Firestone, editors, *NASA Technical Memorandum 104566*, NASA Goddard Space Flight Center, Greenbelt, Maryland, 26 pages, (May 1993).

To actually view the Moon, SeaWiFS had to flip over. Properly oriented, the sensor viewed the Earth; only if was in the upside-down orientation that had endangered the mission after launch could it scan the surface of the moon. The flip required just a quick momentum transfer using the spacecraft's momentum wheels. Yet again, such a maneuver, to be performed routinely, still meant that the satellite had to perform like an Olympic gymnast, accomplishing a complicated gyration without apparent effort, and emerging still smiling (or in the case of SeaWiFS, still scanning).

It might be more exciting to history to describe a dangerous circumstance, an unexpected yaw or pitch that put the mission into peril, but this was not the case. The first lunar calibration maneuver happened without a hitch on November 14, 1997, SeaWiFS viewed the Moon, the band radiances were calculated, and the mission continued.[198]

SeaWiFS continued to collect data through the end of December 1997. A major mission milestone was passed in December when NASA certified ("acceptance") that the sensor met its specifications, and NASA paid OSC a large installment on the contract.[199] Soon after, there were a few days of angst; the SeaWiFS Project began to see strange trends in the derived products, specifically affecting the vital chlorophyll concentration data product. It was ascertained that the inherent difficulty of imaging the ocean through the turbulent atmosphere was once again a key factor: in this case, the global mean atmospheric correction parameters had commenced an unexpectedly rapid degradation trend. Previous experience didn't help to analyze the problem: the Project had not expected the near-IR bands to degrade so quickly, because in the CZCS case history (which was the only one available) the degradation had been most severe in the blue, with very little degradation observed in the red. Robert Barnes jumped on this problem with alacrity and came up with the first correction scheme, based on the lunar calibration data. McClain put it succinctly, "It saved our bacon." [200] Barnes' analysis was one of the motivations for the first major data reprocessing, which required the Goddard DAAC to make sure all the Authorized Research Users wanted to get the first three months of data again – most of them did – and once this had been completed, the imaging mission began in earnest.[201]

---

[198] NASA Ocean Biology Processing Group, "SeaWiFS Lunar Calibration Schedules", http://seawifs.gsfc.nasa.gov/plankton2/d3/userweb/luncal/, (accessed 19 May 2009).

[199] NASA GSFC Office of Public Affairs, "SeaWiFS Team Wins Pecora Award", Release 02-144, 3 October 2002, http://www.gsfc.nasa.gov/news-release/releases/2002/02-144.htm, (accessed 20 May 2009).

[200] Robert A. Barnes, Robert E. Eplee, Jr., Frederick S. Patt, and Charles R. McClain, "Changes in the radiometric sensitivity of SeaWiFS".

[201] There is no official written statement that the SeaWiFS Project performed a global reprocessing at the end of the first three months of the mission (completed in January 1998). A few sources indicate implicitly that this is what happened; in the author's experience, this was the actual timeframe, and it was partly due to a planned refinement of the at-launch algorithms based on analysis of the actual data from the first three months of the mission at the time of data acceptance, as well as the discovery of the degradation trends. Greg Janée and James Frew, "Preserving the Context of Science Data", http://www.alexandria.ucsb.edu/~gjanee/archive/2008/agu.pdf, (accessed 20 May 2009); Ocean Biology Processing Group, "SeaWiFS Reprocessing #1", http://oceancolor.gsfc.nasa.gov/REPROCESSING/SeaWiFS/R1/, (accessed 20 May 2009).

At just the right time.

El Niño of the Century: The Sequel

Although not known at the time, the OCTS viewed an important period in the ebb and flow of the winds and currents of the Pacific Ocean – the preparatory period for an imminent El Niño event. The precursors were detected in early 1997. In May, El Niño arrived in full force, and NASA missions were already watching. One mission, the TOPEX/Poseidon mission, had arrived in orbit in 1992, and began another unprecedented data set collection, this one of the small changes in the height of the ocean. TOPEX/Poseidon accomplished these measurements by accurately measuring the length of time it took for a radio signal to travel from the satellite, bounce off the ocean surface, and return to the satellite, a remote-sensing method called *radar altimetry*. By making repeated measurements, adjusting for the irregular shape and gravity field of the Earth, correcting for the tidal cycle and absorption by water in the atmosphere, this rigorous data processing scheme extracted images of the centimeter-high hills and valleys of the ocean surface, which in turn indicated where currents flowed and how water moved.[202]

In May 1997, TOPEX/Poseidon detected what appeared to be a warm mountain range moving across the Pacific (its warmth was indicated by the dutiful AVHRR sensors). This was not a small ridge a few centimeters high – it was up to 30 centimeters high, a massive volume of water, as it smashed into the coast of South America and began to spread north and south. More and more warm water flooded along the Equator into the eastern Pacific in a succession of Kelvin waves.[203] The 20th century's second "El Niño of the century" arrived with a vengeance.

Fires erupted over Indonesia and Australia as the cooler waters in the western Pacific led to quick, severe drought. The high ridge of warm water stretched nearly two-thirds of the way across the Pacific when SeaWiFS began its imaging operations in September.[204]

Thus, something was clearly missing from the expected pattern of color in the Pacific Ocean when SeaWiFS began compiling its clear monthly views of the world. Now that a true synoptic basin-wide view was possible, it was utterly obvious that not just the Galapagos Islands were affected; the entire Pacific equatorial upwelling zone, an expected band of light green, had *disappeared.* And in the western Pacific, even despite the compilation of multiple SeaWiFS observations of Indonesia and Southeast Asia, parts of the ocean were still obscured due to the pervasive pall of smoke emanating from smoldering tropical rain

---

[202] Jet Propulsion Laboratory, "TOPEX/Poseidon Fact Sheet", http://topex-www.jpl.nasa.gov/mission/tp-fact-sheet.html, (accessed 20 May 2009).
[203] Tony Phillips, "A Curious Pacific Wave", http://science.nasa.gov/headlines/y2002/05mar_kelvinwave.htm, (accessed 20 May 2009).
[204] Jet Propulsion Laboratory, "TOPEX/Poseidon image shows El Niño is still strong," September 25, 1997, http://topex-www.jpl.nasa.gov/elnino/970925.html, (accessed 20 May 2009).

forests.[205]

Unlike the massive El Niño event of 1982-1983, this mission had several sensors from space watching it, as well as the in-water TOGA-TAO (Tropical Atmosphere – Global Ocean – Tropical Atmosphere-Ocean) array of buoys. So the scientists watched as the warm water spread north and south along the coasts of Central and South America. The warm water covered the cold upwelling along the Peruvian coast, limiting the extent of the nutrient-enriched productivity in the classic diagnostic indicator of an El Niño. Along the Pacific coast of Central America, three zones of productivity generated by intense winter winds rushing through passes of the cordillera spine, also virtually disappeared. Powerful storms lashed the California and Pacific Northwest coasts, again causing flooding and landslides, though damage estimates were considerably less than in the 1982-1983 event.[206] Starving California sea lions were observed with their ribs showing, indicating the difficulty of finding fish.[207]

As El Niño progressed into 1998, satellite oceanographers watched with considerable interest as the event began to evolve. The warm flood of water from the western Pacific began to dissipate, and the sea surface slowly lowered down to its normal state. Biological oceanographers watched the daily SeaWiFS images for indications that the equatorial upwelling had resumed. Meanwhile, *Discover* magazine put the early success of SeaWiFS on its pages in February, giving the science-minded public an update on the new sensor in the sky watching the dramatic changes taking place in the world's largest ocean.[208]

La Niña Takes over the Pacific

In a matter of six-eight weeks from late April to early June 1998, the equatorial upwelling zone reappeared with a remarkable burst of phytoplankton activity. Daily images clearly showed high chlorophyll concentrations perched on the peaks of waves proceeding across the Pacific basin. One SeaWiFS image of this event was later to appear on the cover of the December 10, 1999 issue of *Science*.[209] Around the Galapagos, the transition from nearly clear water to waters bursting with fat phytoplankton occurred in less than three

---

[205] NASA Scientific Visualization Studio, "SeaWiFS Indonesian Smoke," http://www.nasaimages.org/luna/servlet/detail/NSVS~3~3~11898~111898:SeaWiFS-Indonesian-Smoke, (accessed 20 May 2009).

[206] Jan Null, "El Niño, La Niña, and California flooding."

[207] Don Terry, "Battered Sea Lions Find Refuge from El Niño," *New York Times*, 16 February 1998; Philip Colla, "California sea lion pup starving during 1997-8 El Nino event, Coronado Islands," http://www.oceanlight.com/info.php?img=02417, (accessed 20 May 2009).

[208] "Ocean view," *Discover,* 19(2), 22 February 1998.

[209] SeaWiFS Project, NASA GSFC, and Orbimage, "SeaWiFS satellite composite of Earth's biosphere (ocean chlorophyll and land vegetation), acquired between 12 and 19 July 1998," *Science*, 286(5447), (10 December 1999).

weeks, to the gustatory delight of zooplankton, the local seabirds, and sea lions.[210]

As the oceanographers watched, they began to realize that they were seeing a somewhat different pattern of events than had apparently occurred in 1982 and 1983. Now, rather than returning to the normal "average" conditions of the Pacific, the world's largest ocean was now playing host to El Niño's alter ego – La Niña. In a La Niña event, the surface waters are cooler than normal, upwelling is intensified, and with SeaWiFS observing closely, it was clear that the productivity of the equatorial Pacific belt was considerably higher than normal.

This major La Niña event was more unprecedented in size and scope than the 1997-1998 El Niño had been, because the 1982-1983 El Niño provided one other example of a large ENSO event. The new La Niña conditions persisted throughout the rest of 1998 and into mid-1999. Thus, it was only after the first year and a half of SeaWiFS orbital operations that the Pacific finally returned to its average state – nicely bracketing the highs and lows of Pacific productivity for oceanographers exulting in the remarkable flow of data from the little satellite that could. Early in 1999, with La Niña still controlling the Pacific, Ray Sambrotto described the eventful first year-and-a-quarter of SeaWiFS data in *Nature*.[211] This short article nearly summarized the difficulty of ocean color science for *Nature*'s readers when he referred to the CZCS: "The expectations for the original CZCS project were meagre because several difficulties presented themselves. The sensor needed to fish out spectral changes in sunlight that enters, and then reflects back from surface water, a signal that is dwarfed by visible radiation from the atmosphere." Sambrotto also noted with a touch of humor, "In an apt demonstration that one man's noise is another's signal, the SeaWiFS visible and near infrared bands have been used to track aeolian dust. Such aerosols are thought to transport trace metals (such as iron) needed for phytoplankton growth from land to open ocean regions, where they occur in very low concentrations. Ironically, the aerosol signal is part of the atmospheric correction made before estimating chlorophyll."

SeaWiFS the Global Observer

The important addition of a bilinear gain to SeaWiFS meant that the sensor, which had originally been intended to be strictly an ocean-observing instrument, had become a sensor capable of considerably more. Now – rather than simply covering over bright targets like lands and clouds with a black mask (which was still employed for chlorophyll data when the radiances indicated that the pixels were brighter than allowed for the normal oceanic range of reflectivity) – SeaWiFS could return true color images of a wide range of global phenomena.

In June 1998, SeaWiFS provided a dramatic demonstration of how useful an instrument

---

[210] F. P. Chavez, P. G. Strutton, G. E. Friederich, R. A. Feely, G. C. Feldman, D. G. Foley, and M. J. McPhaden, "Biological and Chemical Response of the Equatorial Pacific Ocean to the 1997-98 El Niño," *Science,* 286 (5447), 2126 – 2131 (10 December 1999).

[211] Raymond Sambrotto, "Oceanography: colour renewal from space," *Nature,* 397, 301-302, doi:10.1038/16808, (28 January 1999).

providing daily global views of the world could be. That summer, in Florida, firefighters were contending with an outbreak of resilient wildfires that produced remarkably heavy amounts of smoke. The smoke was so heavy, in fact, that firefighters on the ground could not find the actual location of the fires (which were occurring in central Florida near Ocala, sometimes with smoke so thick it was necessary to close sections of Interstate 75). Looking for assistance, one of the fire captains asked NASA if they had any instruments in space that could help them find the fires. The SeaWiFS Project was contacted, and Feldman made a quick call to Orbimage to release an image of the Florida peninsula and send it to the firefighters on the ground. The image did the job; fire-fighting forces were able to zero in on the location of the fires with the aid of the image, and finally managed to bring them under control.[212] Through the first years of the mission, SeaWiFS was able to view a remarkable number of events, in the ocean, on land, and in the atmosphere. In April 1998, a dust storm swept out of the Gobi desert and proceeded to cross the Pacific Ocean. SeaWiFS was able to track this Asian dust cloud all the way to the Pacific coast of California.[213]

In February 2000, SeaWiFS observed a massive dust cloud depart from the Saharan desert with a distinctive "hammerhead" shape, akin to that of the hammerhead jets observed in upwelling zones.[214] Portions of this cloud traveled to Brazil, the United States, and even to Britain. This image – which notably is not primarily about ocean biology – is one of the most widely-distributed and well-known SeaWiFS images. It was published as a stand-alone photograph in the New York Times on March 7, 2000, with the caption: "On Feb. 26, a satellite captured what a federal researcher called "probably the largest dust event in the past 100 years" – a sandstorm billowing from Africa's northwest coast 1,000 miles into the Atlantic."

SeaWiFS proved adept at imaging hurricanes and typhoons. In its first year, it watched hurricanes Bonnie and Danielle. In 1999, SeaWiFS observed the development of the drought-busting Hurricane Dennis, which brought desperately needed rains to the mid-

---

[212] NASA, "A Unique Vantage for Tempests", http://www.gsfc.nasa.gov/topstory/20010327page3.html, (accessed 21 May 2009); Robert Lee Hotz, "Satellites hottest idea in battling wildfires,"*Los Angeles Times*, (18 July 1998).

[213] Rudolf B. Husar, D.M. Tratt, B.A. Schichtel, S.R. Falke, F. Li, D. Jaffe, S. Gass´o, T. Gill, N.S. Laulainen, F. Lu, M.C. Reheis, Y. Chun, D. Westphal, Brent N. Holben,, C. Gueymard, I. McKendry, Nornan Kuring, Gene C. Feldman, C. McClain, Robert J. Frouin, J. Merrill , D. DuBois,
F. Vignola, T. Murayama , S. Nickovic, W.E. Wilson, K. Sassen, N. Sugimoto, and W. C. Malm, "The Asian Dust Events of April 1998," *Journal of Geophysical Research*, 106(D16), 18,317–18,330, 2001; NASA GES DISC, "SeaWiFS Observes Asian Dust Transport over the Pacific Ocean", http://daac.gsfc.nasa.gov/oceancolor/scifocus/oceanColor/asian_dust.shtml, (accessed 21 May 2009).

[214] NASA Visible Earth, "Dust Storm Sweeps from Africa into Atlantic," http://visibleearth.nasa.gov/view_rec.php?id=157, (accessed 21 May 2009); "A sandstorm the size of Spain," *New York Times* (7 March 2000), stand-alone image with caption, http://www.nytimes.com/2000/03/07/science/a-sandstorm-the-size-of-spain.html (accessed 21 May 2009).

Atlantic states, particularly Virginia.[215] A few weeks later, SeaWiFS observed Hurricane Floyd as it churned up the east coast of the United States, and in its wake it saw the entire ocean bottom on the continental shelf churned into a shade of slate gray.[216] A week later, the turbid flood waters spawned by Floyd in North Carolina were observed mixing into the Gulf Stream and subsequently were transported hundreds of kilometers into the Atlantic Ocean.[217]

SeaWiFS quickly proved that it could monitor outbreaks of toxic phytoplankton blooms, an operational use of the data which had been highly anticipated, even though it was generally unable to tell if the blooms were toxic or benign. In March 2001, SeaWiFS detected a bloom, diagnosed in the water as the phytoplankter *Chattonella*, which was responsible for killing at least 700 tons of farm-raised salmon off of Norway.[218] Scientists watching the Gulf of Mexico began to determine the conditions when high chlorophyll concentrations detected by SeaWiFS indicated the presence of the detested *Karenia brevis*. And in 2002, oceanographers examining SeaWiFS data observed a strange, nearly black patch of water close to Florida Bay and the southwestern coast of Florida. Quickly organized reconnaissance cruises discovered that the bloom was initially a *K. brevis* red tide, which evolved into a standard, but abnormally rich, phytoplankton bloom. The local fish populations abandoned the area. The black water patch also severely stressed the nearby coral reefs.[219]

SeaWiFS also added new information to oceanographer's knowledge of the Benguela upwelling zone off South Africa and Namibia. For years, the local population had occasionally reported a strange greenish-yellow occurrence of water associated with a noxious smell, dead fish, and flocks of seabirds attracted to the dead fish. SeaWiFS allowed researchers to observe how often these phenomena took place, and discovered that due to the low population living on the arid coast, they happened far more often than reported. In this region, the intense productivity means that a large amount of dead phytoplankton falls on the sea floor. Oxygen is rapidly used up by bacteria decomposing the organic material, and the job of digesting the rich soup is taken over by bacteria that use sulfur for energy, rather than oxygen. These bacteria generate immense amounts of hydrogen sulfide gas (H_2S,

---

[215] Emma Kelly, "SeaStar captures Hurricane Dennis Development," *Flight International*, 39, (September 8, 1999).
[216] NASA GES DISC, "Sedimentia," http://disc.sci.gsfc.nasa.gov/oceancolor/additional/science-focus/ocean-color/sedimentia.shtml, (accessed 29 October 2009).
[217] NASA Earth Observatory, "Hurricane Floyd: Fearing the Worst," http://earthobservatory.nasa.gov/Features/FloydFear/, (accessed 21 May 2009); "NASA technology tracks consequences of Hurricane Floyd," *Science Daily*, 3 November 1999, http://www.sciencedaily.com/releases/1999/11/991103080450.htm, (accessed 21 May 2009).
[218] NASA GSFC, "SeaWiFS Sensor Marks Five Years," http://www.gsfc.nasa.gov/topstory/20020801seawifs.html, (accessed 21 May 2009).
[219] Katie Greene, "Black Water Mystery," *Science* (1 April 2002); Noreen Parks, "Black Water Mystery Solved," *Science* (23 April 2003), http://sciencenow.sciencemag.org/cgi/content/full/2003/423/3, (accessed 21 May 2009); Chuanmin Hu., K. E. Hackett, M. K. Callahan, Serge Andréfouët, J. L. Wheaton, J. W. Porter, and Frank E. Muller-Karger, "The 2002 ocean color anomaly in the Florida Bight : a cause of local coral reef decline?" *Geophysical Research Letters* 30(3), 1151-1154, doi:10.1029/2002GL016479 (15 February 2003).

"rotten egg gas") as they consume the organic feast. So much H_2S builds up that it can suddenly erupt from the sea floor and rise to the surface, first becoming a white sulfur mineral and then slowly turning into bright yellow elemental sulfur. The toxic gas causes fish kills, while the suspended minerals result in the unusual greenish patches observed by SeaWiFS.[220]

SeaWiFS data also provided indicators of the fate of penguin populations in Ross Sea of Antarctica, during the period 2000-2003 when large drifting tabular icebergs broke loose and dramatically affected the ecology of this remote region.[221] The icebergs drifted in the Southern Ocean near the Ross Sea and occasionally became stuck on the sea floor, which blocked the normal movements of sea ice out of the sea during austral summer. As a result, the normally open waters during summer were covered, which markedly curtailed summer blooms of phytoplankton in the Ross Sea. Because the important zooplankton *Euphausia superba*, otherwise known as krill, feeds on these blooms, and because penguins consume huge quantities of krill, the curtailment of primary production led Kevin Arrigo and colleagues to expect effects on the indigenous Adelie and Emperor penguins, because the penguins "… time their reproduction so that their chicks fledge in the early summer, at the time of maximum food availability. These organisms will be particularly sensitive to any environmental perturbation that shifts temporally the availability of their food source."

Ocean color data was even used by the international sporting community. In 2002, Volvo Ocean Race yachts received SeaWiFS data from the SeaWiFS Project, which tracked their progress around the world. In return, the yachts were outfitted with radiometers, but the exigencies of open ocean racing precluded any scientific use of the data. However, when the Volvo racers were in port, they and the Volvo corporation promoted awareness of the ocean environment through a series of public presentations and an associated Web site that allowed daily race updates combined with ocean color imagery.[222]

The Scientific Returns

While SeaWiFS images were busy attracting publicity, the main task of the mission, serious and accurate science, was also taking place. SeaWiFS continued its regular habit of

---

[220] Scarla J. Weeks, Bronwen Currie, and Andrew Bakun, "Massive emissions of toxic gas in the Atlantic," *Nature*, 415, 493-494 (31 January 31, 2002); NASA GES DISC, "A Bloom By Any Other Name", http://disc.gsfc.nasa.gov/oceancolor/scifocus/oceanColor/sulfur_plume.shtml, (accessed 21 May 2009).

[221] Kevin R. Arrigo, Gert L. van Dijken, David G. Ainley, Mark A. Fahnestock, and Thorsten Markus, "Ecological impact of a large Antarctic iceberg," *Geophysical Research Letters*, 29(7), 1104-1107, doi:10.1029/2001GL014160; "Giant iceberg blocks Ross Sea Food Chain", *Ameriscan*, 3 October 2003, http://www.ens-newswire.com/ens/oct2003/2003-10-03-09.asp#anchor5, (accessed 30 October 2009).

[222] IOCCG, "SeaWiFS' Contribution to Volvo Ocean Race," IOCCG News, (October 2001), http://www.ioccg.org/news/Oct2001/Octnews.html, (accessed 21 May 2009); NASA GSFC, "NASA Joins the Volvo Ocean Adventure in Oceanography Education Events in Chesapeake Area," http://www.gsfc.nasa.gov/topstory/04052002volvoocean1.htm, (accessed 21 May 2009).

flipping over monthly as it passed over Antarctica to take a look at the moon. This repeated observational routine provided the Project engineers a continuous record of the changes in the sensitivity of every SeaWiFS band, allowing calibration adjustments to be applied to the lengthening data set.[223] MOBY returned data each time SeaWiFS observed the waters off Lanai, providing another data set for instrument calibration and revisions that could be applied to the radiance data. Cruises, including the JGOFS Southern Ocean process studies and a wide variety of research campaigns around the world, provided more data to the SeaWiFS Project, which was added to a growing observational database called SeaBASS (described in more detail in Chapter 8).[224] The SIRREX series continued, becoming even more rigorous than thought possible.[225] The SIRREX collaboration with NIST resulted in the SeaWiFS transfer radiometer, an instrument that allowed calibrated radiance measurements from an instrumental standard to be transferred to instruments that

---

[223] Robert E. Eplee, Jr., Robert A. Barnes, and Charles R. McClain, "SeaWiFS detector and gain calibrations: four years of on-orbit stability," *Proceedings SPIE,* 4814, 282-288, (2002).
[224] Paul J. Werdell, Sean Bailey, Giulietta Fargion, Christophe Pietras, Kirk Knobelspiesse, Gene Feldman, and Charles McClain, "Unique data repository facilitates ocean color satellite validation," *Eos, Transactions of the American Geophysical Union,* 84(38), 379, (2003).
[225] James L. Mueller, "Volume 14, The First SeaWiFS Intercalibration Round-Robin Experiment, SIRREX, July 1992," in S.B. Hooker and E.R. Firestone, editors, *NASA Technical Memorandum 104566*, NASA Goddard Space Flight Center, Greenbelt, Maryland, 60 pages, (September 1993); James L. Mueller, B. Carol Johnson, Christopher L. Cromer, John W. Cooper, James T. McLean, Stanford B. Hooker, and Todd L. Westphal, "Volume 16, The Second SeaWiFS Intercalibration Round-Robin Experiment, SIRREXB, June 1993," in S.B. Hooker and E.R. Firestone, editors, *NASA Technical Memorandum 104566*, NASA Goddard Space Flight Center, Greenbelt, Maryland, 121 pages, (May 1994); James L. Mueller, B. Carol Johnson, Christopher L. Cromer, Stanford B. Hooker, James T. McLean, and Stuart F. Biggar, "Volume 34, The Third SeaWiFS Intercalibration Round-Robin Experiment.(SIRREX-3), 19-30 September 1994," in S.B. Hooker E.R. Firestone, and James G. Acker, editors, *NASA Technical Memorandum 104566*, NASA Goddard Space Flight Center, Greenbelt, Maryland, 78 pages, (March 1996); B. Carol Johnson, Sally S. Bruce, Edward A. Early, Jeanne M. Houston, Thomas R. O'Brian, Ambler Thompson, Stanford B. Hooker, and James L. Mueller, "Volume 37, The Fourth SeaWiFS Intercalibration Round-Robin Experiment (SIRREX-4), May 1995," in S.B. Hooker and E.R. Firestone, editors, *NASA Technical Memorandum 104566*, NASA Goddard Space Flight Center, Greenbelt, Maryland, 65 pages, (May 1996); T. Riley and Sean Bailey, "The Sixth SeaWiFS/SIMBIOS Intercalibration Round-Robin Experiment (SIRREX-6) August–December 1997," in S.B. Hooker and E.R. Firestone, editors, *NASA Technical Memorandum 1998-206878*, NASA Goddard Space Flight Center, 26 pages, (1998); B. Carol Johnson, Howard W. Yoon, Sally S. Bruce, Ping-Shine Shaw, Ambler Thompson, Stanford B. Hooker, Robert E. Eplee, Jr., Robert A. Barnes, Stephane Maritorena, and James L. Mueller, "The Fifth SeaWiFS Intercalibration Round-Robin Experiment (SIRREX-5), July 1996," in S.B. Hooker and E.R. Firestone, editors, *NASA Technical Memorandum 1999--206892*, Volume 7, NASA Goddard Space Flight Center, Greenbelt, Maryland, 75 pages, (1999); Stanford B. Hooker, Scott McLean, Jennifer Sherman, Mark Small, Gordana Lazin, Giuseppe Zibordi, and James W. Brown, "The Seventh SeaWiFS Intercalibration Round-Robin Experiment (SIRREX-7), March 1999," in S.B. Hooker and E.R. Firestone, editors, *NASA Technical Memorandum 2002-206892*, Volume 17, NASA Goddard Space Flight Center, Greenbelt, Maryland, 69 pages, (2002); Giuseppe Zibordi, Davide D'Alimonte, Dirk van der Linde, Jean-François Berthon, Stanford B. Hooker, James L. Mueller, Gordana Lazin, and S. McLean, 2002: The Eighth SeaWiFS Intercalibration Round-Robin Experiment (SIRREX-8), September--December 2001," in S.B. Hooker and E.R. Firestone, editors, *NASA Technical Memorandum 2003--206892*, Volume 21, NASA Goddard Space Flight Center, Greenbelt, Maryland, 39 pages, (2003).

were subsequently deployed at sea.[226] It also produced the SeaWiFS quality monitor, a "field-deployable stable light source", which was used to quantity the stability of the sensors use for in-water or at-the-surface measurements. The NIST collaboration also provided calibration support to the MOBY ground operations in Hawaii.

One of the reasons that the SeaWiFS calibration and validation program was so successful was that McClain had considerable autonomous control over the budget for these activities. If necessary, he could designate funds for specific investigators (including Howard Gordon, Ken Carder, Dave Siegel and Greg Mitchell) to refine the atmospheric correction algorithm and provide *in situ* data sets. McClain also tasked Stan Hooker to initiate an internal field program so that NASA would have its own presence in the field, address *in situ* data quality issues, and work on improved instrument designs.[227] McClain stated that "He [Hooker] ran with this ball far beyond my dreams"; these activities are described further in Chapter 8.

All of this successful science meant one thing; a lot more people should know about it. A publication about the 1997-1998 El Niño and the subsequent La Niña provided NASA with a way to get the SeaWiFS project more media attention.[228] In the publication, Behrenfeld and his co-authors used SeaWiFS chlorophyll data and models of primary productivity based on chlorophyll (as well as land productivity) to calculate how big the shift in the Earth's production of carbon had been between the El Niño period (when primary production was markedly reduced) to the La Niña period (when primary production was markedly elevated). This paper prodded Mary Cleave, who had moved to NASA HQ in 2000 (while Chuck McClain took over as SeaWiFS Project Manager) to organize NASA's first ever televised **Earth Science Update.**[229] The Earth Science Update, broadcast live, took place on March 29, 2001, with Cleave (now the Deputy Associate Administrator for Earth Science) interviewing Behrenfeld, co-author Paul Falkowski, Gene Feldman, and carbon cycle modeler Jorge Sarmiento of Princeton University.[230]

Subsequently, in 2002, André Morel and David Antoine summarized the progress that had been made to that point in the estimation of net primary productivity from

---

[226] B. Carol Johnson, J.B. Fowler, and Christopher L. Cromer, "The SeaWiFS Transfer Radiometer (SXR)," in S.B. Hooker and E.R. Firestone, editors, *NASA Technical Memorandum 1998-206892*, Volume 1, NASA Goddard Space Flight Center, Greenbelt, Maryland, 58 pages, 1998; NIST, "Calibration Support for Ocean Color Science," http://physics.nist.gov/Divisions/Div844/facilities/emet/ocean.html, (accessed 21 May 2009).
[227] Stanford B. Hooker and Charles R. McClain, "The calibration and validation of SeaWiFS data," *Progress in Oceanography*, 45, 427-465, (2000); Charles R. McClain, Gene C. Feldman, and Stanford B. Hooker, "An overview of the SeaWiFS Project and strategies for producing a climate research quality global ocean bio-optical time series," *Deep-Sea Research* II, 51(1-3), 5-42, (2004).
[228] Michael J. Behrenfeld, J. T. Randerson, Charles R. McClain, Gene C. Feldman, Sietse O. Los, Compton J. Tucker, Paul G. Falkowski, C.R. Field, C.B. Frouin, Wayne E. Esaias, Dorota D. Kolber, and Nathan H. Pollack, "Biospheric Primary Production During an ENSO Transition," *Science,* 291, 2594-2597, (2001).
[229] NASA Johnson Space Center Oral History Project, "Mary Cleave".
[230] NASA Headquarters, "NASA's First Earth Science Update – New Insights Into the Biological Record of Earth", http://www.gsfc.nasa.gov/news-release/releases/2001/n01-14.htm, (accessed 21 May 2009).

phytoplankton.[231] Their paper discussed how phytoplankton productivity was calculated and now these numbers could be integrated in models to better characterize the ocean carbon cycle.

The routine task of distributing SeaWiFS data to the oceanographic community continued at the Goddard DAAC. The most notable events during this period where the data reprocessings, when the entire archive of data was reanalyzed with modified algorithms, regenerated, and retransmitted to the DAAC, and subsequently re-sent to researchers who wanted to receive the newest and best data. During the third reprocessing in April-May 2000, so much data was received and delivered that a few of the DAAC's tape recorders burnt out, and the DAAC operations staff which mailed the tapes worked at their maximum capacity.[232] The user community for SeaWiFS data had grown from the initial count of over 300 to number in the thousands. The mailing list generated from the SeaWiFS Authorized User registrations exceeded 2,000, although not everyone on the list actually received and analyzed data, because according to the terms of the Authorized User Agreement, scientists who shared the data, such as co-authors, were also required to register as Authorized Users.

Other organizations around the world also acquired and distributed SeaWiFS data. In the United Kingdom, Peter Miller, Steve Groom, and Samantha Lavender inaugurated a data archive, the Remote Sensing Data and Analysis Service (RSDAS), facilitating distribution to UK and European researchers.[233]

Appendix 3 discusses some of the salient research results obtained by researchers using SeaWiFS data.

Selling the data

While SeaWiFS was proving itself as a remarkable scientific success, there was another side to the mission: the Orbimage side of selling the data. Orbimage created an entire system to provide mapped ocean color data to commercial interests, primarily targeting fishery operations.[234] Orbimage had also expected to sell the data to oil and gas offshore drilling operations for detection of potentially hazardous currents, and to commercial shipping companies to be used for fuel conservation, by following (or avoiding) currents

---

[231] André Morel and David Antoine, "Small Critters – Big Effects," *Science*, 296(5575), 1980-1982, (14 June 2002).

[232] James G. Acker, Suhung Shen, Gregory Leptoukh, George Serafino, Gene Feldman, and Charles McClain, "SeaWiFS Ocean Color Data Archive and Distribution System: Assessment of System Performance".

[233] Peter Miller, Jamie Shutler, Rory Hutson, Peter Land, Tim Smyth, and Steve Groom, "Recent Developments at RSDAS," Proceedings of Sensing and Mapping the Marine Environment 2, Institute of Physics, London, April 5, 2005, http://www.research.plym.ac.uk/geomatics/sensemap2/SMap05-abstracts.pdf, (accessed 20 May 2009); Samantha Lavender and Steve Groom, "The SeaWiFS Automatic Data Processing System (SeaAPS)," *International Journal of Remote Sensing*, 20, 1051-1056, (1999).

[234] "Orbimage begins SeaStar fisheries information service," *Space Daily*, 24 November 1997, http://www.spacedaily.com/news/orb2-97b.html, (accessed 21 May 2009).

indicated by the remote-sensing data.

All of these potential markets had been outlined in the EOSAT SeaWiFS report.[235] When OSC undertook the mission, they proceeded, as any good company should, to examine the markets and then price the data according to what they thought the market would pay for it.

Orbimage aggressively marketed the data and provided software to use it; they retained the name SeaStar for the software and the data products derived from the satellite. In 1997, they recorded $1.3 million dollars in SeaStar sales; in 1998 the revenues increased substantially to $11.7 million.[236] The company astutely realized that ocean color data was not enough; they combined SeaWiFS data with AVHRR SST to produce their fish-finding products.[237] Other fish-finding companies, such as ROFFS (Roffer's Ocean Fishing Forecasting Service) use similar strategies.[238] The Goddard DAAC occasionally received queries to find out if the SeaWiFS science data could be obtained in real-time; these queries were referred to Orbimage.

The fish-finding map service proved to be the most sucessful of Orbimage's efforts to market SeaStar data. They also had discovered that their cost models had been way too high, and as a result, the mission, while it brought the fledgling company considerable attention, was a loss-leader. An analysis of the SeaWiFS data-buy indicated that the commercial market failed to mature as had been forecast, so it was unclear if the 14-day data embargo for research users was significant. It was also stated that for future contracts, it would be better to have a longer time-frame, so that partners in the private sector can eventually make a profit. In this analysis, the effect of overestimating commercial profit was evident; it was stated that there was "...a need to price the product to reflect the investment, based on market analyses that reflect the true costs of providing data both to government-sponsored users and to commercial users." [239]

Orbimage's fortunes would rise and fall in the late 1990 and the earliest years of the 21st century. One of the main events was the anticipated launch of Orbview-4, a high resolution Earth imager, along with the QuikTOMS satellite. After Midori had failed in orbit, NASA decided quickly that the two NASA instruments lost on the satellite were vital to continuing Earth observations. Ball Aerospace successfully built and launched the QuikSCAT satellite carrying the SeaWinds scatterometer on June 19, 1999 on a reliable Titan

[235] Joint EOSAT/NASA SeaWiFS Working Group, "System Concept for Wide Field-of-view Observations of Ocean Phenomena from Space."
[236] "Orbimage 10-K for 12/31/98", http://www.secinfo.com/dsvRq.613f.htm, (accessed 24 June 1999); "Orbimage Holdings Inc/DE 10-K for 12/31/99", http://www.secinfo.com/dsvRq.51Aa.htm, (accessed 24 June 1999).
[237] "Orbimage to launch new information products in 2002," *World Fishing Magazine*, http://www.worldfishing.net/news/news_story.ehtml?o=171, (accessed 21 May 2009).
[238] Roffer's Ocean Fishing Forecasting Service, http://www.roffs.com/about_cc.html, (accessed 21 May 2009).
[239] Space Studies Board, National Academy of Sciences, "Toward Successful Public-Private Partnerships", Chapter 3 in *Toward New Partnerships in Remote Sensing: Government, the Private Sector, and Earth Science Research*, National Academies Press, 82 pages, (2002).

II – one of the fastest design-build-launch efforts in the history of the agency.[240] QuikSCAT is still in operation (in mid-2009) and scientists and oceanographers wish that a replacement was at least in the planning stages. [QuikSCAT scanning failed in November 2009.][241]

In similar fashion, NASA also built the QuikTOMS satellite in collaboration with OSC. though the urgency was not as great due to the operating EP-TOMS satellite. Originally slated to fly on a Russian satellite, QuikTOMS became a free-flyer when the Russian satellite mission fell through.[242] OSC stepped up, put the ozone-monitoring instrument on a satellite bus, and created a dual-satellite launch plan, partnering QuikTOMS with Orbimage's Orbview-4 on OSC's Taurus launch vehicle. On September 21, 2001, the Taurus carrying the two satellites lifted off from Vandenberg, but approximately a minute-and-a-half into the launch, the second stage apparently cut off early and released the third stage with the satellites. Lacking sufficient velocity to achieve orbit, the two satellites plunged into the Indian Ocean.[243]

Orbimage subsequently declared bankruptcy soon after the loss of Orbview-4, but successfully emerged from the Chapter 11 state in December 2003, a few months after the successful launch of the high-resolution Orbview-3 satellite.[244] Part of the reason that Orbimage emerged from bankruptcy was that it received a successful insurance settlement for the loss of Orbview-4.[245] Orbimage subsequently acquired the Ikonos high-resolution satellite company, which was operating two 1-meter resolution satellites, and renamed itself GeoEye.[246] GeoEye landed several military remote-sensing contracts and partnered with Google, rapidly expanding its revenue sheet. Its next satellite, GeoEye-1 (which was launched on September 6, 2008) offered somewhat astounding 41-cm resolution data, but software problems that caused pointing errors initially plagued the data release. GeoEye-1

---

[240] Jet Propulsion Laboratory, "Mission – SeaWinds on QuikSCAT", http://winds.jpl.nasa.gov/missions/quikscat/index.cfm, (accessed 21 May 2009); Jet Propulsion Laboratory, "Publications - QuikScat Team Wins American Electronics Achievement Award," http://winds.jpl.nasa.gov/publications/qs_award.cfm , (accessed 21 May 2009).
[241] Jet Propulsion Laboratory, "Publications - NASA Assessing New Roles for Ailing QuikScat Satellite," http://winds.jpl.nasa.gov/publications/quikscatStatus.cfm, (accessed 13 January 2010).
[242] NASA, "QuikTOMS Mission," Press Kit, September 2001, http://www.gsfc.nasa.gov/gsfc/earth/quiktoms/QT_PRESSKIT1.PDF, (accessed 21 May 2009).
[243] Orbital Sciences Corporation, "Orbital's Launch of Taurus Rocket Unsuccessful," http://www.orbital.com/NewsInfo/release.asp?prid=336, (accessed 21 May 2009); "Taurus Rocket Fails to Deliver QuikTOMS to Orbit," http://www.spacedaily.com/news/ozone-01f.html, (accessed 21 May 2009).
[244] Fran Keating, "Orbimage Files for Bankruptcy," *Geospatial Solutions*, 12(5), page 19, (May 2002); "Orbimage Officially Emerges from Chapter 11," *SpaceRef*, (5 January 2004), http://www.spaceref.com/news/viewpr.html?pid=13326, (accessed 21 May 2009); "Orbimage's Orbview-3 Satellite Successfully Reaches Orbit," *SpaceRef*, (27 June 2003), http://www.spaceref.com/news/viewpr.html?pid=11972, (accessed 21 May 2009).
[245] Eric Tegler, "Eye in the Sky," *Smart CEO*, http://www.smartceo.com/content/eye-sky, (accessed 21 May 2009).
[246] "Orbimage completes acquisition of Space Imaging: Changes brand name to GeoEye," *GISCafe*, http://www10.giscafe.com/nbc/articles/view_article.php?articleid=234676, (accessed 21 May 2009).

began delivering (and selling) data in February 2009.[247]

The Struggle to Renew

NASA rediscovered the uncertainties of dealing with a private company for vital ocean remote-sensing data at the end of 2002, when it was dealing with an Orbimage mired in Chapter 11 to renew the contract for SeaWiFS data. Weeks of nearly fruitless negotiations culminated in Orbimage leaving the table in early December, dimming hopes for a new contract. Somehow lured back, Orbimage finally agreed to a $1.1 million dollar extension of the contract for the next year.[248] NASA officials indicated that they hoped this would give Orbimage more time to develop its commercial sector for the SeaStar data, but Orbimage vice-president Tim Puckorius countered that the main market for the data was to scientific researchers. Orbimage had been hoping for $5.1 million dollars. Chuck Trees, the manager of the Ocean Biology program at NASA HQ who had started his tenure in September 2001, described the circumstances at the end of 2002 as one of serendipitous fortuity. When reviewing his program budget, knowing that several large programs had come to completion, he discovered that he had an additional $1 million dollar surplus. He dutifully asked his superior, Jack Kaye, what to do with the extra money, hoping that Kaye would grant him about $500,000 for the next three years, which would be more useful for funding science. Kaye told him to save it, and they would know where it was if they needed it. Because SeaWiFS had been slated as a five-year mission, NASA had not procured any funds for a continuation past 2002. When the SeaWiFS contract extension negotiations began, Kaye suggested using the surplus, and Trees, acting for NASA offered the $1 million he had – which was all NASA could offer, in addition to the control of the broadcast wavelengths used by the satellite (see below).[249] Trees recalls that this was about "20 times less than the figure that they were talking about", and added that without the fortuitous surplus, he is not sure that SeaWiFS would have accomplished its long operational record.

During 2003, a groundswell of scientific support grew for SeaWiFS, including an online poll with nearly 600 names (in June of that year) prodding for another continuation of the

---

[247] "GeoEye-1 gains success with VAFB launch," *SatNews Publishers*, http://launch.geoeye.com/LaunchSite/assets/documents/GeoEye-1_Gains_Success_with_VAFB_Launch.pdf, (accessed 21 May 2009); Peter B. DeSelding, "Software Glitches Delay GeoEye Satellite Use," *Space News*, (17 November 2008), http://www.space.com/businesstechnology/081117-business-monday-satellite.html, (accessed 21 May 2009); Turner Brinton, "GeoEye begins selling imagery from its newest satellite," *Space News*, 9 February 2009, http://launch.geoeye.com/LaunchSite/assets/documents/Space_News_02_09_09.pdf, (accessed 21 May 2009).

[248] Brian Berger, "Orbimage disappointed with NASA contract to extend Orbview-2 deal," *Space News*, 14 January 2003, http://www.space.com/spacenews/archive03/orbimagearch_011303.html, (accessed 21 May 2009).

[249] Charles Trees initially commented informally on this topic at the Ocean Color Collaborative Historical Workshop, January 13-14, 2009, St. Petersburg, FL. Confirmed in email message received 24 June 2009.

SeaWiFS data buy.[250] Orbimage held out even longer at the end of 2003, resulting in a week where NASA did not receive data, but eventually signed another contract extension; the data acquired during the lost week was subsequently transmitted to NASA to fill the gap.[251] 2004 did not go well, either, and Gene Feldman had to issue an announcement regarding the end of the contract before the new contract was finalized.[252] In all of these negotiations, one item in particular that kept Orbimage negotiating was a big carrot; NASA held the rights to the broadcast license frequencies for the direct downlink stations. Without a contract with NASA, the satellite could not legally broadcast its data to the NASA receiving stations.[253] Negotiations continued in early 2005, and eventually a contract was signed in April.[254] Orbimage again supplied the data that they had retained while negotiating the contract, to maintain the continuity of the data set.

Subsequent contract renegotiations went more smoothly, but with one major loss; the 2005 contract did not include a provision for the continued acquisition and transmittal of the high-resolution HRPT station data, which had provided some of the most dramatic images produced by the satellite. NASA would now only receive the 4km GAC data to continue the global ocean color data record. In the United States, NOAA stepped in and provided funds for the continued collection of the high-resolution SeaWiFS data over U.S. waters, partly to provide a well-known data source for its Gulf of Mexico red tide forecast system, which is described subsequently.[255]

The Ocean Biology Processing Group Becomes an Archive

In tandem with data archive decisions made concerning MODIS data, in late 2003, Deputy Administrator Mary Cleave decided to call upon her former colleagues on the SeaWiFS Project to pioneer a new concept in the archive and distribution of Earth remote sensing data, called "Missions to Measurements". Rather than have the operations to collect, calibrate, and produce the data in one location, with the "finished" data then sent to the NASA DAAC system for archive and distribution, Cleave directed the Goddard DAAC to halt its archive and distribution activities for SeaWiFS and MODIS ocean data, and to

---

[250] Brian Berger, "Scientists hopeful government will renew Orbview-2 support," *Space News*, (7 July 2003), http://www.space.com/spacenews/archive03/orbviewarch_070703.html, accessed 15 March 2010.
[251] IOCCG, "New contract for SeaWiFS?" *IOCCG News*, http://www.ioccg.org/news/Dec2003/news.html, (accessed 21 May 2009).
[252] Gene Feldman, "Clarification of SeaWiFS data access," http://oceancolor.gsfc.nasa.gov/forum/oceancolor/topic_show.pl?tid=401, (accessed 21 May 2009).
[253] Gene Feldman, comments recorded at the Ocean Color Collaborative Historical Workshop, January 13-14, 2009, St. Petersburg, FL.
[254] IOCCG, "SeaWiFS is back!" *IOCCG News*, May 2005, http://oceancolor.gsfc.nasa.gov/forum/oceancolor/topic_show.pl?tid=532, (accessed 21 May 2009).
[255] NASA, "The Ocean Chromatic: SeaWiFS enters its second decade," http://www.nasa.gov/vision/earth/lookingatearth/seawifs_10th_feature.html, (accessed 21 May 2009); IOCCG, "Tenth IOCCG Committee Meeting," http://www.ioccg.org/reports/ioccg_meeting10.html, (accessed 21 May 2009).

turn them over to the SeaWiFS Project, which was renamed the Ocean Biology Processing Group (OBPG) in February 2004.[256] The OBPG created a new system to process Authorized User registrations and to handle data requests; as of May 2009, the OBPG had distributed over 55,000 Gigabytes of SeaWiFS data.[257]

Phytoplankton Need Iron to Grow

The resumption of NASA's ocean color remote sensing program with SeaWiFS coincided with a significant advance in scientific understanding of the interactions between ocean biology (represented by phytoplankton) and ocean chemistry (represented by nutrients). For many years, oceanographer John Martin had investigated the cycling of carbon in the oceans and the chemistry of trace metals, first at Stanford's Hopkins Marine Station and subsequently at the Moss Landing Marine Laboratory near Monterey Bay.[258] Martin's trace metal work, involving meticulous cleaning of sampling equipment (frequently with washes of acid), a large amount of Teflon, and rigorous sampling protocols, indicated that previous measurements of iron in ocean waters had been too high due to contamination, and actual seawater iron concentrations were very low. Martin theorized that iron was likely a required nutrient for phytoplankton growth, and now that it was clear how low seawater iron concentrations actually were, it could explain previous enigmas in the patterns of ocean productivity.

These enigmas were called high nutrient, low chlorophyll (HNLC) regions of the global ocean.[259] The main three HNLC regions are the Southern Ocean, the equatorial Pacific, and the high latitudes of the North Atlantic. Martin suspected that the reason for the low chlorophyll concentrations (indicative of low phytoplankton productivity) was the lack of iron in the surface waters, as the only significant source of iron away from the coast would be dust from desert regions. The three main HNLC regions were too far from significant sources of dust to get sufficient iron for robust primary productivity.

Martin thus proposed what came to be known as "The Iron Hypothesis", i.e., that seawater iron could be a limiting nutrient for phytoplankton growth. "Limiting nutrient" simply means the nutrient that is in short supply for the physiological requirements of biological growth. For coastal populations of phytoplankton, sufficient iron is supplied by rivers, so commonly nitrate (and less commonly phosphate) will be the limiting nutrients. In the HNLC regions of the open ocean, nitrate and phosphate concentrations are high

---

[256] Gene C. Feldman and NASA OBPG, "From Missions to Measurements: an Ocean Color Experience", 8 January 2007, presentation for the SeaDAS Training Session, University of Maryland – Baltimore County, http://oceancolor.gsfc.nasa.gov/DOCS/Presentations/seadas_training_jan2007.ppt, (accessed 19 May 2009).

[257] NASA OBPG Data Distribution Statistics, http://oceancolor.gsfc.nasa.gov/cgi/ocdist_stats.cgi, statistics current as of 21 May 2009.

[258] John Weier, "John Martin (1935-1993)", NASA Earth Observatory, http://earthobservatory.nasa.gov/Features/Martin/martin.php, (accessed 22 May 2009).

[259] John Weier, "John Martin (1935-1993)".

enough to keep the phytoplankton happy, but Martin thought that there wasn't enough iron.

Breakthrough ideas such as Martin's aren't immediately accepted by the scientific community; they have to be tested. So Martin proposed an additional process study to the JGOFS program early in the 1990s, beyond the planned studies in the equatorial Pacific, Arabian Sea, and Southern Ocean. The proposal was straightforward; go to an HNLC area, dump in some iron (in a form that would be assimilated by phytoplankton; obviously dumping in rusty railroad tracks wouldn't work) and see what happens. JGOFS accepted the new process study and decided to perform it in the equatorial Pacific near the Galapagos islands; one of the cited reasons was Gene Feldman's 1986 CZCS study.[260] (Feldman confirmed that he discussed the iconic *Science* cover image of the Galapagos chlorophyll plume with Martin, because it "lent credence to the whole idea".)[261] JGOFS dubbed the process study "Iron-Ex" (for Iron Enrichment Experiment).[262] Despite Martin's untimely death from prostate cancer in June 1993, Iron-Ex I proceeded in the equatorial Pacific waters near the Galapagos Islands in October 1993. The experiment appeared to work, as chlorophyll concentrations increased by a factor of 3, although other results were not as clear-cut; larger increases in chlorophyll and greater consumption of other nutrients were expected. This initial study also examined the high chlorophyll region near the Galapagos, and the paper published subsequently in *Nature* used the CZCS image of the Galapagos pigment plume that had previously appeared on the cover of *Science*.[263] Sampling of the plume showed higher iron concentrations than in the central equatorial Pacific, indicating that iron-rich volcanic ash released from the Galapagos islands and marine sediments alleviated the iron deficiency in the waters around the archipelago.

Because of the uncertainties in the results of Iron-Ex I, a second experiment was proposed, and this became Iron-Ex II. Iron-Ex II featured three additions of iron, rather than one, taking place in May and June 1995.[264] This time the results were unequivocal;

---

[260] Gene C. Feldman, "Patterns of phytoplankton production around the Galapagos Islands," in J. Bowman, M. Yentsch, and W.T. Peterson, editors, Tidal *Mixing and Plankton Dynamics; Lecture Notes on Coastal and Estuarine Studies*, Volume 17, 77-106, (1986).

[261] Gene Feldman, comments recorded at the Ocean Color Collaborative Historical Workshop, January 13-14, 2009, St. Petersburg, Florida.

[262] U.S. JGOFS Planning and Coordination Office, "Iron Enrichment Experiment", U.S. JGOFS Planning Report Number 15, 28 pages, (February 1992).

[263] John H. Martin, Kenneth H. Coale, Kenneth, S. Johnson, Steve E. Fitzwater, R. Michael Gordon, Sara J. Tanner, C.N. Hunter, V.A. Elrod, J.L. Nowicki, T.L. Coley, Richard T. Barber, S. Lindley, A.J. Watson, K. Van Scoy, C.S. Law, M.I. Liddicoat, R. Ling, T. Stanton, J. Stockel, C. Collins, A. Anderson, Robert Bidigare, Michael Ondrusek, M. Latasa, Frank J. Millero, K. Lee, W. Yao, J.Z. Zhang, G. Friederich, Carole Sakamoto, Francisco Chavez, K. Buck, Zbigniew Kolber, R. Greene, Paul Falkowski, S.W. Chisholm, Frank Hoge, Robert Swift, James Yungel, S. Turner, P. Nightingale, A. Hatton, Peter Liss, and Neil W. Tindale, "Testing the Iron Hypothesis in Ecosystems of the Equatorial Pacific Ocean," *Nature*, 371(6493), 123-129, (1994).

[264] Kenneth H. Coale, Kenneth S. Johnson, Steve E. Fitzwater, R. Michael Gordon, Sara Tanner, Francisco P. Chavez, Laurie Ferioli, Carole Sakamoto, Paul Rogers, Frank Millero, Paul Steinberg, Phil Nightingale, David Cooper, William P. Cochlan, Michael R. Landry, John Constantinou, Gretchen Rollwagen, Armando Trasvina, and Raphael Kudela, "A massive phytoplankton bloom induced by an ecosystem-scale iron fertilization experiment in the equatorial Pacific Ocean," *Nature*, **383**, 495 – 501, (10 October 1996).

chlorophyll concentrations soared as phytoplankton bloomed profligately. Intriguingly, the results utilizing chlorophyll fluorescence did not only indicate that the phytoplankton grew faster; the results also indicated that the phytoplankton became "healthier", just as any animal or person deprived of vitamins or food will get healthier when their nutrition improves. For the phytoplankton, the improved health was indicated by improved photosynthetic efficiency – they made more chlorophyll in their cells, and used this chlorophyll to manufacture more carbon.[265]

Iron-Ex I and II demonstrated that Martin had been correct, but the blooms stimulated by the iron supplementation were not witnessed by remote-sensing instruments. The experiments, however, set the stage for larger and more ambitious research projects. Essentially, the scientists needed to determine if the warm and sunny tropical Pacific waters were unique in their response to iron. Would adding iron to the cold and windswept waters of the Southern Ocean induce the same response?

To find out, an oceanographic project based in New Zealand, with international collaborators (including many representatives from Australia and the UK) set out to add iron to the Southern Ocean. This experiment took the name SOIREE, for Southern Ocean Iron RElease Experiment. In February 1999, with SeaWiFS watching from orbit, the SOIREE party of oceanographers started adding iron to the ocean, and repeated the process on the 3rd, 5th, and 7th day of the experiment. The cold-water phytoplankton living near Antarctica reacted just as happily to the iron infusion as their Pacific brethren, and a strong bloom with elevated chlorophyll concentrations developed. The bloom drifted in a semi-circular pattern, so as SeaWiFS spied it over successive days, it traced out a three-quarter circle in the SeaWiFS 8-day data product, clearly visible south of Australia.[266]

Subsequent experiments with iron addition took place in the northern Pacific (SERIES, the Subarctic Ecosystem Response to Iron Enrichment Study) and SOFeX, the Southern Ocean Iron Enrichment Experiment. ("Fe" is the chemical symbol for iron in the hybrid acronym.) In SERIES, the area was frequently covered by clouds, but the phytoplankton were not dissuaded by lower light levels, and bloomed well – SeaWiFS was able to acquire one image of the SERIES bloom on July 29, 2002, which was 19 days after the iron had been added.[267] SOFeX, which preceded SERIES during January and February

---

[265] Michael J. Behrenfeld, Anthony J. Bale, Zbigniew S. Kolber, James Aiken, and Paul G. Falkowski, "Confirmation of iron limitation of phytoplankton photosynthesis in the equatorial Pacific Ocean," *Nature* 383, 508 – 511, (10 October 1996).

[266] Philip W. Boyd, Andrew J. Watson, Cliff S. Law, Edward R. Abraham, Thomas Trull, Rob Murdoch, Dorothee C.E. Bakker, Andrew R. Bowie, K.O. Buesseler, Hoe Chang, Matthew Charette, Peter Croot, Ken Downing, Russel Frew, Mark Gall, Mark Hadfield, Julie Hall, Greg Jameson, Julie LaRoche, Malcolm Liddicoat, Roger Ling, Maria T. Maldonado, R. Michael McKay, Scott Nodder, Stu Pickmere, Rick Pridmore, Steve Rintoul, Karl Safi, Philip Sutton, Robert Strzepek, Kim Tanneberger, Suzanne Turner, Anya Waite, and John Zeldis, "A mesoscale phytoplankton bloom in the polar Southern Ocean stimulated by iron fertilization," *Nature*, 407, 695-702, (12 October 2000); Edward R. Abraham, Cliff S. Law, Philip W. Boyd, Samantha J. Lavender, Maria T. Maldonado, and Andrew R. Bowie, "Importance of stirring in the development of an iron-fertilized phytoplankton bloom", *Nature*, 407, 727-730 (12 October 2000).

[267] Philip W. Boyd, Cliff S. Law, C.S. Wong, Yukihiro Nojiri, Atsushi Tsuda, Maurice Levasseur, Shigenobu Takeda, Richard Rivkin, Paul J. Harrison, Robert Strzepek, Jim Gower, R. Mike McKay,

2002, added iron in two separate regions of the Southern Ocean, one region where concentrations of silicic acid are elevated, and another region where silicic acid concentrations are low.[268] Silicic acid is an important nutrient because diatoms, frequently the dominant form of phytoplankton in ocean waters, need silicon to form their glittering assembly of frustule shapes. The question was: would other phytoplankton species that didn't need silicon take over the role of primary producers when the diatoms ran out of silicon?

The answer was a definite yes. Where silicic acid concentrations were low, a mixed population of diatoms and non-siliceous phytoplankton grew in response to iron enrichment. MODIS-Terra viewed this bloom, which had been stretched into a narrow ribbon by oceanic currents, 28 days after iron had been added. In the southern region of the experiment, SeaWiFS viewed a large patch of high chlorophyll 20 days after iron was added. As expected, diatoms were the dominant blooming phytoplankton where there was sufficient silicic acid. The SOFeX results did show that the phytoplankton utilized different forms of nitrogen in oceanic waters to keep their blooming healthy.

Red Tide Predictions

In 2005, Rick Stumpf of NOAA described the newly-operational National Ocean Service Harmful Algal Bloom Forecast System.[269] This system used ocean color data in conjunction with wind and SST data to predict the outbreak and movement of *Karenia brevis* red tides on the shores of the Gulf of Mexico. This system realized one of the predicted

---

Edward Abraham, Mike Arychuk, Janet Barwell-Clarke, William Crawford, David Crawford, Michelle Hale, Koh Harada, Keith Johnson, Hiroshi Kiyosawa, Isao Kudo, Adrian Marchetti, William Miller, Joe Needoba, Jun Nishioka, Hiroshi Ogawa, John Page, Marie Robert, Hiroaki Saito, Akash Sastri, Nelson Sherry, Tim Soutar, Nes Sutherland, Yosuke Taira, Frank Whitney, Shau-King Emmy Wong and Takeshi Yoshimura, "The decline and fate of an iron-induced subarctic phytoplankton bloom," *Nature*, 428, 549-553, (1 April 2004).

[268] Kenneth H. Coale, Kenneth S. Johnson, Francisco P. Chavez, Ken O. Buesseler, Richard T. Barber, Mark A. Brzezinski, William P. Cochlan, Frank J. Millero, Paul G. Falkowski, James E. Bauer, Rik H. Wanninkhof, Raphael M. Kudela, Mark A. Altabet, Burke E. Hales, Taro Takahashi, Michael R. Landry, Robert R. Bidigare, Xiujun Wang, Zanna Chase, Pete G. Strutton, Gernot E. Friederich, Maxim Y. Gorbunov, Veronica P. Lance, Anna K. Hilting, Michael R. Hiscock, Mark Demarest, William T. Hiscock, Kevin F. Sullivan, Sara J. Tanner, R. Mike Gordon, Craig N. Hunter, Virginia A. Elrod, Steve E. Fitzwater, Janice L. Jones, Sasha Tozzi, Michal Koblizek, Alice E. Roberts, Julian Herndon, Jodi Brewster, Nicolas Ladizinsky, Geoffrey Smith, David Cooper, David Timothy, Susan L. Brown, Karen E. Selph, Cecelia C. Sheridan, Benjamin S. Twining, and Zackary I. Johnson, "Southern Ocean Iron Enrichment Experiment: Carbon Cycling in High- and Low-Si Waters," *Science,* **304** (5669), 408-414 (16 April 2004); Ken O. Buesseler, John E. Andrews, Steven M. Pike, and Matthew A. Charette, "The Effects of Iron Fertilization on Carbon Sequestration in the Southern Ocean," *Science,* **304** (5669), 414-417 , (16 April 2004).

[269] Richard P. Stumpf, Richard Patchen, Mary Culver, Debra Payton, and Mark Vincent, "Red tide prediction to benefit public health and coastal economies," Proceedings of the 14th Biennial Coastal Zone Conference, New Orleans, Louisiana, July 17-21, 2005. http://www.csc.noaa.gov/cz/2005/CZ05_Proceedings_CD/pdf%20files/Stumpf.pdf, (accessed 22 May 2009).

benefits of routine ocean color observations from space, the ability to detect and forecast HABs to protect both humans and aquaculture.[270] SeaWiFS also returned striking images of noxious phytoplankton blooms in the Baltic Sea which became increasingly widespread as the mission progressed, due to the documented increase in eutrophication in the Baltic.[271] Researchers from Poland, Germany, Belgium, Denmark, and Estonia routinely used SeaWiFS data to observe the Baltic and North Seas.[272]

The Green Tide of Cholera

Conventional "green" tides – phytoplankton blooms – also have a potential connection to human health, due to the dreaded disease of cholera. The life cycle of the bacterium which causes cholera, *Vibrio cholerae*, includes a survival strategy of dormancy where the bacteria will hitch a ride on the backs of copepods, and also live inside their guts. This connection was established by Dr. Rita Colwell, who founded the University of Maryland Biotechnology Institute and who also served as the Director of the National Science Foundation. Colwell determined that one way to help rural populations to avoid cholera was simply to have them strain raw freshwater through several layers of sari cloth.[273] She also determined that following the fates of zooplankton was key to determining where cholera outbreaks were likely to occur and recur.

Colwell also collaborated with researchers to examine the causes of cholera outbreaks in cities in India, with an eye toward developing a predictive capability. Because copepods feast on phytoplankton, and because certain conditions contribute to phytoplankton blooms, in the 1990s the research team examined sea surface height data from TOPEX/Poseidon

---

[270] Heidi Dierrsen, James Acker, Stewart Bernard, and Grant Pitcher, "Hazards: Natural and Man-Made," Chapter 9 in Trevor Platt, Nicholas Hoeppffner, Venetia Stuart, and Christopher Brown, editors, *Why Ocean Colour? The Societal Benefits of Ocean Colour Techology*, IOCCG Report Number 7, 83-102 (2008).

[271] NASA GES DISC, "Bad Bloom Rising", http://daac.gsfc.nasa.gov/oceancolor/scifocus/oceanColor/bad_bloom.shtml, (accessed 22 May 2009).

[272] M. Darecki, S. Kaczmarek, and J. Olszewski, "SeaWiFS ocean colour chlorophyll algorithms for the southern Baltic Sea," *International Journal of Remote Sensing*, **26,** 247-260, 2005; M. Darecki, A. Weeks, S. Sagan, P. Kowalczuk, and S. Kaczmarek, "Optical characteristics of two contrasting case 2 waters and their influence on remote sensing algorithms," *Continental Shelf Research* , **23(3-4),** 237-250 (2003); Thomas Ohde and Herbert Siegel, "Correction of bottom influence in ocean colour satellite images of shallow water areas of the Baltic Sea," *International Journal of Remote Sensing*, 22(2&3), 297-313, (2001); Kevin G. Ruddick, Fabrice Ovidio, and M. Rijkeboer, "Atmospheric Correction of SeaWiFS Imagery for Turbid Coastal and Inland Waters," *Applied Optics*, 39(6), 897-912 (2001); P. V. Jørgensen, "SeaWiFS data analysis and match-ups with *in situ* chlorophyll concentrations in Danish waters," International Journal of Remote Sensing, 25(7-8), 397-1402, (2004);
Anu Reinart and Tiit Kutser, "Comparison of different satellite sensors in detecting cyanobacterial bloom events in the Baltic Sea," *Remote Sensing of Environment*, 102 (1-2), 74-85, (30 May 2006).

[273] Rita R. Colwell, "Global climate and infectious disease: the cholera paradigm," *Science*, 274 (5295), 2025 – 2031, doi: 10.1126/science.274.5295.2025 (20 December 1996).

mission in conjunction with SST data to determine conditions that were conducive to phytoplankton blooms in the Bay of Bengal. The addition of chlorophyll data, desired and anticipated in 1997 in an article published a day after the launch of SeaWiFS, enabled a direct examination of chlorophyll concentrations.[274] The initial use of SeaWiFS chlorophyll data in conjunction with other data types was described in a 2000 publication.[275] In 2008, a large research team published a study describing a combination of chlorophyll, precipitation and SST data to establish the remotely-sensed fingerprints of cholera outbreaks.[276]

SeaWiFS reaches 10 years

September 2007 marked the 10th anniversary of the first month of data collection from SeaWiFS. To commemorate the event, Ocean Biology and Biogeochemistry Program Manager Paula Bontempi convened a NASA TV "Science Update" production at Goddard Space Flight Center, featuring Gene Feldman, Chuck McClain, James Yoder, Marlon Lewis, and Mike Behrenfeld. The panelists evaluated the impact of the 10 years of SeaWiFS data collection and warned that a mission that could provide a similar record for the next 10 years was not currently being planned.[277]

Bontempi noted that due to the prevailing political winds at the time, it was necessary to get clearance from the NASA Office of Public Affairs to hold the 10-year SeaWiFS anniversary Science Update. When she was in the process of getting this approval, she discovered that the head of the NASA Public Affairs office had never actually heard of SeaWiFS. To determine its scientific impact, they conducted a publications search using the same criteria as that used for scientific contributions from the Hubble Space Telescope – and SeaWiFS got the same number of hits as the Hubble. According to Bontempi, the Public Affairs office was "blown away by that, because nobody had ever heard of SeaWiFS". And despite this demonstration of scientific prowess, SeaWiFS was still deemed "too politically sensitive" for the Science Update to be held at NASA HQ. Bontempi speculated that one reason was potential misconstrual of interpretations of ocean color data; she said "… people don't connect the biology and chemistry of the ocean, and the optics of the ocean, with climate – if they do, it might bring up something negative, some impact of

[274] John Travis, "Spying diseases from the sky," *Science News*, 152(5), 72-74, (2 August 1997).
[275] Brad Lobitz, Louisa Beck, Anwar Huq, Byron Wood, George Fuchs, A. S. G. Faruque, and Rita Colwell, "Climate and infectious disease: Use of remote sensing for detection of *Vibrio cholerae* by indirect measurement," *Proceedings of the National Academy of Sciences*, 97(4), 1438-1443, (15 February 2000).
[276] Guillaume Constantin de Magny, Raghu Murtugudde, Mathew R. P. Sapiano, Azhar Nizam, Christopher W. Brown, Antonio J. Busalacchi, Mohammad Yunus, G. Balakrish Nair, Ana I. Gil, Claudio F. Lanata, John Calkins, Byomkesh Manna, Krishnan Rajendran, Mihir Kumar, Bhattacharya, Anwar Huq, R. Bradley Sack, and Rita R. Colwell, "Environmental signatures associated with cholera epidemics," *Proceedings of the National Academy of Sciences, 105*(46), 17676-17681 (18 November 2008).
[277] NASA, "The Ocean Chromatic: SeaWiFS enters its second decade".

change that they don't want to hear." [278]

The Stroke of Midnight 2008

The 2007 NASA TV show highlighted the reliability and long-lived survivability of the plucky SeaWiFS instrument and Orbview-2 satellite. But as 2007 turned into 2008, something went wrong. SeaWiFS lost its ability to transmit necessary telemetry to accompany the reliable radiance measurements – telemetry which allowed the images to be navigated, i.e., to allow the data processing to determine where the satellite was observing.[279] Without this telemetry, the vast open spaces of the ocean viewed from space became truly uncharted; one patch of deep blue or light green water looks pretty much like any other.

The remarkable reliability of Orbview-2 meant that Orbimage had not required a very large staff to monitor and maintain the satellite. As a result, when the telemetry stream disappeared and the satellite was commanded into "safe hold" while the problem was analyzed, the working knowledge of the satellite required to fix it was no longer immediately available. After several attempts to revive the satellite were unsuccessful or only partially successful, Orbimage found it necessary to contact the engineering staff at Orbital Sciences Corporation for assistance. This initial contact went slowly due to somewhat strained relations between the two companies dating back to the loss of Orbview-4 and a subsequent lawsuit filed by Orbimage against OSC. [280] The engineering staff finally decided to rev up a backup computer system that had never been required before. When the B-side computer turned on and began operating the satellite, the engineers discovered that it was "dumb" – because it had not received the many patches and operating system updates that had been sent to the prime A-side computer over the course of the mission. So it took additional time to identify and send the necessary software updates to get the B-side up to speed.[281] After several months, stretching into July 2008 (even though partial imaging operations had resumed in April, the satellite did not operate for an entire month in April, May, or June), the satellite and sensor returned to full operational status in August 2008.[282]

One other operational factor affected SeaWiFS data quality over the course of the mission. After several years, the operations staff became reluctant to turn on the hydrazine thruster system to maintain the orbital position of the satellite. Because the satellite was now drifting, it no longer maintained its timing with the solar day, and so it began to drift away from its ideal noon equatorial crossing time. The initial drift was slow, but began to

[278] Paula Bontempi, comments recorded at the Ocean Color Collaborative Historical Workshop, January 13-14, 2009, St. Petersburg, Florida.
[279] Fred Patt, email announcement from the SeaWiFS Project, dated 3 January 2008; Gene Feldman, email announcement from the SeaWiFS Project, dated 3 January 2008.
[280] "Orbimage CEO Matt O'Connell," *Earth Imaging Journal*, http://www.eijournal.com/Interview_O%27Connell.asp, (accessed 22 May 2009); Eric Tegler, "Eye in the Sky".
[281] Mary Cleave and Gene Feldman recorded interview, 21 November 2008.
[282] Gene Feldman, email announcement from the SeaWiFS Project, 8 August 2008.

accelerate in 2006 and 2007. In late 2008, SeaWiFS was crossing the Equator at about 1:40 PM local time. This change in the observational geometry affects the global normalized water-leaving radiances measured by the sensor.[283]

SeaWiFS - The Epilogue:

Much of this history was completed in 2009, but SeaWiFS and Orbview-2 continued to operate – albeit with other extended hiatuses due to safehold incidents, which lasted for weeks to months. The equatorial crossing time problem was finally deemed to be a crucial problem, so NASA, GeoEye, and OSC collaborated to raise the orbit of Orbview-2 to 781.5 kilometers using 13 thruster burns, and this allowed the satellite to starting drifting back toward a noon equatorial crossing time. The orbit-raising maneuver, completed in July 2010, was one of the last remarkable success stories of the precedent-setting SeaWiFS Project.[284]

Unfortunately, the word "final" applies. In late December 2010, Orbview-2 stopped transmitting to Earth. After several weeks of trying to reestablish contact, the mission was declared over on Valentine's Day, February 14, 2011.[285]

Summary: The SeaWiFS mission surmounted both remarkable odds that had defeated numerous other attempts to get a NASA CZCS follow-on mission into space, as well as many unplanned and unexpected dilemmas created by its unique identity as a public-private partnership. That it succeeded so well is a tribute to a large number of people who took risks, worked long hours, cut corners where possible and didn't cut them when necessary, and who did everything possible to increase the scientific returns from this mission. SeaWiFS benefited in many ways from being a small and "agile" mission for which the scientific staff could make decisions quickly, and get or deploy resources efficiently, to augment the mission. SeaWiFS also benefited by having a larger, better-funded partner in MODIS, by sharing some aspects of the MODIS mission, such as the sea-truth data collected by the MOBY instrument and expertise of some of the MODIS science team members. Ultimately, SeaWiFS may not have received all the recognition it might have deserved because it was a satellite and instrument that was owned by a private company and not by the U.S. government and NASA. However, there is no doubt that the data collected by SeaWiFS currently constitute the longest well-calibrated ocean color data set ever obtained, a data set that will yield oceanographic insight for years to come.

---

[283] NASA Ocean Biology Processing Group, "SeaWiFS Calibration Drift", http://oceancolor.gsfc.nasa.gov/SeaWiFS/On_Orbit/lcal/cal_drift.html, (accessed 22 May 2009).
[284] Gene Feldman, "seawifs is back after a successful orbit-raising campaign," email message, 13 July 2010.
[285] Gene Feldman, "the end of an incredible era," email message, 14 February 2010.

6

PUSHING THE ENVELOPE

Imagine this: $17 billion dollars for a program of advanced Earth remote sensing satellites. A satellite carrying 10 (or more) separate sensors, all designed to look at the Earth in various ways and with various bands – at the same time – scanning the atmosphere, the oceans, and the land, collecting enormous volumes of daily data to describe the fundamental processes of Earth's ecosystems and climate. And not just *one* such satellite; actually three separate sets of satellites, each with the same sensors, to be launched in five-year intervals.

That very brief description is the original concept of the **Earth Observing System (EOS)**.

Actually, a previous system, "System Z", predates EOS. System Z dates back to the early days of the Reagan administration, and it was designed conceptually to utilize the envisioned capabilities of the Space Shuttle.[1] System Z was envisioned at the time when NASA was planning to use both the Atlantic (Canaveral) and Pacific (Vandenberg) Shuttle launch facilities. According to Dixon Butler, the name "System Z" was coined by Dr. Bert Edelson, who was the Associate Administrator for Space Science and Applications[2]. Edelson realized that large satellites in low Earth orbit could be used for Earth science. Butler was the Executive Secretary of a group organized by Edelson to study the idea. Butler describes that he came up with the idea of a set of satellites equipped with Earth observing instruments designed to examine the role of water in Earth's environment.[3]

One of the key ideas for System Z was that a large low-earth orbit satellite would allow servicing by the Space Shuttle astronauts, much like the Hubble or Solar Max. Thus, putting a large payload of complex instruments on a satellite was not conceived as vulnerable because astronauts could replace a faulty instrument if necessary. This would also allow instruments to be replaced with new and better versions as technology improved. After various meetings and design concepts, System Z was renamed the Earth Observing System or *Eos* (now EOS, after a trademark conflict was cleared up).[4] System Z was one of the first

---

[1] Dixon Butler, "The Early Beginnings of EOS: "System Z" Lays the Groundwork for a Mission to Planet Earth," *The Earth Observer*, Volume 20(5), EOS Project Science Office, NASA Goddard Space Flight Center, 4-7, September-October 2008.

[2] Dixon Butler, "The Early Beginnings of EOS: "System Z" Lays the Groundwork for a Mission to Planet Earth."

[3] Dixon Butler, "The Early Beginnings of EOS: "System Z" Lays the Groundwork for a Mission to Planet Earth."

[4] Dixon Butler, "The Early Beginnings of EOS: "System Z" Lays the Groundwork for a Mission to Planet Earth."

multi-instrument platform concepts, and several aspects of the eventual EOS are first described for System Z, including the use of the Tracking and Data Relay Satellite System (TDRSS) to for system communications. The TDRSS satellites were already in orbit to facilitate Space Shuttle communications, and they also enable communications with many other satellite systems.[5]

The next step was a formalization of the goals of the System Z concept. This was done in the Bretherton reports, based on the work of a committee led by climate scientist Francis Bretherton.[6] The two Bretherton reports formulated the official agency basis for NASA's Mission to Planet Earth (MTPE) and EOS.[7] The initial EOS was set up with a $17 billion dollar, ten-year budget and the system goals and mission were described in a 1984 NASA report, *Earth Observing System, Science and Mission Requirements Working Group Report.*[8] The mission payload included a "Surface Imaging and Sounding Package" (SISP), which consisted of four instruments. The two instruments of most interest to the ocean color community were the Moderate Resolution Imaging Spectrometer (MODIS) and the High Resolution Imaging Spectrometer (HIRIS).[9]

A 1987 report authored by astronaut Sally Ride advocated that MTPE be one of NASA's four major thematic programs.[10]

The End of "System Z" and the new EOS

The Challenger disaster marked the end of plans to utilize Vandenberg as a Shuttle launch facility.[11] President Reagan redefined the Space Shuttle's role, eliminating the polar-

---

[5] "TDRSS: Tracking and Data Relay Satellite System", http://msl.jpl.nasa.gov/Programs/tdrss.html, (accessed 6 August 2009).

[6] John M. Logsdon, Stephen J. Garber, Roger D. Launius, and Ray A. Williamson, editors, *Exploring the Unknown: Selected Documents in the History of the U.S. Civil Space Program*, NASA SP-2004-4407, *Volume VI: Space and Earth Science*, page 448.

[7] John M. Logsdon, Stephen J. Garber, Roger D. Launius, and Ray A. Williamson, editors, *Exploring the Unknown: Selected Documents in the History of the U.S. Civil Space Program*, NASA SP-2004-4407, *Volume VI: Space and Earth Science.* **Note:** The two Bretherton reports from the NASA Advisory Council were entitled *Earth System Science, A Program for Global Change*, report of the Earth System Sciences Committee, 1986; and *The Crisis in Space and Earth Science, A Time for New Commitment*, report of the Space and Earth Science Advisory Committee, 1986.

[8] John M. Logsdon, Stephen J. Garber, Roger D. Launius, and Ray A. Williamson, editors, *Exploring the Unknown: Selected Documents in the History of the U.S. Civil Space Program*, Volume VI; NASA, "Earth Observing System: Science and Mission Requirements, Working Group Report, Volume I", *NASA Technical Memorandum 86129*, 1984.

[9] Raymond E. Arvidson, Dixon M. Butler, and Richard E. Hartle, "Eos: The Earth Observing System of the 1990s," *Proceedings of the IEEE*, 73(6), 1025-1030 (1 June 1985).

[10] Sally Ride, *NASA Leadership and America's Future in Space: A Report to the Administrator*, NASA, August 1987, cited in Volume VI, John M. Logsdon, Stephen J. Garber, Roger D. Launius, and Ray A. Williamson, editors, *Exploring the Unknown: Selected Documents in the History of the U.S. Civil Space Program.*

[11] John M. Logsdon, Stephen J. Garber, Roger D. Launius, and Ray A. Williamson, editors, *Exploring the Unknown: Selected Documents in the History of the U.S. Civil Space Program*, Volume VI, page 456.

orbiting option, which was only possible from Vandenberg. So the Challenger explosion ended plans for the second Shuttle launch facility at Vandenberg, and forced EOS to return to the use of launch vehicles. This change required substantial changes and a conceptual redesign of the entire EOS program.[12]

Based on these new requirements, the EOS mission design evolved into a two-satellite concept, with two multi-instrument platforms, EOS-A and EOS-B. EOS-A was the morning satellite, and EOS-B the afternoon satellite, with the instrument payloads designated for different observational conditions. Land and ocean observations were known to be favored in the morning, when the warmth of the Sun has not yet caused the extensive buildup of clouds over land and oceans.[13] Cloud properties, radiative balance, and precipitation (among numerous other variables) were therefore designated for the afternoon satellite.[14] EOS was conceived as a fifteen-year program with new copies of the satellites, i.e., another satellite with the same instrument payload, to be launched every five years.[15]

Ocean Color in the Original EOS

The success of CZCS, and the continuing parallel frustration in realizing a CZCS follow-on mission, was known to the EOS planners. Therefore, soon after EOS was initiated, the capabilities of the instruments necessary for ocean color observations were considered. The instrument in the original payload design that was designated for ocean color was MODIS.

Recognition of the particular needs of ocean color led to the development of a two-instrument MODIS concept: **MODIS-N** (with the N standing for "nadir") and **MODIS-T** (with the T standing for "Tilt").[16] The need to tilt an ocean color observing instrument extended back to the CZCS mission.[17] The MODIS Science Team was initially formed under the leadership of Vince Salomonson in 1987, a position that was formalized in 1988.[18] The plan for MODIS was to have the skilled GSFC engineering staff create MODIS-T "in-house", while MODIS-N was designated to be built under a contract to the Hughes Santa

---

[12] John M. Logsdon, Stephen J. Garber, Roger D. Launius, and Ray A. Williamson, editors, *Exploring the Unknown: Selected Documents in the History of the U.S. Civil Space Program*, Volume VI.
[13] John M. Logsdon, Stephen J. Garber, Roger D. Launius, and Ray A. Williamson, editors, *Exploring the Unknown: Selected Documents in the History of the U.S. Civil Space Program*, Volume VI, page 458; NASA, "The First EOS Satellite: EOS AM-1," http://terra.nasa.gov/Publications/AM-1_brochure.pdf, (accessed 6 August 2009).
[14] NASA, "Mission to Planet Earth: Earth Observing System (EOS) Spacecraft and Instruments", http://www.gsfc.nasa.gov/gsfc/service/gallery/fact_sheets/earthsci/fs-96(07)-13.htm, (accessed 6 August 2009).
[15] John M. Logsdon, Stephen J. Garber, Roger D. Launius, and Ray A. Williamson, editors, *Exploring the Unknown: Selected Documents in the History of the U.S. Civil Space Program*, Volume VI, page 458.
[16] Thomas J. Magner and Vincent V. Salomonson, "Moderate resolution imaging spectrometer-tilt (Modis-T)," *International Journal of Imaging Systems and Technology*, 3(2), 121-130, (1991).
[17] Warren Hovis, "The Nimbus-7 Coastal Zone Color Scanner (CZCS) Program".
[18] Vincent V. Salomonson recorded interview, 2 December 2008.

Barbara Research Center (SBRC).[19] Thomas Magner of GSFC was the MODIS-T instrument manager. There was a crucial difference in the two instruments as well; MODIS-N, though originally termed a spectrometer, was actually going to be a spectro-radiometer, collecting radiance measurements in specific bands, initially 36.[20] (Salomonson indicated that the initial number of MODIS-N bands was actually 42.)[21] MODIS-T was designed to be an actual spectrometer, which rather than collecting radiance measurements at specific points in the spectrum, as a radiometer does, would collect data over the entire spectrum, creating a full ocean spectrum; MODIS-T thus was designed with 64 bands, spaced so closely that there were essentially no gaps in coverage from the IR to the UV and at all points in between (particularly the useful visible range).[22] Such full-spectrum data provides considerably more analyzable diagnostic information than data collected at discrete, separate points in the spectrum.

A third instrument, HIRIS, was also somewhat in the plans of the ocean color community.[23] While MODIS was designed to gather data at about one kilometer resolution, HIRIS would collect much more data for a much smaller area, 30 meters, equivalent to the Landsat ETM, but with full spectral resolution, with 10 nanometer spectral bands from 0.4-2.5 microns. The swath width for HIRIS was only 30 km; thus, MODIS was indicated to be the global "big picture" instrument, and HIRIS the microscope (from space), targeting unusual or perplexing features of interest. HIRIS data was of particular interest to ocean color scientists seeking to deconvolute the perplexities of Case II coastal waters; Kendall Carder of the University of South Florida received funding for this research effort.[24] HIRIS was also recognized very early on to be one of the most expensive instruments that EOS would carry.[25]

So, starting in 1988 and moving into the very early 1990s, scientists watched with interest as the MODIS engineering teams at GSFC and SBRC started creating the subsystems that would be merged into instruments that were some of the most ambitiously complex Earth observing instruments ever conceived.

---

[19] Vincent V. Salomonson recorded interview, 2 December 2008.

[20] Vincent V. Salomonson, "The Moderate Resolution Imaging Spectrometer: An EOS Facility Instrument Candidate for Application of Data Compression Methods,", *Proceedings of the Space and Earth Science Data Compression Workshop*, SEE N92-12425 03-59, 13-23, IEEE, (1991).

[21] Vincent V. Salomonson recorded interview, 2 December 2008.

[22] Vincent V. Salomonson, William L. Barnes, Peter W. Maymon, Harry E. Montgomery, and Harvey Ostrow, "MODIS : Advanced facility instrument for studies of the Earth as a system," *IEEE Transactions on Geoscience and Remote Sensing*, 27(2), (March 1989).

[23] Jeff Dozier and Alexander F.H. Goetz, "HIRIS - EOS instrument with high spectral and spatial resolution," in *Proceedings of the 21st International Symposium on Remote Sensing of Environment*, Ann Arbor, MI, Oct. 26-30, 1987, Volume 1 (A89-10926 01-43), Ann Arbor, MI, Environmental Research Institute of Michigan, 31-39, (1987).

[24] Kendall L. Carder, "Curriculum Vitae," http://www.marine.usf.edu/faculty/documents/carder-kendall-cv-10-2007.pdf, (accessed 6 August 2009).

[25] Jeff Dozier, "Planned EOS Observations of the land, ocean and atmosphere," *Atmospheric Research*, 31, 329-357, (1994).

The MODIS Science Team: Selection and Commitment

With the MODIS instrument build initiated, the next step was to select the science teams that would support them; the scientists who would create the algorithms which would convert the radiance measurements into useful geophysical variables. NASA had a good model to follow for this effort; the NETs of the Nimbus 7 mission. But in the case of MODIS, and the long-term commitment that being on the science team would entail, there was a crucial difference; the science team members would actually sign contracts that would have schedules and deliverables, and the length of the contracts were unprecedented; with NASA expecting the first EOS satellite launch in 1998, the researchers were thus contemplating a decade-long commitment, which Salomonson said was "unconscionable, in retrospect".[26]

The MODIS Ocean science team included some familiar names with CZCS heritage, and several new faces as well. Wayne Esaias was selected as the Ocean Team Leader, and also took a very big role in algorithm development; creating algorithms for ocean primary productivity based on MODIS data.[27] One of his primary collaborators on this effort was Janet Campbell of the Bigelow Laboratory for Ocean Sciences.[28] Howard Gordon took responsibility for the advanced atmospheric correction algorithms. Along with Dennis Clark, who was working on MOBY development and open ocean (Case I) algorithm development, these were the two members with direct connections to the NET team. Robert Evans of RSMAS had a key role, which would be to create a database matching at-sea observations (either from ships or buoys) with MODIS data for direct instrument calibration.[29] William Balch of Bigelow would be working with Gordon on algorithms for coccolithophore calcite concentration.[30] Frank Hoge of the NASA Wallops Flight Facility, whose specialty was determining phytoplankton pigment concentrations with laser-induced fluorescence, would be developing algorithms to produce the concentration of these non-chlorophyll pigments.[31] Mark Abbott of Oregon State University also joined the team; Abbott planned to develop algorithms for data products derived from specialized new bands in the MODIS instrument. His chief deputy at OSU would be Richard Letelier.[32] Kendall Carder of the University of South Florida proposed to develop advanced

[26] Vincent V. Salomonson recorded interview, 2 December 2008; Kendall L. Carder, "Curriculum Vitae"; NASA GSFC, "Minutes of the MODIS Science Team Meeting, September 24-26, 1990," http://modis.gsfc.nasa.gov/sci_team/meetings/199009/minutes.pdf, (accessed 7 August 2009); NASA GSFC, "Minutes of the MODIS Science Team meeting, February 1991", http://modis.gsfc.nasa.gov/sci_team/meetings/199102/minutes.pdf, (accessed 7 August 2009).
[27] Wayne E. Esaias, Mark R. Abbott, Ian Barton, Otis B. Brown, Janet W. Campbell, Kendall L. Carder, Dennis K. Clark, Robert H. Evans, Frank E. Hoge, Howard R. Gordon, William M. Balch, Richard Letelier, and Peter J. Minnett, "An Overview of MODIS Capabilities for Ocean Science Observations," *IEEE Transactions on Geoscience and Remote Sensing*, 36(4), 1250-1265, (July 1998).
[28] Wayne E. Esaias, et al., "An Overview of MODIS Capabilities for Ocean Science Observations."
[29] Wayne E. Esaias, et al., "An Overview of MODIS Capabilities for Ocean Science Observations."
[30] Wayne E. Esaias, et al., "An Overview of MODIS Capabilities for Ocean Science Observations."
[31] Wayne E. Esaias, et al., "An Overview of MODIS Capabilities for Ocean Science Observations."
[32] Wayne E. Esaias, et al., "An Overview of MODIS Capabilities for Ocean Science Observations."

algorithms, called "semi-analytical" algorithms, designed to be applied to the difficult Case II water problem.[33]

These investigators were the team members designated to look at the ocean color or biological side, but there was another facet to MODIS, the SST or physical side. In addition to the ocean color bands, MODIS would be equipped with bands for infrared wavelength detection, which for the oceans would generate SST products. Otis Brown of RSMAS was the SST algorithm lead, and he would be collaborating with Ian Barton of CSIRO (Australia), and Peter Minnett of RSMAS, as well as Evans.[34]

The Budget Axe Swings

$17 billion dollars is a lot of money for anything, especially in 1990 dollars. At times when the entire budget for NASA comes under scrutiny, $17 billion dollars invites special scrutiny. Starting in 1990, the EOS Program began to face a series of reviews and a succession of budget cuts, causing major changes to the entire program. The pressure began with significant cuts to the NASA budget imposed during the Presidency of George H.W. Bush, which drastically reduced the EOS Program budget from its lofty $17 billion to a still substantial, but $6 billion dollars smaller, budget of $11 billion.[35] The budget cuts to NASA and to EOS forced a massive revamping of the entire program. One particular target was the project cost of the data system being developed to support the EOS program, EOSDIS (EOS Data and Information System). [36]

Edward Frieman was made chairman of the EOS Engineering Review Committee, which had to consider the program and recommend where the budget cuts would fall. In 1991, the committee recommended abandoning the huge 14-instrument EOS-A and EOS-B satellites in favor of smaller satellites with complementary sensors targeting specific aspects of the Earth's land-atmosphere-ocean system. The committee also recommended examining and redesigning the EOSDIS.[37]

In light of the Frieman Report – and also as a result of a meeting of the Investigators Working Group (IWG) in the same year in Seattle – at the MODIS Science Team meeting held October 1-3, 1991, the continuance of MODIS-T as an EOS payload instrument was under serious reconsideration.[38] Salomonson recommended that the

---

[33] Wayne E. Esaias, et al., "An Overview of MODIS Capabilities for Ocean Science Observations."
[34] Wayne E. Esaias, et al., "An Overview of MODIS Capabilities for Ocean Science Observations."
[35] John M. Logsdon, Stephen J. Garber, Roger D. Launius, and Ray A. Williamson, editors, *Exploring the Unknown: Selected Documents in the History of the U.S. Civil Space Program*, Volume VI, page 458.
[36] John M. Logsdon, Stephen J. Garber, Roger D. Launius, and Ray A. Williamson, editors, *Exploring the Unknown: Selected Documents in the History of the U.S. Civil Space Program*, Volume VI, page 464 - 466.
[37] Edward Frieman, "Report of the Earth Observing System Engineering Review Committee", in John M. Logsdon, Stephen J. Garber, Roger D. Launius, and Ray A. Williamson, editors, *Exploring the Unknown: Selected Documents in the History of the U.S. Civil Space Program*, Volume VI, pages 566-575.
[38] The Investigators Working Group was composed of members of the original EOS Interdisciplinary Investigations, which were projects composed of several team members intended to use data from multiple EOS instruments. The original team projects and leaders are described in a paper by Jeff

MODIS science team develop "compelling scientific arguments" in support of MODIS-T, otherwise known as "credible, rigorous, and consistent arguments" to present to the EOS Payload Panel, which was to meet later that month.[39] In the same Science Team meeting minutes, Steve Running presented evaluation of an alternative under consideration, which was favored by the MODIS disciplines, to have a MODIS-N on both the morning and afternoon satellites. The Ocean group, not surprisingly, still strongly favored MODIS-T. At the same time, the Ocean team had to consider the alternatives, and modeling studies performed by Watson Gregg indicated that two MODIS-N instruments provided about the same amount of ocean coverage as one MODIS-T (as well as the tiltable SeaWiFS).[40] There were certainly deficiencies; the viewing angles at low latitudes (i.e., the tropics) for some areas would be unfavorable, and the tilting instrument was therefore a better option.

Unfortunately, it was also a very expensive option; too expensive for the slashed EOS budget profile. The Payload Panel dropped MODIS-T from the instrument manifest in favor of the dual MODIS-N deployment option.[41] Mark Abbott said that they could "blame me" for this decision (he was on the panel); he stated that they were "trying to balance the costs".[42] Mary Cleave noted that MODIS-T was a convenient target, as it was an in-house instrument, which meant there wasn't a contract with a private corporation to build it, as was the case for MODIS-N being built at the time by Hughes SBRC.[43] Dennis Clark commented that a lot of the MODIS team members had signed on to work with MODIS-T; when it was dropped, they had to decide if it was worth the effort for them to continue.[44]

The ink had barely dried on the Frieman report, with little time available for NASA to respond, when NASA Administrator Daniel Goldin imposed *another* budget cut on EOS, an additional 30% cut, so that the total EOS budget running from late 1992 to the year 2000

Dozier, "Looking Ahead to EOS: The Earth Observing System," *Computers in Physics*, May/June 1990, pp. 248-259, http://www.ssrc.ucsc.edu/PaperArchive/dozier-cip90.pdf, (accessed 7 August 2009). The members of the interdisciplinary investigations convened at Investigators Working Group meetings.

[39] NASA GSFC, "Minutes of the MODIS Science Team Meeting, October 1-3, 1991," http://modis.gsfc.nasa.gov/sci_team/meetings/199110/minutes.pdf, (accessed 7 August 2009).

[40] NASA GSFC, "Minutes of the MODIS Science Team Meeting, October 1-3, 1991."

[41] The Payload Panel report, by Berrien Moore and Jeff Dozier, "Adapting the Earth Observing System to the Projected $8 Billion Budget: Recommendations from the EOS Investigators" is the first explicit statement that MODIS-T was dropped. The report appears in "John M. Logsdon, Stephen J. Garber, Roger D. Launius, and Ray A. Williamson, editors, *Exploring the Unknown: Selected Documents in the History of the U.S. Civil Space Program*, Volume VI, 628-639. In the NASA report "Report to Congress on the Restructuring of the Earth Observing System", 9 March 1992, MODIS-T is no longer represented. The report also includes the first mention of EOS Color: Greg Mitchell (comments recorded at the Ocean Color Collaborative Historical Workshop, January 13-14, 2009, St. Petersburg, FL) said that he was able to get an initial commitment to EOS Color, which appears on a reproduced display figure. Mitchell said "I had to get the guy that made the charts... to make sure it's on the chart!" The report to Congress also appears in John M. Logsdon, Stephen J. Garber, Roger D. Launius, and Ray A. Williamson, editors, *Exploring the Unknown: Selected Documents in the History of the U.S. Civil Space Program*, Volume VI, pages 586-613.

[42] Mark Abbott recorded interview, 26 November 2008.

[43] Mary Cleave and Gene Feldman recorded interview, 21 November 2008.

[44] Dennis Clark, interview transcript notes.

was down to $8 billion.[45] These impressive cuts forced further reconsideration of the satellite instrument payloads, a task undertaken by the EOS Payload Advisory Panel. It was also clear that the instrument capabilities themselves would face budget pressures. Even though MODIS-T had been terminated, the instruments being built (now two of them) were still very remarkable sensors.

MODIS Capabilities for Ocean Observations

MODIS was designed to be an instrument capable of acquiring observational data for the land, atmosphere, and oceans. To fully describe its land and atmospheric capabilities is beyond the scope of this history. In brief, MODIS had bands dedicated to land processes, including thermal bands for fire detection, visible-wavelength bands for land cover research (which included broader bandwidth bands with higher spatial resolution, 500 meters and 250 meters) as well as a suite of 1 km resolution bands. The IR channels also had land applications, similar to the IR bands on Landsat. The MODIS atmospheric bands allowed observation of clouds, atmospheric aerosols, water vapor, and atmospheric temperature (Table 1).[46]

For the oceans, MODIS had nine bands, with the "reddest" bands (centered at 748 and 869.5 nm reserved for atmospheric correction). MODIS still had the classic 443 nm band for chlorophyll absorption, and heritage bands (at slightly different locations than CZCS or SeaWiFS) centered at 412.5, 488, 551, and 667 nm. The other two bands, centered at 531 nm and 678 nm, provided new information. The 531 nm band allowed better detection of the accessory pigments called phycobilins which were the objects of Frank Hoge's attention, while the 678 nm band was designed specifically to examine something quite interesting: the fluorescence of chlorophyll.[47] When the Sun shines on the chlorophyll molecule, the light-stimulated molecule re-emits light. The amount of light emitted is related to the physiological state of the organism containing the chlorophyll. Stronger fluorescence indicates vibrant, healthy, actively photosynthesizing organisms. Weaker fluorescence indicates that the photosynthetic activity has diminished; lack of nutrients to power biological processes or even bleaching caused by extended exposure to bright sunlight can cause this to happen. So chlorophyll fluorescence offered a window into the living room of ocean's phytoplankton; rather than just indicate how much chlorophyll was in the water, fluorescence could indicate how active the phytoplankton were, which was significant to primary productivity estimates. Furthermore, the fluorescence signal could be detected in Case II waters, potentially improving estimates of chlorophyll concentration and primary productivity in turbid coastal waters.

---

[45] John M. Logsdon, Stephen J. Garber, Roger D. Launius, and Ray A. Williamson, editors, *Exploring the Unknown: Selected Documents in the History of the U.S. Civil Space Program*, Volume VI, page 460.
[46] Table adapted from http://modis.gsfc.nasa.gov/about/specifications.php.
[47] Wayne E. Esaias, et al. "An Overview of MODIS Capabilities for Ocean Science Observations."

Orbit:	705 km, 10:30 a.m. descending node or 1:30 p.m. ascending node, sun-synchronous, near-polar, circular
Scan Rate:	20.3 rpm, cross track
Swath Dimensions:	2330 km (across track) by 10 km (along track at nadir)
Telescope:	17.78 cm diam. off-axis, afocal (collimated), with intermediate field stop
Size:	1.0 x 1.6 x 1.0 m
Weight:	250 kg
Power:	225 W (orbital average)
Data Rate:	11 Mbps (peak daytime)
Quantization:	12 bits
Spatial Resolution:	250 m (bands 1-2)
(at nadir):	500 m (bands 3-7), 1000 m (bands 8-36)
Design Life:	5 years

Primary Use	Band	Bandwidth[1]	Spectral Radiance[2]	Required SNR[3]
Land/Cloud Boundaries	1	620-670	21.8	128
	2	841-876	24.7	201
Land/Cloud Properties	3	459-479	35.3	243
	4	545-565	29.0	228
	5	1230-1250	5.4	74
	6	1628-1652	7.3	275
	7	2105-2155	1.0	110
Ocean color/ Phytoplankton/ Biogeochemistry	8	405-420	44.9	880
	9	438-448	41.9	838
	10	483-493	32.1	802
	11	526-536	27.9	754
	12	546-556	21.0	750
	13	662-672	9.5	910
	14	673-683	8.7	1087
	15	743-753	10.2	586
	16	862-877	6.2	516
Atmospheric Water Vapor	17	890-920	10.0	167
	18	931-941	3.6	57
	19	915-965	15.0	250

Primary Use	Band	Bandwidth[1]	Spectral Radiance[2]	Required NEΔT(K)[3]
Surface/Cloud Temperature	20	3.660-3.840	0.45	0.05
	21	3.929-3.989	2.38	2.00
	22	3.929-3.989	0.67	0.07
	23	4.020-4.080	0.79	0.07
Atmospheric Temperature	24	4.433-4.498	0.17	0.25
	25	4.482-4.549	0.59	0.25
Cirrus Clouds	26	1.360-1.390	6.00	150[4]
Water Vapor	27	6.535-6.895	1.16	0.25
	28	7.175-7.475	2.18	0.25
	29	8.400-8.700	9.58	0.05
Ozone	30	9.580-9.880	3.69	0.25
Surface/Cloud Temperature	31	10.780-11.280	9.55	0.05
	32	11.770-12.270	8.94	0.05
Cloud Top Altitude	33	13.185-13.485	4.52	0.25
	34	13.485-13.785	3.76	0.25
	35	13.785-14.085	3.11	0.25
	36	14.085-14.385	2.08	0.35

[1]Bands 1 to 19, nm; Bands 20-36, μm
[2](W/m^2-μm-sr)
[3]SNR=Signal-to-noise ratio; NEΔT=Noise-equivalent temperature difference } Performance goal is 30%-40% better than required
[4]SNR

Table 1. MODIS Specifications.

The MODIS bands were narrower than for CZCS or SeaWiFS; the larger scan mirror and light collection aperture allowed these stringent optics while still having acceptable signal-to-noise ratios. These narrower bands offered particular advantages; as an example, the 10 μm – wide atmospheric correction band at 748 nm avoided the interference of an atmospheric oxygen spectral feature located a few nanometers away.[48]

The MODIS capabilities in the IR were also significant. The long-running history of the AVHRR virtually required that MODIS have an SST band at the classic wavelength of 11 μm; MODIS added a 4 μm band to provide a second SST data product. Other IR bands were designated to improve cloud and aerosol detection which could interfere with the SST determination, allowing either corrections or the emplacement of data "flags" for conditions conducive to errors.[49]

MODIS had a number of design aspects that would allow better calibration and characterization of the instrument as it operated in orbit. It was equipped with a solar diffuser with both a low and high (bright) calibration setting, and had three separate doors: one to view the Earth, one to view the Sun, and one to view deep space.[50]

In the early 1990s, the MODIS science teams all knew what the design goals for MODIS were; the problem was, it still had to be built, in an era with declining budgets,

[48] Wayne E. Esaias, et al. "An Overview of MODIS Capabilities for Ocean Science Observations."
[49] Wayne E. Esaias, et al. "An Overview of MODIS Capabilities for Ocean Science Observations."
[50] Xiaoxiong Xiong, William L. Barnes, Bruce Guenther, and Robert E. Murphy, "Lessons Learned from MODIS," *Advances in Space Research*, 32(11), 2107-2112, (2003).

while at the same time the engineers at SBRC began to discern where the most difficult aspects of the instrument design would be.

Science Team Meetings: Making MODIS in the Midst of "Descoping"

The Science Team meeting minutes for each of the biannual science team meetings always included an update on the instrument status. At the meeting in April 1992, the complexities and trade-offs of building the complex sensor came under discussion.[51]

One of the primary concerns was the filter specifications. State-of-the-art filters were what allowed the narrow bandwidths desired by the scientists; the problem was that the state-of-the-art necessary to make the filters was apparently a couple of years in the future. So the instrument manufacturer was requesting, if possible, relaxation of the filter requirements. So SBRC provided a list of changes to the filters that they were requesting. Both the budget for instrument fabrication and the schedule for getting it done would be affected by making the filter requirements easier; even making one band easier to do would be significant.[52]

The problem was that there were specific science reasons for having bands with narrow bandwidths at the specific selected locations. Relaxing the filter requirements and broadening the bandwidth would change how the algorithms interpreted the data. So it was not just a simple matter of the science team agreeing to the suggested changes; their impact had to be considered, not just on paper, but in modeling simulations run on computers.[53]

But there was more. The calibration team noted that the critical scan mirror already had a layer of contamination built up on it, 50 angstroms thick.[54] The layer of contamination resulted from a problem which occurred during thermal/vacuum testing; one of the protocols of testing was to have a "hot nadir door" to simulate solar exposure in space. During the course of the testing, the door was overheated, releasing volatiles that were deposited on the scan mirror. In the words of the ocean color scientists, the door was "smoking".[55]

Having two missions in development was an ongoing area of confusion; with SeaWiFS and MODIS both underway, there were competing budgets and conflicting responsibilities. Robert Evans noted that there was tension because the overlap of budgets and responsibilities meant that it wasn't clear who was the ultimate budget authority, particularly with regard to the important contract obligations of the MODIS team

---

[51] NASA GSFC, "Minutes of the MODIS Science Team meeting, April 13-16, 1992," http://modis.gsfc.nasa.gov/sci_team/meetings/199204/minutes.pdf, (accessed 7 August 2009).
[52] NASA GSFC, "Minutes of the MODIS Science Team meeting, April 13-16, 1992."
[53] NASA GSFC, "Minutes of the MODIS Science Team meeting, April 13-16, 1992."
[54] NASA GSFC, "Minutes of the MODIS Science Team meeting, April 13-16, 1992."
[55] Howard Gordon and other attendees, comments recorded at the Ocean Color Collaborative Historical Workshop, January 13-14, 2009, St. Petersburg, FL.

members.[56] The Ocean team also had to evaluate the impact of the loss of MODIS-T on the list of science products they were obligated to produce.

The next stage of MODIS development took place in the era of "descoping". Checking with Merriam-Webster, "descoping" is not a word that has entered the popular parlance; in non-technical terms, it means reduced capabilities for less cost. So "descoping" appear to be a term particular to NASA. What it meant was that EOS was working with a budget of about $8 billion, necessitating budgetary balancing acts between instruments. One way to deal with the lower budget was to move up the launch date to June 1998.

The MODIS Oceans science team had to deal with changing product requirements, a problem because the algorithms for many of their products were interdependent with other measurements. The next stage of concern was all the bands that MODIS had; deleting a few bands would save money but hurt science.[57]

The entire MODIS Science Team was also dealing with the requirement from the Project and instrument management that the algorithms for data products be peer-reviewed. Each Science team member was responsible for their own set of Algorithm Theoretical Basis Documents (ATBDs), and they were also intending to publish papers in peer-reviewed journals and also in technical reports.[58]

In 1993, the overall EOS plan briefly included the EOS Color mission as a follow-on to SeaWiFS, which was still expected to be launched in late 1993 or 1994 (Figure 6.1).[59] Charles McClain had just been selected as the EOS Color Project Scientist.[60]

One of the other key aspects of EOS was EOSDIS and the EOSDIS core system (ECS). EOSDIS would be the distribution and archive element of EOS. MODIS Oceans team member Robert Evans expressed concern that there really had been no communications between the instrument teams and EOSDIS. Evans stressed the need for open communications, and also wanted to know where the processing capabilities for the science team member's algorithms would operate.

The early state of EOSDIS was one of considerable uncertainty. NASA struggled with how to create an unprecedentedly large data system for a prodigious amount of data, trying to balance the reliability of current data systems with the need for modernization as new technologies matured. Costs were always a concern as the EOS budgets were slashed. At the same time, there was some expectation within the realm of computer technology that the needs of EOSDIS would provide an impetus to develop much better data technologies. The challenges, struggles, and uncertainties permeating the developmental atmosphere of the early EOSDIS were described in *Science* in 1993, published on a Friday the 13th, for anyone

---

[56] NASA GSFC, "Minutes of the MODIS Science Team meeting, April 13-16, 1992."
[57] NASA GSFC, "Minutes of the MODIS Science Team meeting, April 13-16, 1992."
[58] NASA GSFC, "Minutes of the MODIS Science Team meeting, April 13-16, 1992."
[59] G. Asrar and D.J. Dokken, editors, *Earth Observing System (EOS) Reference Handbook,* National Aeronautics and Space Administration, Earth Science Support Office, Washington DC, Document Resource Facility, (1993).
[60] NASA GSFC, "Minutes of the MODIS Science Team meeting, March 24-26, 1993."

superstitiously concerned about the eventual success of EOSDIS.[61]

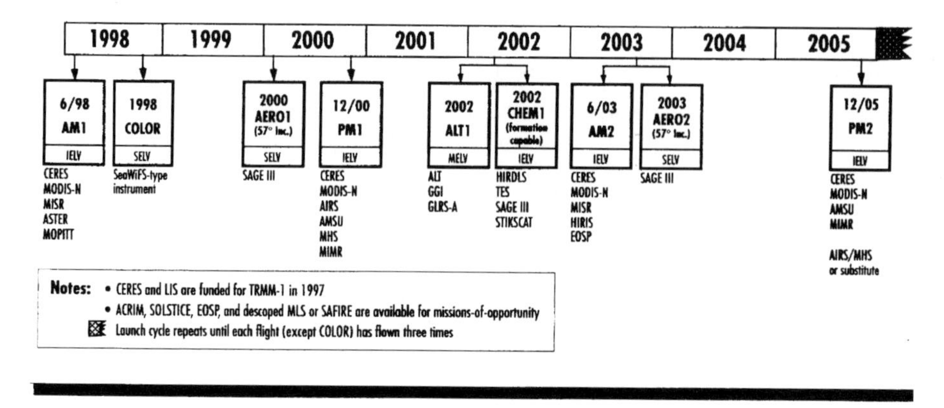

Figure 6.1. Restructured EOS Mission Profile, with EOS-AM-1, EOS-PM, and COLOR.

During 1993, stray light effects in MODIS had been identified as a problem, just as Howard Gordon had speculated it might be. But the instrument team had already addressed it; Frank Muller-Karger from NASA HQ noted on the record that their handling of the stray light problem had been impressive. Steve Wharton, head of the recently-created Global Change Data Center at GSFC, stated that his top priority was improving communications between EOSDIS and the EOS science community.[62]

The problem of "ghost images" in MODIS had been analyzed thoroughly. They occurred due to reflections from the focal planes, and they were being fixed by tilting the focal planes and using intermediate filters. Testing had indicated that the visible range and near IR range focal planes would not be significantly affected.[63]

EOS Color was still an active project in 1993. EOS Color was expected to have a mission team consisting of only have six or seven members, and no associated science team. Also of particular concern to the MODIS Ocean team were interactions with the other sensors being built: the Japanese OCTS and Global Imager (GLI), and the European MERIS.[64]

May 1995 was marked by a MODIS Science Team meeting at GSFC, which featured an overview of the status of ocean color. The overview was presented by Robert Frouin, from Scripps, who was now at NASA HQ as the Ocean Biology program manager and also the MODIS co-program scientist. Due to the launch delay for SeaWiFS (they were still

---

[61] Eliot Marshall, "Fitting planet Earth into a user-friendly database," *Science*, 261(5123), pp. 846 and 848 (13 August 1993).
[62] NASA GSFC, "Minutes of the MODIS Science Team meeting, March 24-26, 1993."
[63] NASA GSFC, "Minutes of the MODIS Science Team meeting, March 24-26, 1993."
[64] NASA GSFC, "Minutes of the MODIS Science Team meeting, March 24-26, 1993."

hoping and planning for the launch of SeaWiFS at earliest opportunity), Charles Kennel, NASA Associate Administrator and director of Mission to Planet Earth, had decided not to fly the EOS Color mission with Landsat 7.[65] Rather, he was going to wait to see if SeaWiFS succeeded; if it did, he favored an "enhanced" calibration/validation effort. (This note is one of the seeds for NASA's eventual successful international collaboration project, the Sensor Intercomparison for Marine Biological and Interdisciplinary Ocean Studies (SIMBIOS) Project, described in Chapter 7.) A three-day workshop had already convened in Miami to discuss how to use ocean color data from the vast multitude – actually six missions at the time, but others were already being discussed in other countries – of ocean color-capable sensors. The working plan was to initiate a project which would facilitate international and interagency data sharing and interactions. (Frouin's narrative indicates where SeaStar was residing during the long wait for launch, at a clean storage facility in Germantown, Maryland.)

Politics and government, and the funding thereof, intersected in November 1995. There had been a MODIS Science Team meeting scheduled for that month, but the U.S. government budget crisis that resulted in the furlough of all non-essential employees caused the cancellation of the official meeting.[66] However, there were still science interest group discussions held during this period, clearly only between the *essential* personnel who were still coming to work. As might be expected under those particular circumstances, Michael King reported that NASA didn't have a budget for fiscal year 1996 yet. Bill Barnes briefly discussed plans for the "next generation" of MODIS instruments to fly on the subsequent EOS platforms after the AM and PM satellites – this was still the plan for EOS in 1995. At the unofficial convocations, the Ocean group presented their data validation plan, which was now going to be conducted in conjunction with SIMBIOS. There was substantial concern that the bright target recovery effects would not be fully addressed at launch. Simulated data would be delivered to allow testing of the data system and algorithm processing.[67]

In 1996, Janet Campbell was introduced as a new member of the Ocean science team, with her main focus on the development and validation of regional "site-specific" algorithms. Also in 1996, SBRC delivered the MODIS protoflight model, just in time for Christmas – and SBRC verified that the first MODIS already met the specifications for radiometric accuracy (Figure 6.2).[68] The MODIS Ocean Team noted that though their plan for data validation had been developed in 1990, it was still valid, relying on MOBY in combination with ship-based data initialization cruises.[69]

---

[65] NASA GSFC, "Minutes of the MODIS Science Team meeting, March 24-26, 1993."
[66] NASA GSFC, "MODIS Science Interest Group Minutes, November 15-17, 1995," http://modis.gsfc.nasa.gov/sci_team/meetings/199511/minutes.pdf, (accessed 11 August 2009).
[67] NASA GSFC, "MODIS Science Interest Group Minutes, November 15-17, 1995."
[68] Herbert J. Kramer, "Terra Mission (EOS/AM-1)," http://directory.eoportal.org/presentations/129/7312.html, (accessed 5 March 2010).
[69] Herbert J. Kramer, "Terra Mission (EOS/AM-1)."

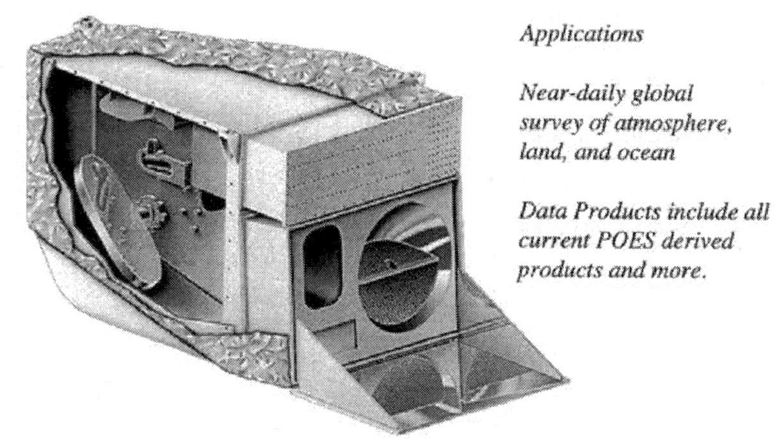

Figure 6.2 Cutaway diagram of MODIS.

The Oceans Team in 1996 remained very concerned that remaining stray light issues were being ignored, and they agreed that airborne observation campaigns were vital to characterize atmospheric variability.[70] The need for ships to do at-sea validation – which had always been a concern for NASA – had to be considered, with various options, particularly considering the possibility (prominent on their minds due to the SeaWiFS delays) of launch delays.[71]

The EOSDIS science processing segment fell six months behind schedule in 1996, partly due to underestimation of how much software was required for the data server. There had been some difficulty in recruiting sufficient programmers skilled with the necessary software languages, so the ECS had failed its Test Readiness Review in August. This key software was also necessary to support the Tropical Rainfall Measuring Mission (TRMM) and Landsat 7.[72]

In a wonderful example of how time informs us that we were naïve, during one of the 1996 MODIS Science Team meetings, Mark Abbott asked about how EOS data would be accessed, and the answer was one of Timmy Turner's (the central character in the cartoon series *Fairly Oddparents*) fallback lines: "Internet".[73] Abbott was concerned that reliance on the Internet would reduce data throughput, particularly for users interested in global products.[74] (In retrospect, this concern was justified, as the advances in the Internet could

---

[70] NASA GSFC, "MODIS Science Team Meeting Minutes, May 1-3, 1996, Section 7.0, Ocean DisciplineGroup Meeting."
[71] NASA GSFC, "MODIS Science Team Meeting Minutes, May 1-3, 1996, Section 7.0, Ocean DisciplineGroup Meeting."
[72] NASA GSFC, "MODIS Science Team Meeting Minutes, May 1-3, 1996, Section 7.0, Ocean DisciplineGroup Meeting."
[73] Internet Movie Database, "The Fairly OddParents (2001)", http://www.imdb.com/title/tt0235918/, (accessed 11 August 2009).
[74] NASA GSFC, "MODIS Science Team Meeting Minutes, October 10-11, 1996."

not be envisioned. The Goddard DAAC ceased sending any data on physical media in 2005, relying instead wholly on electronic network distribution.)[75]

The Ocean team was still very concerned about polarization sensitivity in MODIS. Esaias even suggested a launch slip to get a better characterization of the protoflight model.[76]

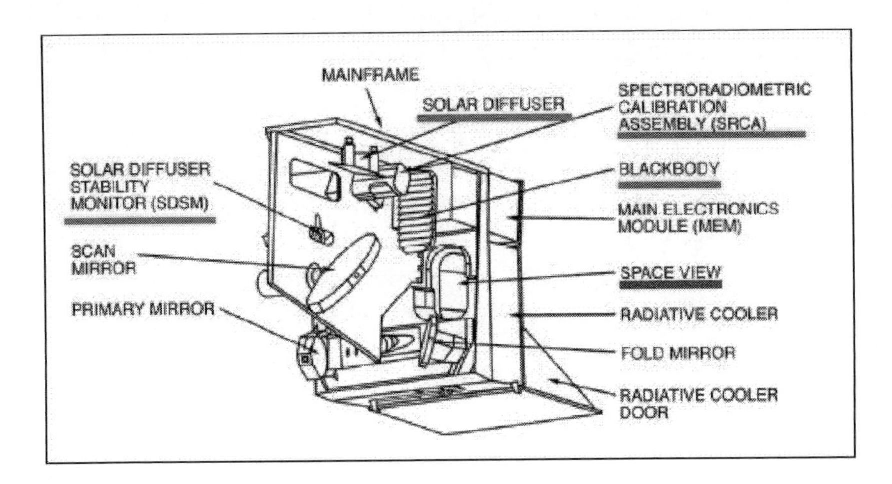

Figure 6.3. Schematic diagram of MODIS, showing the various calibration systems.

Regarding the polarization issue, Vince Salomonson said that prior to launch, there was still uncertainty about the response vs. scan angle in the protoflight (AM) model; because high bay testing had been conducted with the flight (PM) model, it was much better characterized (Figure 6.3).[77] Salomonson knew the protoflight MODIS wasn't well characterized, and even wanted to have it shipped back to Goddard for more testing, but with the instrument already on site awaiting spacecraft integration, the schedule slip was deemed unacceptable by Ghassem Asrar at HQ (NASA Associate Administrator for Earth Science), and so the slip was nixed. Salomonson said "I knew there would be problems."[78]

Despite the problems, the MODIS Science Team agreed to an on-orbit spacecraft maneuver to view the moon, despite apprehensions. The idea of flipping SeaWiFS over had not initially been met with enthusiasm, but its necessity had been realized and ultimately included in the mission profile. Flipping over the EOS-AM-1 spacecraft – a much larger spacecraft with a long solar panel boom, and five state-of-the-art, expensive instruments, was not going to be done without some concern, but despite those concerns, this risky maneuver would still be done.[79]

Piling up frequent flyer miles to Hawaii in 1996, Dennis Clark was in the initial

---

[75] NASA GES DISC, "End of Media Distribution", http://disc.sci.gsfc.nasa.gov/gesNews/end-of-media-distribution, (accessed 24 December 2009).
[76] NASA GSFC, "MODIS Science Team Meeting Minutes, October 10-11, 1996."
[77] Herbert J. Kramer, "Terra Mission (EOS/AM-1)".
[78] Vincent V. Salomonson recorded interview, 2 December 2008.
[79] Vincent V. Salomonson recorded interview, 2 December 2008.

stages of MOBY deployment and testing. Even though the community was eager to get MOBY data, Clark indicated that he did not want to initiate routine deployments until each of the MOBYs had been deployed at-sea at least once.[80]

Early 1997 was marked with several noteworthy circumstances: SeaWiFS launch was impending, and OCTS was delivering ocean color and SST data. MODIS-AM was being reviewed for shipping readiness at SBRC before it would be shipped to Valley Forge, Pennsylvania for spacecraft integration. The second MODIS would be carried on the EOS-PM platform, and the spacecraft was behind schedule due to a contract award protest, but was now in the 13th month of a 54-month development period. The feasibility of conducting one deep space calibration maneuver with EOS-PM was being considered; the main difference between the maneuver with the AM and PM spacecraft was that the PM spacecraft would carry the Advanced Microwave Scanning Radiometer (AMSR), which featured a large rotating scanning antenna. Because it was not feasible to stop the mirror rotation periodically during the mission to perform a calibration maneuver, the current plan was to perform one maneuver early in the mission before the scanning antenna was turned on. This would allow MODIS (and also CERES) one, and one only, on-orbit calibration attempt.[81]

EOSDIS was behind schedule, which was fairly routine; the ECS core system, originally schedule for delivery in October or November 1996, was now hoping for May 1998. That release was intended to allow "all the capabilities necessary to support early mission calibration and algorithm testing and refinement." The MODIS Oceans Team even considered having an emergency backup plan for their data system.[82] At this point in time, the software and algorithms for the Ocean team were on schedule.[83] Vince Salomonson was encouraged to see the continuance of data with the heritage of CZCS being acquired by OCTS.[84]

Because EOSDIS continued to be an area of concern for the science teams, and because the vast size of the ocean would produce immense data volumes, a plan was floated to reduce the data volume by going to 5-km resolution data sets, rather than 1 km. The Oceans team also considered what the collective "next step" might be beyond MODIS, because it was felt there was little chance MODIS would be on the next EOS satellites after PM and CHEM (as it turned out, there were not to be any second-generation EOS satellites at all).[85]

---

[80] NASA GSFC, "MODIS Science Team Meeting Minutes, October 10-11, 1996," Section 7.3.
[81] NASA GSFC, "MODIS Science Team Meeting Minutes, May 14-16, 1997," http://modis.gsfc.nasa.gov/sci_team/meetings/199705/minutes.pdf, (accessed 11 August 2009).
[82] NASA GSFC, "MODIS Science Team Meeting Minutes, May 14-16, 1997."
[83] NASA GSFC, "MODIS Science Team Meeting Minutes, May 14-16, 1997."
[84] NASA GSFC, "MODIS Science Team Meeting Minutes, May 14-16, 1997."
[85] NASA GSFC, "MODIS Science Team Meeting Minutes, May 14-16, 1997," Section 2.0.

MOBY at Sea, Sometimes Adrift

The adventures of the Marine Optical Buoy (MOBY) became much more critical to the MODIS Science Team and the SeaWiFS Project following the launch of SeaWiFS and as the launch of MODIS on EOS-AM-1 drew closer and closer. Dennis Clark and NOAA had established an operational base on Lanai which allowed maintenance and repair to the alternate MOBYs while one operated in the ocean. MOBY operations were conducted under the Marine Optical Characterization Experiment (MOCE) umbrella, and required several cruises a year for MOBY recovery and deployment, as well as many additional expeditions to maintain, upgrade, and calibrate the instruments on land.[86] The MOBY deployments required ship support: two of the primary ships that provided support were the University of Hawaii's R/V *Moana Wave* and the R/V *Ka'imikai O Kanaloa*.

It is difficult enough to operate a well-calibrated, sophisticated, accurate spectrometer in a land-based laboratory; the MOCE personnel had to maintain and calibrate three spectrometers capable of operating autonomously at sea (which meant thoroughly cleaning off the accumulated bio-fouling after each recovery), as well as keeping the supporting electronics and communications equipment on the buoy operating. MOBY was a challenge to recover and deploy; the main instrument mast was 12 meters long, with a heavy instrument bay at one end and the above-water part of the buoy at the other, which was festooned with antennas and solar panels. During the deployments, they would release the upper end of the buoy, and the "headache ball" (the big ball above the crane hook, frequently utilized in cartoons) would swing free, while divers were in the water attaching the crane to the buoy. During one of the first deployments, one crew member was caught between the deck crane and the buoy itself, and was nearly crushed. Another time, when retrieving the first prototype, which was made entirely of aluminum, a lifting point broke, and one diver was seriously endangered until additional straps could be put on the buoy.[87]

In 1996 and 1997, anticipating the launch of SeaWiFS and MODIS, MOBY operations increased in frequency. After Clark had a first look at MOBY's data, he made some design changes, including a new spectrometer equipped with charge-coupled devices (CCDs) to increase the instrument's critical signal-to-noise ratio. (He also switched from an aluminum buoy to a foam buoy, which provided better stability in waves.) The CCDs had to be cooled to -30° C using pumped refrigerant – and everything had to fit into the previous instrument casing. He contracted the work to a company called American Holographic in Massachusetts, and for several months nothing happened. So Clark took a team to the company, and they allowed the team to use their equipment to custom-build the spectrometer and re-machine the parts. The ultimate instrument had 4-position mirrors to measure upwelling and downwelling radiances, LEDs for stray light checks, and a stable incandescent lamp for calibration. The instrument also had a dichroic mirror, so that the light was split and directed to a red spectral range spectrometer and a blue spectral range

---

[86] NOAA, "Marine Optical Characterization Experiment", http://www.star.nesdis.noaa.gov/sod/orad/mot/moce/, (accessed 11 August 2009).
[87] Dennis Clark, interview transcript notes.

spectrometer.[88] This arrangement significantly reduced stray light from one side of the visible spectrum to the other.

The CCDs also had a downside – they were cold, and they had to be kept in a vacuum. The CCD "space" in the instrument was purged with nitrogen before pumping down, but occasionally the vacuum had to be restored. Following one of the re-establishments of the vacuum, the instrument started returning bad data, and this was determined to be due to excess humidity – something to be expected in an oceanic environment. The actual problem was traced to a batch of bad desiccant, the chemical that absorbs water. The cold of the CCDs caused condensation on the optical surfaces, leading to the bad data.[89]

Clark also initiated a collaboration with the National Institute of Standards and Technology for MOBY calibration, and NIST was already collaborating on SeaWiFS and MODIS calibrations. The Center for Hydro-Optics and Remote Sensing (CHORS) in San Diego, under the direction of Jim Mueller and Chuck Trees, became the focal point for MOBY calibrations. Clark noted, "I got trained by Ros [Austin] that the calibration process is not very glamorous, but it's critical!" [90]

As Clark had noted to the MODIS Ocean science team, he wanted to have each of the three MOBYs in and out of the water at least once before commencing a routine deployment schedule. So during 1997, calibration and deployment activities continued. And as the MOBYs stayed in the water longer and longer, incidents and accidents also increased. It turned out that MOBY was one of the best FADs (fish accumulation devices) in the calm waters near Lanai, and fishermen soon realized this, so MOBY became a prime fishing spot.[91] Unfortunately, this meant that fishing boats might collide with the buoy. Furthermore, some of the incidents when the mooring line was cut were apparently due to vandalism. The FBI had established a field office in Hawaii due to incidents of vandalism regarding NOAA buoys, and after the second incident of vandalism, Clark contacted the bureau. Despite the fact that the FBI didn't do anything, the fishing community in Maui is "tight", and word got out that they FBI had gotten involved. Subsequently, MOBY vandalism incidents were markedly curtailed.[92]

The last months of 1997 demonstrated the inherent difficulties of maintaining the MOBY system. The initial SeaWiFS validation cruise from September 22 to October 4 was cut short inauspiciously because one of the research vessels (the R/V *Kila*) ran aground on its first day at sea.[93] A couple of weeks later, just before Halloween, MOBY was cut loose and made a ghostly disappearing act, but was recovered off Penguin Bank, where the long

---

[88] Dennis Clark, interview transcript notes.
[89] Dennis Clark, interview transcript notes.
[90] Dennis Clark, interview transcript notes.
[91] Dennis Clark, interview transcript notes.
[92] Dennis Clark, interview transcript notes.
[93] NOAA, "MOCE Team Expedition Log, MOBY L22-M206OB", http://www.star.nesdis.noaa.gov/sod/orad/mot/moce/personnel/moby_l22.html, (accessed 13 August 2009).

mast fortunately ran aground.[94] "Flopper stoppers" were added to the mooring in early November; "flopper stoppers" helped reduce the resonant frequency of MOBY in the waves; without them, MOBY would occasionally submerge and then pop to the surface quite abruptly, and quite high, which wasn't good for the delicate electronics and optical sensors.[95] The month was not even over when MOBY was cut loose a second time, and again recovered on Penguin Bank.[96] Yet despite its nautical excursions, MOBY was working and delivering data, while the MODIS science team grappled with elevated concerns and encroaching deadlines.

It is worth pointing out, given the difficulties, complexities, and expense of sustaining the MOBY deployments off Hawaii, that Clark's original MOBY proposal had been to deploy MOBYs at five different "high quality" sites, to fully characterize the oceanic optical environment. As Clark stated clearly, "Thank God there was only one MOBY." [97]

Back on Land, with a Sensor in Orbit

At the October 1997 MODIS science team meeting, there were likely a few quietly jubilant ocean color scientists, with the shock of the sudden loss of Midori replaced by the gratefulness for the successful launch and initial data acquisition from SeaWiFS. At this meeting, Esaias said that they were expecting to produce all the Level 2 data products at 1km resolution, though they could scale back to 2 km resolution at launch and reprocess to 1km resolution later. Bob Evans was concerned with how rapidly the Level 1B data would be available from the Goddard DAAC; he was anticipating the possibility of doing all of the data processing at Miami if there were problems at the DAAC.[98]

Moving into 1998, technical glitches forced a delay in the EOS-AM-1 launch (which had been moved forward as a cost-saving message years earlier). The main problem was not the spacecraft, as the integration of instruments on the spacecraft was complete (Figure 6.4 and 6.5). [99] The problem lay with the flight operations software, which was not ready for

---

[94] NOAA, "MOCE Team Expedition Log (MOBY-L23)", http://www.star.nesdis.noaa.gov/sod/orad/mot/moce/personnel/moby_l23.html, (accessed 13 August 2009).
[95] NOAA, "MOCE Team Expedition Log (MOBY-L24), http://www.star.nesdis.noaa.gov/sod/orad/mot/moce/personnel/moby_l24.html, (accessed 13 August 2009); Dennis Clark, interview transcript notes.
[96] NOAA, "MOCE Team Expedition Log (MOBY-L25:: M207SOBP)", http://www.star.nesdis.noaa.gov/sod/orad/mot/moce/personnel/moby_l25.html, (accessed 13 August 2009).
[97] Dennis Clark, comments recorded at the Ocean Color Collaborative Historical Workshop, January 13-14, 2009, St. Petersburg, Florida.
[98] NASA GSFC, "MODIS Science Team Meeting Minutes, October 22-24, 1997", http://modis.gsfc.nasa.gov/sci_team/meetings/199710/minutes_1097.pdf, (accessed 13 August 2009).
[99] Herbert J. Kramer, "Terra Mission (EOS/AM-1)"; Terra Web site, http://terra.nasa.gov/Events/photos/6d-MODIS_mirrorinspect_md.jpg.

launch. During 1998, this critical software was being "cleaned up" and an alternative system was also being developed. Though MODIS-AM was integrated on the spacecraft, it still exhibited remaining electronic crosstalk and thermal problems. The Oceans team didn't expect these to be fixed on MODIS-AM, but they hoped MODIS-PM would have the necessary fixes. The PM instrument had also been equipped with a new reduced scatter scan mirror, which was intended to improve the SST data products.[100]

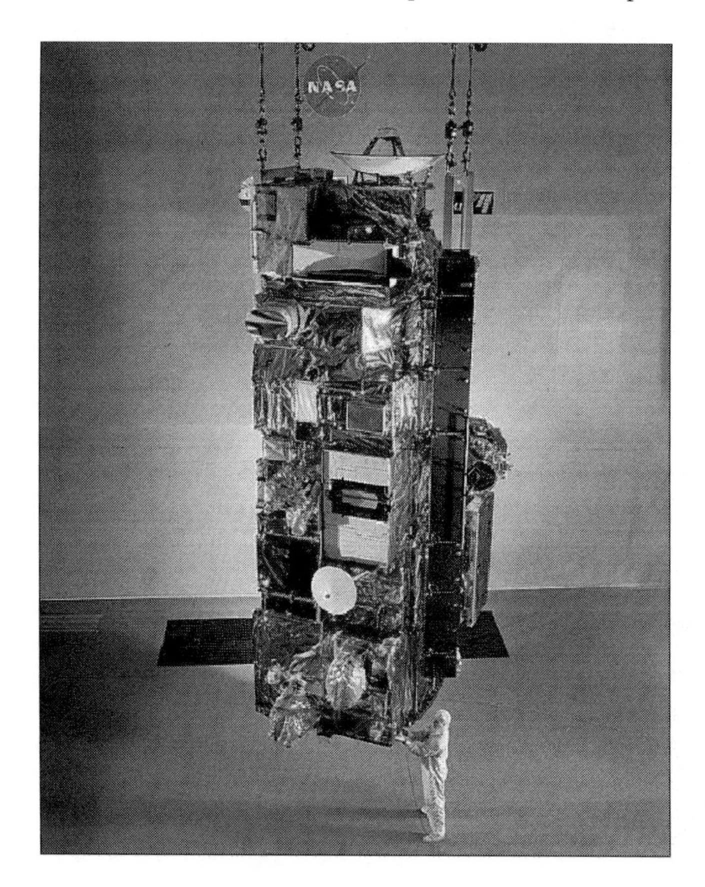

Figure 6.4. The fully-integrated Terra (EOS-AM-1 before launch) in the Valley Forge clean room.

The Marine Optical Characterization Experiment (MOCE) initialization cruise in support of SeaWiFS included Howard Gordon, who was never a fan of at-sea research. According to Gordon, the first MOCE cruise was a "dry run" (rather strange to use that term for oceanographic activities) for MODIS. The research cruise, which took place in January 26 to February 12, 1998 under very sunny skies, had resulted in calibration data for

[100] NASA GSFC, "MODIS Science Team Meeting Minutes, June 22-24, 1998", http://modis.gsfc.nasa.gov/sci_team/meetings/199806/minutes_698.pdf, (accessed 13 August 2009).

all the SeaWiFS bands.[101]

One very unusual idea came out of the Oceans team deliberations in 1998. The team realized that there were still significant problems with the protoflight MODIS (MODIS-AM), which was currently sitting on the spacecraft. Because they now anticipated a launch slip into 1999, they advocated an unusual switch; fully characterize and fix the flight MODIS, pull the protoflight MODIS off the AM satellite and replace it with the flight MODIS, then fix the protoflight MODIS and put it on the PM satellite.[102]

(If it were possible to operate in the present with the benefit of hindsight, this unusual plan likely would have been adopted! However, as it is fraught with risk, it is very rare to disconnect and remove an instrument from a spacecraft when it has already been hardwired and tested. So this plan was never seriously considered. In retrospect, there are certainly regrets that it wasn't.)

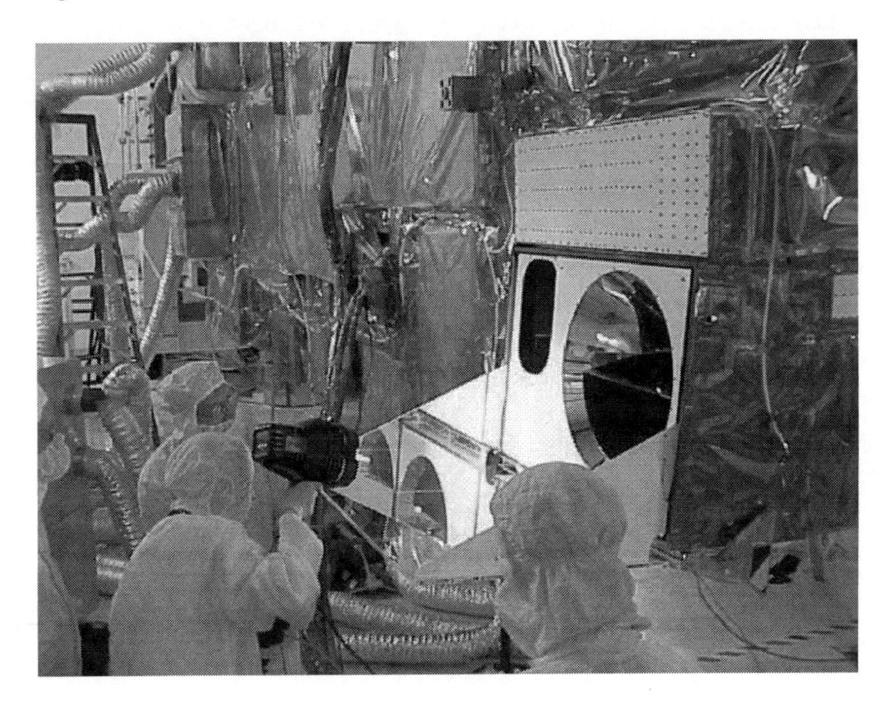

Figure 6.5. Clean room inspection of MODIS-Terra.

The MODIS Oceans team was active in research as they awaited the eventual MODIS launch. Ken Carder, immersed in data validation activities for SeaWiFS, participated in a research cruise off Florida which enabled comparisons of at-sea measurements to SeaWiFS

---

[101] NASA GSFC, "MODIS Science Team Meeting Minutes, June 22-24, 1998," Attachment 9: SeaWiFS Initialization Cruise Results, http://modis.gsfc.nasa.gov/sci_team/meetings/199806/presentations/a9.pdf, (accessed 13 August 2009); NOAA, "MOCE Team Expedition Log (MOCE-4)", (accessed 13 August 2009).
[102] NASA GSFC, "MODIS Science Team Meeting Minutes, June 22-24, 1998," Section 1.18: MOCEAN Group Summary.

data.[103] John Porter had performed aircraft aerosol characterization experiments over Hawaii, including flyovers of Dennis Clark's MOBY initialization cruise.[104] Mark Abbott had experienced the adventure of deploying of several drifting buoys equipped with radiometers into the frigid polar sea during the chilling JGOFS Southern Ocean research campaign. 12 buoys had been deployed; nine of them retrieved radiometric data; and one of the buoys emulated the *Titanic*, disappearing beneath the frigid austral ocean waters after ramming an iceberg.[105] Though Ian Barton hadn't been so adventurous, he was investigating the difference between the "skin" SST, which is the temperature measured exactly at the surface of the ocean, and the "bulk" SST, which is the temperature of the water immediately below the surface skin, and more indicative of the actual temperature of the ocean water.[106] Solar heating caused the skin temperature to be a few tenths of a degree warmer than the bulk temperature, and the difference had to be modeled.

In late 1998, critical deadlines were very close. Because the protoflight MODIS was on the EOS-AM-1 spacecraft, it wasn't possible to do all the testing desired, so the MODIS calibration team was working on strategies to characterize it as well as possible. The flight MODIS destined for EOS-PM was showing improved performance over the protoflight MODIS.[107]

Utilizing MOBY's cellphone connection, direct communication with MOBY allowed the data from the buoy to be posted within 24 hours of collection. MOBY calibration continued, in collaboration with NIST.[108]

Earlier in the year, in March and April 1998, a planned MOBY swap-out had to be scrubbed due to 40-knot wind gusts and 18-foot seas; perhaps ruefully Clark recalled that the Lanai site had been chosen because the seas there were relatively calm.[109] In June, an inspection revealed that the top optical arm on the buoy was missing.[110] Diver inspections indicated when the "flopper stoppers" had come off, such as in August 1998.[111]

---

103 NASA GSFC, "MODIS Science Team Meeting Minutes, June 22-24, 1998," Section 4.4: 'Hot Science' Reports from Team Members.
104 NASA GSFC, "MODIS Science Team Meeting Minutes, June 22-24, 1998."
105 NASA GSFC, "MODIS Science Team Meeting Minutes, June 22-24, 1998."
106 NASA GSFC, "MODIS Science Team Meeting Minutes, June 22-24, 1998," Attachment 28: Skin vs. Bulk SST, http://modis.gsfc.nasa.gov/sci_team/meetings/199806/presentations/a28.pdf, (accessed 13 August 2009).
107 NASA GSFC, "MODIS Science Team Meeting Summary, December 15-17, 1998," http://modis.gsfc.nasa.gov/sci_team/meetings/199812/minutes_1298.pdf, (accessed 13 August 2009).
108 NASA GSFC, "MODIS Science Team Meeting Summary, December 15-17, 1998."
109 NOAA, "MOCE Team Expedition Log (MOBY-L28::M208OBP)," http://www.star.nesdis.noaa.gov/sod/orad/mot/moce/personnel/moby_l28.html, (accessed 13 August 2009).
110 NOAA, "MOCE Team Expedition Log (MOBY-L32)," http://www.star.nesdis.noaa.gov/sod/orad/mot/moce/personnel/moby_l32.html, (accessed 13 August 2009).
111 NOAA, "MOCE Team Expedition Log (MOBY-L36)," http://www.star.nesdis.noaa.gov/sod/orad/mot/moce/personnel/moby_l36.html, (accessed 13 August 2009).

In February 1999, EOS-AM-1 was renamed *Terra*, Latin for "Land".[112]

By May 1999, Terra was now at the Vandenberg launch site. The spacecraft was ready for launch, with a date of June 28, 1999 specified (though a more likely launch in July was "looking solid") Data system testing at the Goddard DAAC was proceeding at this time, under the leadership of Steve Kempler, and the DAAC's Version 1 system was already successfully supporting Tropical Rainfall Measuring Mission (TRMM) data. The Version 2 system would support MODIS in time for the launch. The primary concern for the archive and distribution of MODIS data products was capacity. MODIS produced considerably larger volumes of data than SeaWiFS (the Level 2 products were 1 km resolution rather than 4 km, and the Level 3 global products were 4 km resolution rather than 9 km), and there were many, many more MODIS data products. The SeaWiFS data archive had produced a rather remarkable 8-fold multiplication factor for the amount of data archived versus the amount of data distributed to scientists. For MODIS, at launch, the DAAC was hoping to achieve a 1-for-1 ratio, primarily due to the vastly greater volume of MODIS data. The DAAC was considering subsetting by region to improve this ratio.[113]

At this point in time, Wayne Esaias informed the Oceans Team that the July launch data was "looking solid". That was important for several reasons. One was planning for the deployment of researchers on validation cruises. Another issue of concern to the team was the release schedule for data products. The MODIS Project stated that the data plan was to have an "official release date" for data about 120-150 days after the mission began, but it wasn't clear if this actually meant launch; or the first day of imaging operations; or the first day that it was ascertained that science-quality data was being acquired (i.e., diagnostic tests and calibrations and adjustments would have been essentially completed). Though the data would have plenty of quality "flags" attached, the team was concerned that eager users would ignore the flags and just start trying to do research with the data, regardless of quality!

Following a plan similar to that of the CZCS NET, the team had set up a validation cruise off Baja California, scheduled for October 1999. The schedule for spacecraft maneuvers and door openings following launch was tight and could influence cruise planning. Also of concern was a recent launch failure on May 4, which could have an impact on the launch date.[114] The launch failure of a Delta III rocket attempting to launch the Orion-F3 communications satellite had been due to problems with a second stage rocket motor, resulting in the satellite entering into a useless orbit.[115] This failure did indeed affect the Terra launch data, because it precipitated a stand-down of all launches with vehicles equipped with the RL-10 motor, which had ruptured, and this included the Atlas II-AS that

---

[112] NASA,"Terra/EOS-AM-1".

[113] NASA GSFC, "MODIS Science Team Meeting, 04-05 May 1999," http://modis.gsfc.nasa.gov/sci_team/meetings/199905/MST_mtg_minutes_0599.pdf, (accessed 13 August 2009).

[114] NASA GSFC, "MODIS Science Team Meeting, 04-05 May 1999," MODIS Oceans Discipline Group Meeting, 3-4 May 1999 (section of the May 1999 minutes).

[115] Kevin Forsyth, "History of the Delta Launch Vehicle: Flight Log", http://kevinforsyth.net/delta/log.php, (accessed 13 August 2009).

would launch Terra. The result of this stand-down and relatively short failure investigation was a launch slip to December 16, 1999.[116] (The cause of the rupture was identified as failure of a brazed joint; inspection of Terra's launch vehicle engines did not reveal any defects expected to cause this type of failure.)[117]

The launch slip meant that a MODIS science team meeting slipped in before the launch, in November 1999. Spacecraft readiness was reviewed; the spacecraft had been tested for over 200 hours.[118] Several risk mitigation strategies had been implemented during the delay. Steve Kempler described the Goddard DAAC readiness for the data.[119] The MODIS Ocean team undertook a series of deep breathing exercises, and a few of them headed to the 1999 Fall American Geophysical Union meeting in San Francisco in December.

EOS-AM-1 Launch

Vandenberg prepared to launch the Atlas IIAS with EOS-AM-1 on December 16, 1999. (Figure 6.6) [120] With 40 seconds left in the countdown, computers monitoring the launch vehicle status reported an erroneous condition, and the launch was shut down. Investigation revealed that the report was an error, so the launch was rescheduled for two days later. The launch had to occur in a half-hour launch window; on the 18th, high winds aloft imperiled the attempt, but the decision was made to "go", and the rocket ignited and lifted off with a scant 10 seconds to spare.[121] Initial on-orbit operations went smoothly, included the vital unfolding of the long solar panel power array. Scientists at the AGU meeting watched the launch occurring down the coast live on TV at the NASA exhibition booth. One of the attendees at the AGU meeting who watched the launch with considerable interest was Greg Leptoukh, who had been promoted from the SeaWiFS data management task to the much larger task of managing MODIS atmospheric and ocean data at the Goddard DAAC.[122]

---

[116] David Herring, "Learning to Fly: The Launch and Activation of Terra through the Eyes of Mission Control", May 24, 2000, http://earthobservatory.nasa.gov/Features/LearningToFly/fly.php, (accessed 13 August 2009).
[117] Kevin Forsyth, "History of the Delta Launch Vehicle: Flight Log"; David Herring, "Learning to Fly: The Launch and Activation of TerraThrough the Eyes of Mission Control".
[118] Bruce Guenther, "Spacecraft Status", http://modis.gsfc.nasa.gov/sci_team/meetings/199911/presentations/14-Grady_TerraLaunch.pdf, (accessed 13 August 2009).
[119] Steve Kempler, "GSFC's Earth Sciences (GES) Distributed Active Archive Center (DAAC) Operations Readiness: L1 Production, Archive, Distribution, and User Services", http://modis.gsfc.nasa.gov/sci_team/meetings/199911/presentations/06-Kempler_GDAAC.pdf, (accessed 13 August 2009).
[120] "Terra Launch Photos", http://terra.nasa.gov/Events/launch_photos.php.
[121] David Herring, "Learning to Fly: The Launch and Activation of Terra Through the Eyes of Mission Control".
[122] Steve Kempler, "GSFC's Earth Sciences (GES) Distributed Active Archive Center (DAAC) Operations Readiness: L1 Production, Archive, Distribution, and User Services".

Figure 6.6. Liftoff of TERRA mission at Vandenberg AFB.

The next step was to raise the orbit, similar to the process for SeaWiFS; the final destination was an orbit 705 km above the Earth. A day later, Terra went into safe hold due to electronic disturbances near the troublesome South Atlantic Anomaly, a region in space with higher-than-normal radiation exposure. (SeaWiFS, like many other satellites, occasionally experienced Single Event Upsets (SEUs) near the SAA, sometimes resulting in a safehold condition, but more commonly just inducing a quick computer reset).[123]

Sometimes you get a surprise in space, and the orbit-raising attempt encountered one the first time, due to thruster exhaust hitting the solar panel, which caused the spacecraft to roll unexpectedly and aborted the maneuver.[124] Repositioning the solar array allowed a successful boost to 705 km. At that altitude, doors could begin to open, plastics could be allowed to release volatiles before exposing the sensitive instrument detectors and optics, the instruments could be warmed up and turned on; all routine and necessary operations prior to the actual acquisition of data.

And now the fun began, after more than a decade of planning and developing and

---

[123] David Herring, "Learning to Fly: The Launch and Activation of Terra Through the Eyes of Mission Control: Part 2, The Duck", http://earthobservatory.nasa.gov/Features/LearningToFly/fly_2.php, (accessed 13 August 2009).
[124] David Herring, "Learning to Fly: The Launch and Activation of Terra Through the Eyes of Mission Control: Part 3, The Unplanned Roll", http://earthobservatory.nasa.gov/Features/LearningToFly/fly_3.php, (accessed 13 August 2009).

scoping and descoping and budget cuts and… MODIS was in orbit.

The First Months

With Terra in orbit and going through its pre-ordained list of start-up tasks (a long list; instrument windows and doors had to be opened in the proper order, the sensors themselves had to be warmed up or cooled down, any operating glitches or anomalies had to be addressed immediately), the MODIS science teams had to check and re-check their systems and code to make sure they were ready for the data.

On February 24, 2000, MODIS was ready to go, and it turned its imaging eyes toward the Gulf of Mexico; one of the significant features in this image was the Mississippi Delta and its optically complex waters. The MODIS "first light" image revealed a spectrum of blues and greens and browns (and the white of clouds) in this familiar region, immediately revealing patterns of moving sediments and the flow of green and blackish waters around the delta and along the northern Gulf of Mexico coast. (Figure 6.7.) [125] For the data system and the data archive, the burgeoning flood of MODIS data had started.[126]

[Though it has been mentioned before, Howard Gordon reiterated that MODIS was "probably the most complicated instrument ever to fly in space", due to the extreme wavelength range that it sensed, from the ultraviolet to the infrared. And, Gordon added, "it worked". So some unusual behavior might be expected in such a complex instrument. Gordon described that he had wanted to use the 1.38 μm band to characterize aerosols and cirrus clouds, but he was informed at a MODIS science team meeting that when the satellite had flown over Cuba, they had seen Cuba in the 1.38 μm band – at night, when there was no light. This was due to electronic crosstalk between the infrared detectors in the instrument; Cuba was being detected at night due to its thermal signature at longer IR wavelengths.[127]]

Out in the Pacific Ocean, MOBY was standing by, and collected data on February 13, 2000, when Terra passed overhead. Unfortunately, MODIS wasn't collecting data yet. Still, this proved they could collect MOBY data coincident with a Terra overpass.[128] The 6th Marine Optical Characterization Experiment in April was very successful, collecting data during both SeaWiFS and MODIS overpasses and even an ER-2 (called a U-2 in the log) overflight.[129] The rest of the MOBY support year went fairly uneventfully, except for one

---

[125] Jacques Descloitres, http://visibleearth.nasa.gov/view_rec.php?id=217; MODIS instrument team, http://visibleearth.nasa.gov/view_rec.php?id=160.

[126] NASA Earth Observatory, "MODIS First Light Image: Image of the Day", 10 March 2000, http://earthobservatory.nasa.gov/IOTD/view.php?id=526, (accessed 13 August 2009).

[127] Howard Gordon, comments recorded at the Ocean Color Collaborative Historical Workshop, January 13-14, 2009, St. Petersburg, FL.

[128] NOAA, "MOCE Team Expedition Log (MOBY-L54::M216SOB)," http://www.star.nesdis.noaa.gov/sod/orad/mot/moce/personnel/moby_l54.html, (accessed 14 August 2009).

[129] NOAA, "MOCE Team Expedition Log (MOCE-6)," http://www.star.nesdis.noaa.gov/sod/orad/mot/moce/personnel/moce_6.html, (accessed 14 August 2009).

incident in September when the buoy started transmitting unusual values. A quick trip out discovered that the mooring tether had become entangled with the buoy's upper sensor arm, and MOBY was freed from its self-inflicted snare.[130]

Figure 6.7. MODIS-Terra "first-light" image. Full-scan (left); Mississippi River delta region detail (right).

Early in the mission, members of the SIMBIOS Project examining the new MODIS data discovered a critical error. The calculation of the exact radiances being received by the sensor required an accurate estimation of solar irradiance, and this in turn required an accurate Earth-Sun (F_0) distance, which changes enough to be significant for the data algorithms. Ewa Kwiatkowska observed that there was a seasonal difference between SeaWiFS and MODIS data. Bryan Franz asked her if the difference was approximately 3.5%, and Kwiatkowska confirmed that it was. This particular value led Franz to suspect that the Earth-Sun distance had been double-corrected. Checking the code revealed that Franz was correct; the data processing was therefore using a value assuming that the distance from the Earth to the Sun varied at twice the amount that it usually does. The code was fixed, and the data processing started over. Franz pointed out that there were two important aspects to the discovery of this error; having both an independent team examining the data and data processing assumptions, as well as having a consistent data set for the purpose of comparison.[131]

Bob Evans at the University of Miami was the leader of the science investigator's

---

[130] NOAA, "MOCE Team Expedition Log (MOBY-L61)," http://www.star.nesdis.noaa.gov/sod/orad/mot/moce/personnel/moby_l61.html, (accessed 14 August 2009).

[131] Bryan Franz, email message received 4 January 2010.

computing facility at RSMAS. Changes to the algorithm code – whether they be a variety of correction factors, new values for variables in the calculating equations that were based on the at-sea efforts of the investigators – had to be tested before the new code was sent to EOSDIS to process the entire dataset. EOSDIS would then have to get the code, install the code, and redo the data processing with the new code. So EOSDIS actually had forward and backward data streams – the "forward" stream processed the new data coming down from the sensor in orbit, while the backward stream reprocessed the data that had already been collected. Once the data had been run through the EOSDIS mill, the investigators could then check it out and see if the sensor was returning data with the necessary accuracy and precision.

All of this took considerable time. So when the science team went over the initial results in June 2000, the data were not – as was expected for an instrument with the complexity of MODIS – science quality. The initial data examination generated a long list of issues that had to be addressed. Furthermore, while MODIS had one lunar look in March, the "Deep Space" calibration maneuver had not yet taken place, and was now on the "to do" list for November 2000.[132]

The SST group noted the initial data were affected by mirror side differences: the rotating scan mirror's two sides had different optical properties. For the painstaking level of accuracy required for SST and ocean color, these differences were very significant. There were also differences in the instrument's detectors. It also seemed likely that the instrument was affected by "cross talk" – i.e, one set of detectors was actually detecting a portion of the signal that should only have been detected by a different set of detectors. Despite the problems, initial images of the Central American coast were striking. (Figure 6.8) [133] Nonetheless, the SST team concluded "Until instrumental effects can be sufficiently well corrected, attempts at validating the atmospheric correction and derived SST are futile." [134]

---

[132] MODIS Calibration Support Team, "MODIS Response versus Scan Angle (RVS) Status Report Part I: Reflectance Solar Bands", http://modis.gsfc.nasa.gov/sci_team/meetings/200006/presentations/MCST_WKSHOP_P2062000.pdf, (accessed 14 August 2009); Robert Kozon, "ESMO-MODIS Science Team Presentation", http://modis.gsfc.nasa.gov/sci_team/meetings/200006/presentations/1.02_Ondrus_TerraStatus.pdf, (accessed 14 August 2009).

[133] Image from Robert Evans, Rosenstiel School of Marine and Atmospheric Science.

[134] Otis B. Brown, Peter J. Minnett, Robert H. Evans, Edward J. Kearns, and Richard J. Sikorski, "MODIS Sea-surface temperature", http://modis.gsfc.nasa.gov/sci_team/meetings/200006/presentations/1.08_Brown_SST.pdf, (accessed 13 August 2009).

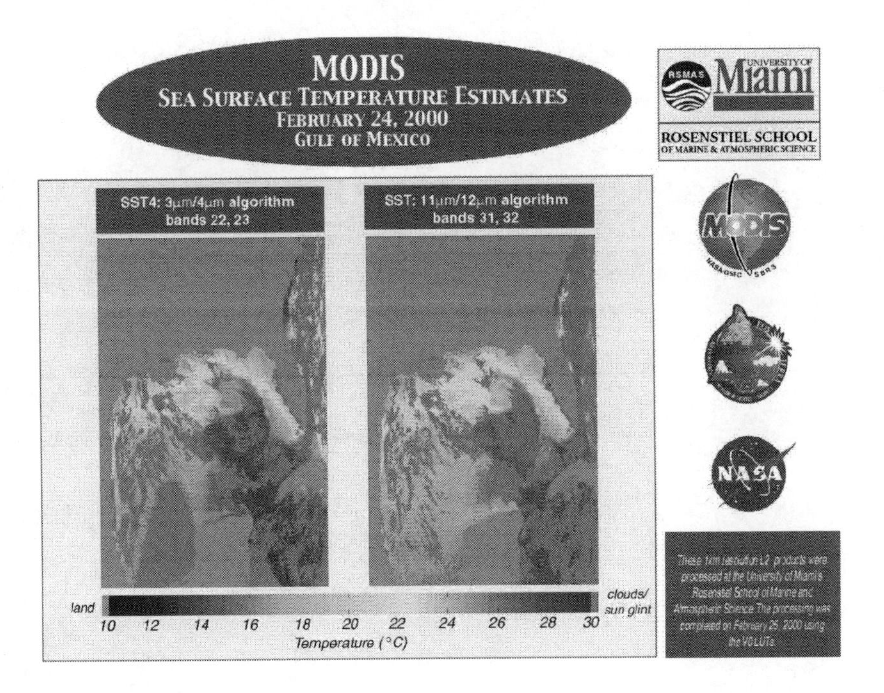

Figure 6.8. MODIS sea surface temperature initial processing for the Gulf of Mexico, from the MODIS SST group at RSMAS.

Howard Gordon had assessed the MODIS atmospheric correction algorithm's performance at this early stage. The basis for comparison was SeaWiFS, and initially, MODIS chlorophyll concentrations and SeaWiFS chlorophyll concentrations were different by roughly a factor of two. That difference was way too high for anybody's satisfaction. There was also severe striping in many of the images (likely due to the mirror side optical disparity), and the water-leaving radiance values were way too high, but calibration efforts had improved the situation significantly from March to April.[135] Gordon summed up his evaluation with sobering, pithy statements:

- Usable Data Requires Incremental/Iterative Resolution of Remaining Problems
- Multiple Processing/Reprocessing Will Be Necessary

Yet at this stage, many (if not most) of the MODIS science team members were familiar with the lessons of the CZCS – getting good data out of a pioneering remote-sensing instrument takes time, and effort.

And more time.

As well as money.

The initial fluorescence data product, being produced by Abbott, Letelier, and

---

[135] Howard R. Gordon, "MODIS Atmospheric Correction Performance: Initial Evaluation", http://modis.gsfc.nasa.gov/sci_team/meetings/200006/presentations/1.09_Gordon_AtmosCorrection.pdf, (accessed 13 August 2009).

Jasmine Bartlett of OSU, looked promising. In fact, five days after MODIS first light, the OSU team produced chlorophyll fluorescence images of the Arabian Sea. (Figure 6.9)[136]

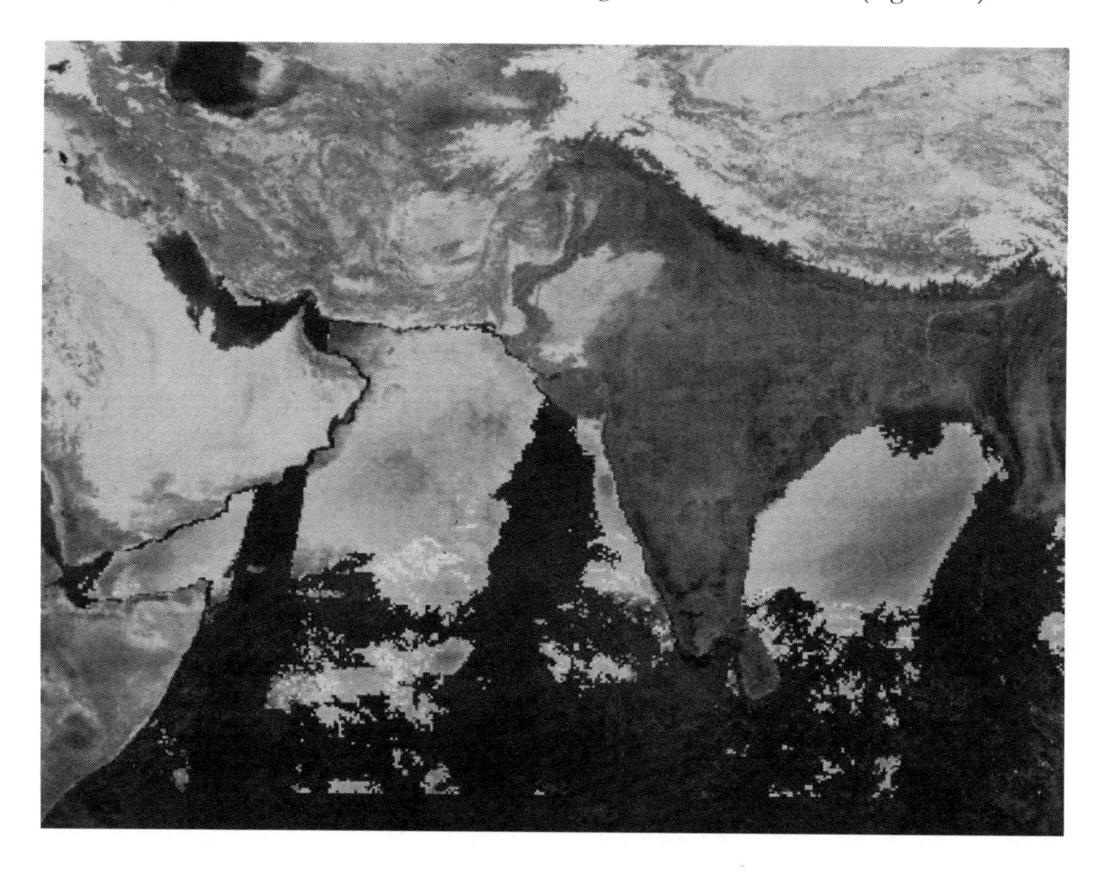

Figure 6.9. MODIS chlorophyll fluorescence "first light" image of the Arabian Sea and Bay of Bengal.

In the Gulf of Maine, Barney Balch from Bigelow was developing the coccolithophorid calcite algorithm, and had conducted a "proof of concept" experiment to see if a much larger algorithm development experiment was feasible.[137] In principle, calculating coccolithophorid calcite concentrations was a tougher tongue-twister to say, and a much simpler data product to produce, than phytoplankton chlorophyll concentrations. The main thing that coccolithophorids do with light is reflect it, as the tiny white biomineral

---

[136] Jasmine S. Bartlett, Mark R. Abbott, and Ricardo M. Letelier, "Phytoplankton Fluorescence from MODIS", http://modis.gsfc.nasa.gov/sci_team/meetings/200006/presentations/1.10_Bartlett_ClhFLH.pdf, (accessed 13 August 2009); NASA Goddard Space Flight Center Scientific Visualization Studio, http://visibleearth.nasa.gov/view_rec.php?id=72.

[137] William Balch, D. Drapeau, B. Bowler, and A. Ashe, "Chalk-Ex: Calibration of the MODIS coccolith algorithm", http://modis.gsfc.nasa.gov/sci_team/meetings/200006/presentations/1.12_Balch_Coccolithophorid.pdf, (accessed 13 August 2009).

microspheres reflect light very effectively. So all that had to be done – in principle – was to dump a sample of this material into the ocean, measure the resulting concentration, and let the satellite see it from space at the same time. The material was easy to get: coccolithophorid calcite is the main material that composes the White Cliffs of Dover in England (and can be mined elsewhere, as more than a few British citizens might object to mining the White Cliffs of Dover).[138] The team had done some small-scale dumps (about a ton) in the Gulf of Maine and imaged the results from an airplane; all that had to be done for MODIS was to scale up the experiment a little, using 25 tons, when MODIS was ready to see it. The timing had to be very good; even slurried chalk sinks pretty fast in the open ocean, so the persistence of the chalk at the surface would not be for an extended period of time.

Data Dependencies

Discussions of data products and algorithms had an underlying theme; the results that scientists were seeking, the data output from the data processing system, were tied very tightly to a chain of data dependencies. The foundation of the process was accurate water-leaving radiances (and the related infrared brightness temperatures that would lead to a sea surface temperature value). For the water-leaving radiances to be accurate, all the characteristics of the instrument that affected them had to be understood, and wherever necessary, corrected. That meant that mirror side differences and detector differences and stray light and anything else in the optical stream had to be characterized *well.* Then the atmospheric correction process could take over and be applied pixel-by-pixel, and hopefully at the end produce sufficiently accurate normalized water-leaving radiances. These, in turn, entered into the processing stream to produce the basic geophysical products, such as chlorophyll concentration, chlorophyll fluorescence line height, coccolithophorid calcite concentration, or sea surface temperature. When these values were available, second generation products – particularly the much-desired primary productivity – could then be calculated. The system was not entirely self-contained, either; other MODIS data products (such as water vapor) or data from other instruments on other satellites (such as ozone concentration) was also required to be in the data processing stream.

At any step in the process, a flaw in the data would lead to very significant errors in the downstream data products. So while a familiar acronym in the realm of computational science is GIGO (Garbage In, Garbage Out), for the processing of data from sophisticated remote-sensing instruments, it might be more appropriately GAGO (Garbage Anywhere, Garbage Out).

So the MODIS Oceans team, along with the instrument team and operators and calibration team – had to clean up the garbage. And that's what they were planning to do.

---

[138] Toby Tyrrell, "Emiliania Huxleyi Home Page", http://www.soes.soton.ac.uk/staff/tt/, (accessed 13 August 2009).

The June 2000 MODIS Science Team meeting also featured a report from Roger Drake on the FM1 MODIS, designated for the EOS-PM satellite. In short summary, the FM1 crew knew about the problems with MODIS-Terra, had already made some necessary fixes, and were working diligently to correct them. They had been given time to make the corrections, but now the FM1 MODIS was on the EOS-PM satellite, which was now called Aqua. The launch of Aqua was set at that point for December 1, 2000. While characterized as "firm", the spacecraft fabricator, TRW, was working triple shifts, 7 days a week in an attempt to make that date.[139]

Forging Forward

During the period of time that elapsed between June 2000 and January 2001, the science team – in particular, the RSMAS oceanographic group seeking to characterize the instrument and then successfully atmospherically correct the data – worked on a myriad of issues in the effort to improve MODIS-Terra data for ocean science. In January 2001, Vince Salomonson characterized the instrument as stable or stabilizing; he characterized this later by saying that MODIS-Terra essentially had a stable configuration a year after launch.[140] Instability was a significant problem when trying to produce science-quality data; each time the instrument was changed (with different gain settings or other correction factors), it entered into what the Ocean team had come to call "epochs". Each epoch required a different application of data processing factors to make data from one epoch be as close as possible to data from a different epoch. Some epochs lasted less than a month.[141]

There had been a troubling event in space in the first year of Terra orbits; in the trackways of Earth remote sensing satellites, another satellite had approached to within 4 km of Terra. That was considered way too close for comfort in the emptiness of space. Furthermore, the Deep Space maneuver had still not been performed; there were concerns that a gyroscope might fail. Unfortunately, the longer they waited to do the maneuver, the more likely it would be for a component to fail.[142]

The DAACs were archiving about 21 Tb of MODIS data per month, but had only distributed about 60 Tb – much less than they were capable of. A survey at the Fall 2000 AGU meeting indicated that the "maturity" of the data from the instrument was still a

---

[139] Roger Drake, "MODIS Protoflight Model (PFM) Instrument Status", http://modis.gsfc.nasa.gov/sci_team/meetings/200006/presentations/1.04_Drake_PFMStatus.pdf, (accessed 13 August 2009).
[140] Vincent V. Salomonson recorded interview, 2 December 2008.
[141] Bob Evans, Ed Kearns, and Kay Kilpatrick, "Latest Terra-MODIS Ocean Color Radiance Corrections," presentation at the MODIS Science Team meeting, Baltimore, MD, July 2004. http://modis.gsfc.nasa.gov/sci_team/meetings/200407/presentations/oceans/Kilpatrick2.ppt, accessed 13 January 2010.
[142] NASA GSFC, "MODIS Science Team Meeting, Wednesday January 24-25, 2001," (Minutes), http://modis.gsfc.nasa.gov/sci_team/meetings/200101/minutes_jan2001.pdf, (accessed 14 August 2009).

concern; other reasons indicated that the multitude of data products was hard to navigate.[143]

Given all that the MODIS science team had contended with, in early January for Wayne Esaias to characterize the MODIS Ocean data effort like this: the "ship was answering the helm, and the seas were calming", seemed a very positive sign. The fluorescence data were particularly exciting and very clean. All of the ocean parameters, including the modeled Level 4 data like primary productivity, were in production and approaching science quality.[144]

One of the particularly intriguing uses of the data was the fact that the 250 m bands, not designed as ocean bands, had been used for red tide detection. The bands had been used as part of a study connecting the deposition of iron-rich dust from the Sahara desert on the Gulf of Mexico, which provided a vital nutrient for the nitrogen-fixing phytoplankton *Trichodesmium*.[145] When *Trichodesmium* bloomed, more nitrogen entered the water column, and this source of nitrogen could trigger harmful algal blooms, including *Karenia brevis*, the flagellated scourge of the Florida coast. Striping had been markedly reduced in the images, and the data showed remarkable "fine structure", chlorophyll concentration variability that exceeded the capabilities of SeaWiFS. MODIS was now producing an unprecedented 1km resolution data set covering nearly the entire global ocean.[146]

In one year, the Miami data processing group led by Bob Evans had made major improvements to the ocean data products. Their list of corrective actions was impressive:

1. Remove the response vs. scan (RVS) problem.
2. Remove the polarization effects (the MODIS rotating mirror design meant that the angle of incidence of light on the mirror changed from east to west on each scan line!); this required a polarization correction for each band.
3. Adjust the detector gains.
4. Remove the sun glint problem
5. "Tweak" the aerosol radiance filter in the atmospheric correction bands to improve the calculation of epsilon, a vital atmospheric correction variable.
6. Finally, evaluate the saturation water-leaving radiance fields down to the sea surface.
7. Also, apply a filter to remove noise associated with different atmospheric optical models used for the atmospheric correction process. [147]

Tediously, painstakingly, this step-by-step process reduced the uncertainties in ocean

---

[143] NASA GSFC, "MODIS Science Team Meeting, Wednesday January 24-25, 2001," section entitled "Data Distribution and User Services".

[144] NASA GSFC, "MODIS Science Team Meeting, Wednesday January 24-25, 2001," section entitled "Oceans Summary."

[145] John J. Walsh and Karen A. Steidinger, "Saharan dust and Florida red tides: The cyanophyte connection". *Journal of Geophysical Research,* 106, 11597-11612 (2001).

[146] NASA GSFC, "MODIS Science Team Meeting, Wednesday January 24-25, 2001," section entitled "Oceans Summary."

[147] NASA GSFC, "MODIS Science Team Meeting, Wednesday January 24-25, 2001," section entitled "Oceans Summary."

data produced by the quirky MODIS-Terra. There was still work to be done; Evans said that there would be new water-leaving radiance and chlorophyll fields for each of the two inhomogenous mirror sides. And as Evans noted, all of the data had been test cases; all of those processing steps and corrections and fixes had not yet been implemented in the full MODIS data processing system.[148]

As for Aqua, TRW's triple-shift efforts had not resulted in a December 2000 launch; the current launch date was set for July 12, 2001, but Aqua Project Scientist Claire Parkinson expected a launch "much later in the year". (Figure 6.11)[149]

Figure 6.11. Aqua in the TRW clean room. MODIS is the instrument nearest to the top of the satellite in this orientation.

[Even while MODIS scientists were grappling with the instrument's quirks, initial work on the instrument which was expected to succeed MODIS was underway. Specifications had been drawn up for the Visible Infrared Imaging Radiometer Suite

---

148 NASA GSFC, "MODIS Science Team Meeting, Wednesday January 24-25, 2001", section entitled "Ocean Validation Update".

149 NASA GSFC, "MODIS Science Team Meeting, Wednesday January 24-25, 2001," section entitled "Aqua Launch Readiness Report"; Aqua Project Science, http://aqua.nasa.gov/reference/viewImage.php?id=131.

(VIIRS), the heir to MODIS, which would fly on the NPOESS satellites.[150]]

9/11 – September 11, 2001: Despite their laser-like focus on the science they are doing, and the tasks they are assigned to make instrument missions like MODIS successful, scientists do not operate in such a rarefied environment that they do not notice real-world events. And it was impossible for any scientist – and indeed for nearly every citizen of the United States – to not be painfully aware of the deadly terrorist attacks on the World Trade Center and the Pentagon using commercial airliners, as well as the fourth hijacked airliner (Flight 93) that crashed in Pennsylvania apparently due to the actions of its passengers, a flight which had been targeted for either the White House or the Capitol building. Both MODIS-Terra and MODIS-Aqua were destined to be utilized in the subsequent response of the United States to the terrorist attacks. The ASTER instrument on Terra was used in the days following the attack to measure thermal hot spots near the World Trade Center site.[151]

Aqua did not make its scheduled date with the launch pad in July, and was still waiting for launch in December. A month before, in November, Barney Balch and crew hauled 26 tons of chalk to the Gulf of Maine and the Atlantic Ocean, pumped it overboard from huge tanks of white slurry, kept measuring the concentrations of the calcite, and waited for MODIS to fly over and view it. The first experimental attempt in the Gulf of Maine (called *Chalk-Ex* in an echo of the JGOFS Iron-Ex fertilizations off the Galapagos) in August 2000 had demonstrated just how chancy this test was; MODIS did not see the patch, but SeaWiFS, following an hour and a half later, did see it – but just barely, because the bright patch was right at the edge of a large cloud bank. Indeed, MODIS missed the first patch due to a large amount of vertical mixing in the water column, which transported the chalk too deep too quickly for MODIS to see it. But the second aquamarine blue patch of chalk stayed at the surface, and MODIS saw it on November 15. (Figure 6.12) [152] This invaluable data provided excellent validation data to produce accurate estimates of coccolithophorid calcite production on a global scale – a significant element of the Earth's carbon cycle.[153]

---

[150] NASA GSFC, "MODIS Science Team Meeting, Wednesday January 24-25, 2001," section entitled "Introduction to VIIRS."

[151] Roger N. Clark, Robert O. Green, Gregg A. Swayze, Greg Meeker, Steve Sutley, Todd M. Hoefen, K. Eric Livo, Geoff Plumlee, Betina Pavri, Chuck Sarture, Steve Wilson, Phil Hageman, Paul Lamothe, J. Sam Vance, Joe Boardman, Isabelle Brownfield, Carol Gent, Laurie C. Morath, Joseph Taggart, Peter M. Theodorakos, and Monique Adams, "Environmental Studies of the World Trade Center area after the September 11, 2001 attack." *Executive Summary, United States Geological Survey Open File Report 01-0429*, http://pubs.usgs.gov/of/2001/ofr-01-0429/#ExecutiveSummary (November 27, 2001) (accessed 14 August 2009).

[152] William Balch, Bruce Bowler, D. Drapeau, E. Booth, Joaquim Goes, Howard Gordon, Robert Evans, and Kay Kilpatrick, "MODIS Calcite Algorithm: Gulf of Maine, Chalk-Ex, and the World". http://modis.gsfc.nasa.gov/sci_team/meetings/200112/presentations/balch.ppt, (accessed 14 August 2009)

[153] NASA GES DISC, "An Enlightened View of Calcite in the Ocean with MODIS," http://disc.sci.gsfc.nasa.gov/oceancolor/additional/science-focus/ocean-color/calcite.shtml, (accessed 14 August 2009); William Balch, Bruce Bowler, D. Drapeau, E. Booth, Joaquim Goes, Howard Gordon, Robert Evans, and Kay Kilpatrick, "MODIS Calcite Algorithm: Gulf of Maine, Chalk-Ex, and the World"; William Balch, A.J. Plueddeman, Bruce C. Bowler, and D.T. Drapeau,

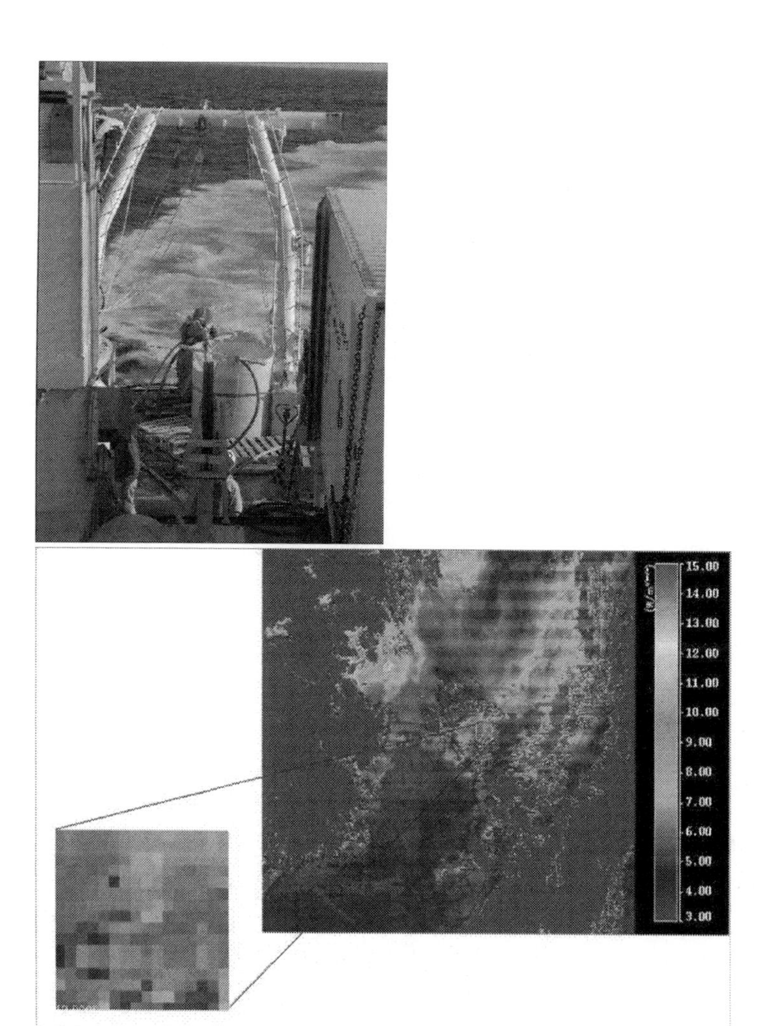

Figure 6.12. Chalk-Ex: Spreading the chalk (top); MODIS image of the patch (bottom); inset image (left) showing the chalk patch pixels as light green.

By the end of 2001, Salomonson was pushing for greater user acceptance of MODIS data by the scientific community, and initiated an "end-to-end" review of the MODIS data processing system. A March 2002 launch data for Aqua was being considered. The milestone for the Oceans teams was that calibration biases had been discovered and corrected for, and for several time periods, the data should be considered validated.[154]

"Chalk-Ex—Fate of $CaCO_3$ particles in the mixed layer: Evolution of patch optical properties, *Journal of Geophysical Research*, 114, C07020, doi:10.1029/2008JC004902, (2009).

[154] NASA GSFC, "MODIS Science Team Meeting, December 17-19, 2001," (Minutes), http://modis.gsfc.nasa.gov/sci_team/meetings/200112/MST2002minutes.pdf, (accessed 14 August 2009).

Hard work had accomplished a great deal, and most of the Oceans team felt that MODIS was ready to provide good data to oceanographers. According to Bob Evans, MODIS data were providing unprecedented view of oceanic features, and the SST data was likely better than the community-standard AVHRR. The global image of 4 µm SST acquired at night was clear and remarkably detailed, demonstrating the quality of this new SST data product.[155]

Howard Gordon was also upbeat on atmospheric correction; the normalized water-leaving radiances were "looking good". However, areas with dust influence, as well as Case 2 waters, still required more effort. Direct comparisons to SeaWiFS data showed one of the expected facts-of-life for MODIS; the sun glint parabola was clearly visible, and it affected data quality adjacent to the glittery pixels, which the tilt of SeaWiFS avoided.[156]

An interesting early result of the synergy between MODIS and SeaWiFS resulted from Janet Campbell's comparisons of the MODIS and SeaWiFS chlorophyll algorithms. Campbell and her colleagues had been able find a clear day in the Atlantic off the U.S. East Coast, when both MODIS and SeaWiFS acquired data, which allowed feature tracking to show the drift of chlorophyll with surface currents, and thus allow estimate of current speeds.

Another area of considerable interest was the research of Frank Hoge, who was validating the phycoerythrin product and other data products, including fluorescence line height, in collaboration with Letelier.[157] Letelier showed fluorescence images of the Arabian Sea featuring eddies with different nutrient concentrations depending on the direction of rotation – the difference in nutrient availability was clearly seen in the different values of chlorophyll fluorescence.[158]

Progress was going forward at a steady pace on the ocean primary productivity product. The validation of this product still awaited valid SST and water-leaving radiance; within a month after the chlorophyll product was validated, they would have a validated primary productivity data product.[159] Thus, chlorophyll was the sticky wicket.

Winds and sharks: In 2001, Marine Optical Characterization Experiments 8 and 9 were conducted; MOCE 8 took place in March and was hindered by high winds and cloud cover, while MOCE 9 took place in November and December and collected more data as MODIS-Terra orbited directly overhead.[160] During this year and previous years, other

---

155 NASA GSFC, "MODIS Science Team Meeting, December 17-19, 2001," section entitled "Calibration of MODIS Ocean Products."
156 NASA GSFC, "MODIS Science Team Meeting, December 17-19, 2001," section entitled "Normalized Water-Leaving Radiances."
157 NASA GSFC, "MODIS Science Team Meeting, December 17-19, 2001," sections entitled "Calcite" and "Phycoerythrin and Other Ocean Color Validation."
158 NASA GSFC, "MODIS Science Team Meeting, December 17-19, 2001," section entitled "Chlorophyll Fluorescence."
159 NASA GSFC, "MODIS Science Team Meeting, December 17-19, 2001," section entitled "Primary Production."
160 NOAA, "MOCE Team Expedition Log (MOCE-8), http://www.star.nesdis.noaa.gov/sod/orad/mot/moce/personnel/moce_8.html, (accessed 14 August 2009); NOAA, "MOCE Team Expedition Log (MOCE-9),

short trips were performed to conduct "diver calibrations" and perform equipment maintenance. During the diver calibration cruise in July (MOBY-L70), the divers encountered aggressive oceanic whitetip sharks, at one time requiring them to fend off a shark with a boat pole. The log notes "If this type of shark behavior becomes the norm, procedural changes will have to be implemented to ensure the safety of the science divers." [161] The sharks weren't as much of a problem the next time the buoy was visited in August.[162] [Dennis Clark indicated that the presence of sharks was partly due to a modified Black-and-Decker brush that had been waterproofed so that it could be taken underwater to clean MOBY's lenses. The brush could only be turned on and couldn't be turned off while underwater. Unfortunately, the whirring electric motor of the brush was an effective shark attractant. When the sharks showed up, the only way to turn off the brush was to get it to the surface without undue delay.][163]

These encounters serve as reminders that remote-sensing oceanography with ocean color data is certainly not just about sitting at a computer terminal looking at colorful images!

Eventful Year: 2002

Though NASA had been hoping to launch Aqua in March, the launch slipped back a little further on the calendar, to May 24, 2002. An early-morning launch of a Delta-II illuminated the skies over Vandenberg, as Aqua was launched into a descending node orbit. (Figure 6.13)[164] Terra orbited in the ascending node: that meant it was moving north as it crossed the Equator. Aqua orbited in the opposite direction, moving south as it crossed the Equator.

Earlier in the year, European Space Agency launched its large and ambitious Envisat Earth observation satellite on March 1, 2002, from Kourou, French Guiana.[165] Envisat was indeed big, carrying eight different instruments, and heavy; it required the heavy-lift Ariane 5 launch vehicle to transport it into Earth orbit. Most importantly for the global ocean color community, Envisat had an ocean color-capable sensor onboard: MERIS, the Medium

---

http://www.star.nesdis.noaa.gov/sod/orad/mot/moce/personnel/moce_9.html, (accessed 14 August 2009).

[161] NOAA, "MOCE Team Expedition Log (MOBY-L70)", http://www.star.nesdis.noaa.gov/sod/orad/mot/moce/personnel/moby_l70.html, (accessed 14 August 2009).

[162] NOAA, "MOCE Team Expedition Log (MOBY-L71)", http://www.star.nesdis.noaa.gov/sod/orad/mot/moce/personnel/moby_l71.html, (accessed 14 August 2009).

[163] Dennis Clark, interview transcript notes.

[164] NASA, "Aqua Spacecraft Launched, Ready To Study Earth's Water Cycle," http://www.nasa.gov/mission_pages/aqua/aqua_era.html, (accessed 14 August 2009); http://aqua.nasa.gov/reference/viewGallery.php?id=11

[165] Mullard Space Science Laboratory, "Spectacular night-time launch places European environmental eye in orbit," http://www.mssl.ucl.ac.uk/general/news/envisat_launch/envisat.html, (accessed 14 August 2009).

Figure 6.13. Early-morning launch of Aqua.

Resolution Imaging Spectrometer. MERIS was an advanced instrument, with several bands capable of 300m resolution, and adjustable band positions. MERIS was equipped with an impressive 15 bands in the visible spectrum.[166] MERIS certified the entry of many countries in Europe into the active collection and analysis of ocean color data from their own instrument, where previously European scientists had relied on data from other missions, particularly CZCS and SeaWiFS. MERIS clearly rivaled the capabilities of MODIS for ocean observations.

By the middle of the year, the instrumental team had to contemplate switching between the A-side data formatter, which was accumulating errors, though not at the expense of the science data. Due to the errors, there would have to be an eventual switch to the B-side formatter. Though it had only been in orbit for a couple of months, Band 6 on MODIS-Aqua was failing. This band, at 1.6 μm, was a band used for land, cloud, and snow properties.[167]

The Oceans team had commenced the reprocessing of their "Collection 4" data, and would finish the reprocessing by October. Collection 4 was going to feature the best and brightest MODIS Ocean data products, and Esaias indicated that overall, they were "very pleased with the data". Some data aspects still caused concern. Mark Abbott reported that the fluorescence line height data had been validated by Frank Hoge's aircraft measurements; there remained confounding areas with high chlorophyll and low fluorescence, possibly influenced by dust aerosols. Dennis Clark was dealing with persistent

---

[166] ESA Earthnet, "The Medium Resolution Imaging Spectrometer Instrument," http://envisat.esa.int/instruments/meris/, (accessed 14 August 2009).
[167] NASA GSFC, "MODIS Science Team Meeting Minutes, July 23-24, 2002", http://modis.gsfc.nasa.gov/sci_team/meetings/200207/MST072002minutes.pdf, (accessed 14 August 2009); L. Wang, J. J. Qu, J. Xiong, X. Hao, Y. Xie, and N. Che, "A new method for retrieving band 6 of Aqua MODIS," *IEEE Geoscience and Remote Sensing Letters*, 3(2), 267-270, doi:10.1109/LGRS.2006.869966, (2006).

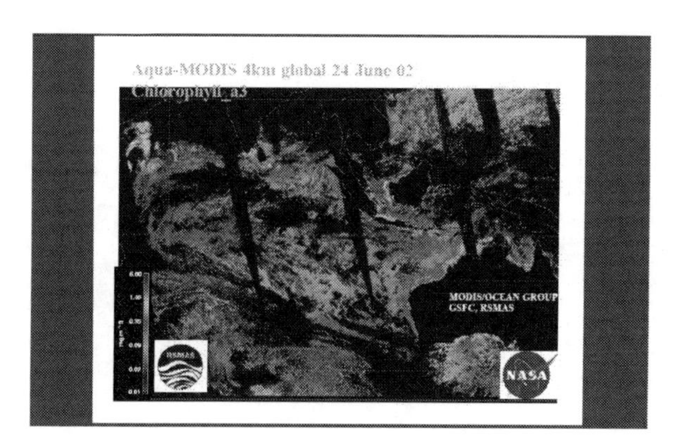

Figure 6.14. MODIS-Aqua first light chlorophyll image, June 24, 2002.

problems reconciling the open water chlorophyll algorithm, partly related to the sensitivity of the product to slight changes in MOBY calibration. In brighter news, the calcite algorithm was working quite well.[168]

The news for MODIS-Aqua was quite good; the experience with the "extreme challenges" of MODIS-Terra was paying off for MODIS-Aqua. The instruments was providing "well-behaved" Level 1 data products, which were now being produced within 24 hours of acquisition.[169]

Over the next months, the Ocean team investigators continued to grapple with the complex process of extracting scientifically-useful data from the puzzle of MODIS-Terra, while simultaneously they were pleased with the relative smoothness of the development of MODIS-Aqua data. Figure 6.14 shows the first light chlorophyll image from MODIS-Aqua. [170]

On December 14, 2002, Japan launched the ADEOS-II satellite, which successfully reached orbit, and was renamed MIDORI-II. Onboard MIDORI-II was the Global Imager (GLI), the Japanese MODIS, a considerably more advanced sensor than the predecessor OCTS. GLI began returning good data early in 2003, and the Japanese once again prepared to participate actively in the global ocean color community.[171]

MOBY continued to function well, with routine maintenance cruises performed throughout 2002. A special MOCE called "Turbid-7" took place in September, with the goal of collecting water samples in "highly turbid" waters to assist with the algorithm

---

[168] NASA GSFC, "MODIS Science Team Meeting Minutes, July 23-24, 2002," section entitled "Ocean Products Validation and Status."

[169] NASA GSFC, "MODIS Science Team Meeting Minutes, July 23-24, 2002," section entitled "Ocean Products Validation and Status."

[170] Wayne Esaias, Kevin Turpie, Donna Thomas, Ron Vogel, and A. Bhatti, "MODIS Ocean Net Primary Production," http://modis.gsfc.nasa.gov/sci_team/meetings/200207/presentations/esaias.pdf, (accessed 14 August 2009).

[171] Japan Aerospace Exploration Agency, "Advanced Earth Observation Satellite – II "Midori-II" ADEOS-II," http://www.jaxa.jp/projects/sat/adeos2/index_e.html, (accessed 14 August 2009).

development effort.[172] This cruise resulted in the collection of data relative to phytoplankton pigments, colored dissolved organic matter, and suspended particulate sediments.

2003: Year of Changes

Terra performed a couple of flips for calibration purposes in March and April of 2003; these deep-space maneuvers (DSMs) were intended to provide calibration data for the SST data product and the visible bands. During the second DSM, the Moon was successfully viewed through the earth-viewing port.[173]

[During this eventful year, the author took part in weekly telephone conferences held by the Principal Investigators of the MODIS Ocean team. Some of the events described here are based on notes taken during those meetings, for which no other documentation is available. The notes provide insight into the week-to-week activities of the team members, and the difficulties that they contended with during a challenging year.]

Operation Iraqi Freedom: Early in 2003, the attention of the United States and the world was drawn to the military operations of the coalition forces in Iraq. A special system to utilize MODIS data, called Satellite Focus, was built to enable the rapid utilization of data from the sensors in the military campaign.[174] In March of 2003, Satellite Focus was an integral part of the military operation when it provided images of a massive dust storm that swept through Iraq, impeding the progress of the campaign by the coalition military forces on the ground.[175] Subsequent to the fall of Baghdad, both MODIS and ASTER observed the black soot plumes of the oil fires lit by the retreating Iraqi forces.[176]

---

[172] NOAA, "MOCE Team Expedition Log (MOCE Turbid-7)," http://www.star.nesdis.noaa.gov/sod/orad/mot/moce/personnel/turbid_7.html, (accessed 14 August 2009).

[173] MODIS Calibration Support Team, "Terra MODIS Instrument Performance History," http://www.mcst.ssai.biz/mcstweb/performance/terra/terra_instrument.html, (accessed 14 August 2009).

[174] http://www.informaworld.com/smpp/section?content=a747979843&fulltext=713240928

[175] Carl Drews, "Sandstorm over Baghdad: The dust storm that stalled the coalition invasion of Iraq," http://acd.ucar.edu/~drews/sandstorm/, (accessed 2 May 2011); Steven D. Miller, Jeffrey D. Hawkins, F. Joseph Turk, Thomas F. Lee, John Kent, Kim Richardson, and Arunas Kuciauskas, "The mission support role played by MODIS during Operation Iraqi Freedom," http://www.nrlmry.navy.mil/sat_training/nexsat/aux_files/SPIE_2004_Denver5548-35_SATFOCUS_MillerEtAl.pdf, (accessed 2 May 2011); Ming Liu, Douglas L. Westphal, Annette L. Walker, Teddy R. Holt, Kim A. Richardson, and Steven D. Miller, "COAMPS Real-Time Dust Storm Forecasting during Operation Iraqi Freedom," *Weather Forecasting*, 22, 192–206, doi: 10.1175/WAF971.1 (2007); Steven D. Miller, J.D. Hawkins, T.F. Lee, F.J. Turk, Kim Richardson, Arunas P. Kuciauskas, J. Kent, R. Wade, C.E. Skupniewicz, J. Cornelius, J. O'Neal, P. Haggerty, K. Sprietzer, G. Legg, J. Henegar, and B. Seaton, "MODIS provides a satellite focus on Operation Iraqi Freedom," *International Journal of Remote Sensing*, 27(7), 1285 – 1296, (2006).

[176] NASA Earth Observatory, "Oil fires in Iraq," http://earthobservatory.nasa.gov/NaturalHazards/event.php?id=11155, accessed May 2, 2011. Two pages from this compilation show a MODIS image,

In May, Chuck Trees (the current and final rotating program head for Ocean Biology and Biogeochemistry) reported that Paula Bontempi, the incoming permanent (civil servant) program head, had reported on the status of the ocean data processing to Ghassem Asrar, and Asrar was said to have been pleased. More intriguingly, the 412 nm band had exhibited a 20% shift since 2003. Clark reported that the last time out to MOBY, there had been a significant puzzle; the readings were 30% low in the blue end of the spectrum. The site had been picked for its usual low chlorophyll concentrations and the rarity of phytoplankton blooms; however, the low values could have been due to a high wind event and a regional bloom. The values measured during this event were large enough to affect SeaWiFS calibration. Later in May, the science team learned that MODIS-Aqua was not showing the same patterns in the Southern Ocean as MODIS-Terra; this difference was a definite conundrum. Later in May, Dennis Clark reported that MOBY had been damaged, losing its mid-depth arm. White paint on the buoy indicated a possible collision with a boat. The replacement couldn't be put into the water until July, so the calibration report would have to be adjusted. The upcoming reprocessing was considered so critical that Chuck Trees said that it could slip to November.[177]

Open sesame? MODIS-Terra threw its controllers a curveball on May 6, when during routine calibration operations, the screen on the solar diffuser failed to open when the solar diffuser door was opened. This anomaly, reported on by *Space News*, attracted some attention to the MODIS-Terra mission.[178] The screen was used to calibrate low radiance levels for the instrument, which were characteristic of ocean values. With the screen in the "closed" position, the higher radiance levels, characteristic of land and cloud reflectivity, could not be calibrated as effectively. The problem was: after the calibration process, the solar diffuser door had been closed. The anomaly with the solar diffuser screen was worrisome, because the screen and the door shared the same motor to open and close. If the solar diffuser door did not open again, the ability to calibrate most of the MODIS bands would be seriously compromised. Consideration of what to do about the solar diffuser screen and the solar diffuser door immediately occupied the primary attention of the MODIS instrument team at GSFC.

On July 2, the command to open the solar diffuser door was sent, and just like the door to the cavern of treasure opened for Aladdin, the MODIS-Terra solar diffuser door opened as commanded.[179] The decision had been made to keep the screen in place, and to not close the door again, due to the anomaly. Rather than a problem for the Ocean team, Esaias saw this anomaly as an advantage, allowing more continuous ocean calibration data to

---

http://earthobservatory.nasa.gov/NaturalHazards/event.php?id=11170, and an ASTER image, http://earthobservatory.nasa.gov/NaturalHazards/view.php?id=11176, both acquired on 31 March 2003.

[177] MODIS Science Team Principal Investigator Meeting Notes: 2 May 2003 , 9 May 2003, 23 May 2003, and 30 May 2003.

[178] Brian Berger, "Instrument Glitch Not a Serious Threat to Terra Mission," http://www.space.com/spacenews/archive03/glitcharch_071403.html, (accessed 14 August 2009).

[179] Brian Berger, "Instrument Glitch Not a Serious Threat to Terra Mission."

be acquired.[180]

In the back of everyone's mind at this time was the salient fact that earlier in the year, NASA had conducted a re-competition for the MODIS science contracts.[181] The MODIS Ocean team had been working on their algorithms for about a decade, and this recompetition would potentially add new voices and ideas to the team – and also potentially mean that some team members who had been together for this entire period of triumph and travail would see an end to their participation. The panels which would consider and choose the new science team members were scheduled to meet in August, and the decisions would be released in October or November.

In late July, after comparisons with SeaWiFS data and evaluation of all the remaining issues, Esaias reluctantly concluded that based on what they knew at that point, they shouldn't proceed with the reprocessing. Both Trees and Bontempi indicated that if radiances "comparable" to SeaWiFS couldn't be achieved, then they shouldn't spin their wheels and waste resources doing it; rather, they should address remaining issues and determine their influence on the reprocessing of MODIS-Aqua data.[182]

Esaias planned to go to NASA HQ in August to describe the status of the critical MODIS-Terra reprocessing.[183] Some of the points in his presentation would be:

- instrument epochs have largely been removed;
- temporal variability is slightly greater;
- chlorophyll agreement good, radiance agreement better;
- disagreement in summer (June) especially in Southern Ocean;
- new BRDF correction;
- improved sunglint correction;
- a number of things have been improved, but no improvements
- to absolute radiances
- only one "bad" fluorescence epoch.

His summary recommendation was to proceed with the reprocessing, in order to provide much better quality science data products to the user community. After the meeting at HQ, the HQ staff was convinced of the need to do the reprocessing, then move on to the Aqua data. But the team did not decide to commence the reprocessing at that time, instead deferring to wait until mid-September, though Chuck Trees left his position at NASA HQ at the end of August. And the MODIS Ocean team also learned that the decisions on the recompetition proposals would also occur in that same timeframe.

In September, Paula Bontempi indicated that she was ready to proceed with the reprocessing "if everyone was in agreement" – the team was still trying to track down

---

[180] Brian Berger, "Instrument Glitch Not a Serious Threat to Terra Mission."
[181] NASA GSFC, "MODIS Science Team Meeting Minutes, July 23-24, 2002"; Steve Graham and Claire Parkinson, "Minutes of the Aqua Science Group Meeting", http://aqua.nasa.gov/doc/pubs/020801_minutes.pdf, (accessed 14 August 2009). (The recompetition process extended well into 2002 and early 2003.)
[182] MODIS Science Team Principal Investigator Meeting Notes: 30 July 2003.
[183] MODIS Science Team Principal Investigator Meeting Notes: 6 August 2003.

discrepancies with matchup comparisons done with SeaWiFS data. At the same time, the uncertainty in the panel decisions was on everyone's mind; Bob Evans wondered what would happen to the current contracts if they were not selected. The decision to proceed with the reprocessing was made in late September.[184]

In October, the MODIS-Terra reprocessing commenced; the team also planned to proceed with MODIS-Aqua data reprocessing in March 2004. The instrument and calibration teams were trying to look at stray light effects that might have cropped up with the open solar diffuser door; while the effects seemed small, on the order of 1-2%, the propagation of errors might explain some of the strangeness still being seen in the MODIS-Terra data. The effect of uncertainty in the recompetition process was particularly acute for MOBY funding, because it took a long time to get money out of the NOAA system!

Bob Evans recommended stopping the reprocessing at March 2003 (the DAAC was reprocessing over a Terabyte a day) because following the problem with the solar diffuser door, everything started to "go way off". Ominously, the MODIS team was also concerned with the termination of the SeaWiFS contract which was scheduled to happen on December 21.[185]

ADEOS redux: Another shock rippled through the ocean color community when Japan's MIDORI-II abruptly ceased communications on October 25. The initial indications were a power failure, again connected to the solar power array. One of the last things that MIDORI-II told its controllers was that the power from the array had abruptly dropped from 6 kilowatts to 1 kilowatt.[186] In an effort to determine the cause of the problem, MIDORI-II was even imaged by radar from the ground, and the images showed that the solar panel was still intact.[187] The sudden loss of power coincided with a strong solar flare while the satellite was over Peru – perilously close to the South Atlantic Anomaly. The final evaluation of the failure blamed the loss on auroral charging, leading to arcing between the power cables, in this unusual zone of the magnetosphere.[188] The loss was not just to Japan; a NASA scatterometer (SeaWinds) on the mission was silenced by the power failure.[189]

The reprocessing was nearly complete in November, and the team members evaluated the new products; Carder was satisfied with the newer results, but the standard chlorophyll product now differed from SeaWiFS by 30%; previously it had been very close.

---

[184] MODIS Science Team Principal Investigator Meeting Notes: 10 September 2003, 17 September 2003, and 1 October 2003.

[185] MODIS Science Team Principal Investigator Meeting Notes: 1 October 2003, 8 October 2003, and 15 October 2003.

[186] Japanese Aerospace Exploration Agency, "Operational Anomaly with Midori-II," http://www.jaxa.jp/press/2003/10/20031025_midori2_e.html, (accessed 17 August 2009).

[187] David Cyranoski, "Satellite loss throws Japan's space programme into disarray," *Nature*, 426, 3 doi:10.1038/426003a, (November 6, 2003).

[188] Hironori Maejima, Shirou Kawakita, Hiroaki Kusawake, Masato Takahashi, Tateo Goka, Tadaaki Kurosaki, Masao Nakamura, Kazuhiro Toyoda, and Mengu Cho, "Investigation of power system failure of a LEO satellite," Paper, *2nd International Energy Conversion Engineering Conference*, Providence, Rhode Island, August 16-19, 2004. (PDF copy acquired from http://www.aiaa.org/content.cfm?pageid=406&gTable=Paper&gID=22021 on 17 August 2009.)

[189] NASA Earth Observatory, "NASA's Newest SeaWinds Instrument Breezes Into Operation," http://earthobservatory.nasa.gov/Newsroom/view.php?id=23070, (accessed 17 August 2009).

The reprocessing had been run at high speed, and that performance could probably be achieved for subsequent reprocessings. In essence, the conclusions of the data analysis were that the differences in chlorophyll and productivity were more than just differences between sensors; the way that the algorithms worked was probably involved. One difference was noted; in the North Atlantic, in summer and fall SeaWiFS had very little variability, while MODIS showed the patterns of a fall bloom. It was speculated that SeaWiFS might be influenced by *Gelbstoff*.

The waiting continued… into December. NASA was waiting to hear from Orbimage on the SeaWiFS contract extension offer; meanwhile, the MODIS Ocean team was waiting to hear who had been selected for the new MODIS science team. While waiting, the reprocessing had been completed, with the results characterized as a "dramatic step in the right direction".[190] Esaias prepared an announcement discussing the period of coverage and the data quality, and the team was turning its attention to the reprocessing of MODIS-Aqua data.[191]

Terra provided a scare for the ground controllers when it suddenly went into Safe mode on December 16; it was carefully reactivated on December 22, and earth observations commenced again on Christmas Eve, making everyone a little more merry.[192]

During 2003, Dennis Clark occasionally participated in the teleconferences from Hawaii, if he was stationed on Lanai working on MOBY or at-sea during calibration and validation cruises which were also conducted out of Honolulu. On a short visit early in January, the sensor windows were obscured by biological gunk – and fish were actually observed swimming in the spectrometer port, which according to the cruise log "*would explain the noisiness observed in the data*".[193] Late in May, it was discovered that MOBY had a broken arm (the middle light sensor arm), which also damaged some of the fiber-optics links.[194] The next MOBY couldn't be rotated into place until July, and it suffered a hard-drive failure following deployment that required a quick fix.[195] The MOBYs had a relatively uneventful year for the rest of 2003.

---

[190] MODIS Science Team Principal Investigator Meeting Notes: 12 November 2003, 19 November 2003, and 26 November 2003.

[191] MODIS Science Team Principal Investigator Meeting Notes: 3 December 2003, 10 December 2003, and 17 December 2003.

[192] Brian Berger, "NASA Officials Search for Explanation to Terra Shutdown," Space News, 20 January 2004. http://www.space.com/spacenews/archive04/terraarch_012004.html, (accessed 17 August 2009).

[193] NOAA, "MOCE Team Expedition Log (MOBY-L88)," http://www.star.nesdis.noaa.gov/sod/orad/mot/moce/personnel/moby_l88.html, (accessed 17 August 2009).

[194] NOAA, "MOCE Team Expedition Log (MOBY-L93)," http://www.star.nesdis.noaa.gov/sod/orad/mot/moce/personnel/turbid_8.html, (accessed 17 August 2009).

[195] NOAA, "MOCE Team Expedition Log (MOBY-L96::M226SB/Oahu-4," http://www.star.nesdis.noaa.gov/sod/orad/mot/moce/personnel/moby_l96.html, (accessed 17 August 2009).

The Results of the Recompetition: 2004 to 2010 (and beyond)

Early in 2004, after considerable delay, the selections resulting from the MODIS science team recompetition were released. For the team that had been devoted to the MODIS Ocean effort since the early 1990s, the selections were tumultuous and somewhat painful, as several dedicated researchers did not receive renewed proposals to continue their research and data product development. The most significant result of the recompetition was the decision to terminate the previous MODIS Ocean data plan, which included a multitude of data products and numerous interconnected algorithms to produce those data products. The new plan was to nearly abandon the attempt to produce scientific-quality data from the difficult-to-characterize MODIS-Terra, which had confounded the MODIS Ocean team from the beginning and caused much more time and effort (and money) than expected to be spent on figuring it out. Instead, the much-better-behaved MODIS-Aqua would become the focus of attention, and the responsibility for producing MODIS-Aqua data products would be transferred from the MODIS Ocean team computing facility to the SeaWiFS Project, which would eventually become the Ocean Biology Processing Group (OBPG). The list of data products was slashed; the OBPG would concentrate on creating a menu of data products very similar to those already being produced for SeaWiFS, with minor differences in identity and data processing scheme due to the differences in band positions between SeaWiFS and MODIS. Some of the data from some of the bands would be archived, but no official data products would be released – in particular, this affected the important fluorescence band data. The production of higher level products, such as a standard ocean primary productivity data product, was also cancelled in favor of the shorter, simpler list of data products.[196]

As a result of this process, the small number of large contracts for science team members was altered to a much larger number of research grants. So when the next MODIS science team meeting took place in July 2004, the full science team (land, atmospheres, and oceans) had increased in size from 28 members to 90 members.[197]

At this meeting, Chuck McClain summarized the elements of the new ocean color data plan, which would concentrate on MODIS-Aqua, which was considered to have the best chance of overlapping with the NPOESS Preparatory Project (NPP) VIIRS mission. McClain indicated that they would use the same calibration/validation plan as SeaWiFS, and planned to continue working with the MODIS calibration team to address MODIS-Aqua's own peculiarities.[198] Following McClain, Gene Feldman discussed how the "missions to measurements" strategy put in place by Mary Cleave would work for MODIS ocean data; the location of a single team responsible for data product development, analysis, and data distribution would facilitate the necessary flexibility to stabilize and improve the ocean color data products. And in his presentation, Feldman provided a very interesting highlight of

---

[196] Paula Bontempi, "MODIS Update Announcement from NASA Head," 30 January 2004, http://oceancolor.gsfc.nasa.gov/forum/oceancolor/topic_show.pl?tid=3, (accessed 17 August 2009).
[197] NASA GSFC, "MODIS Science Team Meeting, July 13, 2004," Minutes, http://modis.gsfc.nasa.gov/sci_team/meetings/200407/minutes.pdf, (accessed 17 August 2009).
[198] NASA GSFC, "MODIS Science Team Meeting, July 13, 2004."

how far the ocean color community had come from the early days of the CZCS, when scientists had waited at the door of the Nimbus communications room for Warren Hovis to bring out a single CZCS scene. Now, the computing facilities of the OBPG allowed a SeaWiFS data reprocessing to take place 3000 times faster than it had taken to acquire the data; this meant the entire SeaWiFS global data set could be reprocessed in roughly a week. The much larger MODIS data volumes slowed things down, so MODIS data reprocessing could be accomplished at a rate about 80 times faster than acquisition; thus, 80 days of 1km global resolution ocean data could be processed in about a day.[199]

McClain also summarized the plans of the new Ocean science team. The main concern was the continuation of calibration and validation efforts. The SIMBIOS Project was ending, MOBY funding was uncertain, and SeaWiFS data continuation was always a concern. [200]

The MOCE squad had a good year in 2004; the MOBYs did not experience any major operational problems throughout the year. During the "Oahu-6" bio-optical algorithm cruise in May 2004, the team even went out to make measurements at night of the fluorescence associated with coral reefs.[201]

In 2005, Paula Bontempi, who was already at NASA HQ as the head of the Ocean Biology and Biogeochemistry Program, also officially became the MODIS Program Scientist and Manager. The actual MODIS mission was now ensconced in the "Science Mission Directorate, Earth-Sun System", which had a familiar personage as acting director: Mary Cleave. In her new dual role, Bontempi encouraged interdisciplinary algorithm development and also applications of MODIS data for decision support.[202]

The general concern of the OBPG this year was that there were many desired data products that could potentially be developed from the satellite data, but the availability of validation data sets (i.e., in-water optical and constituent measurements) was limited, so additional effort was required to get more validation data and enable algorithm choices. Notable activities of the new science team members included Mike Behrenfeld's development of a primary productivity Web site and an analysis of the "best choice" primary productivity algorithm, while Menghua Wang was continuing to work on an improved atmospheric correction method for coastal waters.[203]

MOBY operations continued in 2005, which is the last year for which expedition logs are available. In March, bad weather made it difficult to collect bio-optical data, and in April, during a routine diver calibration check, bad sharks (one of which had to be fended

---

[199] NASA GSFC, "MODIS Science Team Meeting, July 13, 2004."
[200] NASA GSFC, "MODIS Science Team Meeting, July 13, 2004," section entitled "Chuck McClain – Oceans Group Summary and Plans**"**.
[201] NOAA, "MOCE Team Expedition Log (MOBY–L106::M229SOB/Oahu-6)," http://www.star.nesdis.noaa.gov/sod/orad/mot/moce/personnel/moby_l106.html, (accessed 17 August 2009).
[202] NASA GSFC, "MODIS Science Team Meeting, March 22-24, 2005," Minutes, http://modis.gsfc.nasa.gov/sci_team/meetings/200503/minutes.pdf, (accessed 17 August 2009).
[203] NASA GSFC, "MODIS Science Team Meeting, March 22-24, 2005," section entitled "Oceans Group Summary".

off with "anti-foulant tubes") made it difficult to check on the status of the buoy.[204]

2006 marked further expansions in the use of MODIS Ocean data. Bob Arnone and his team at the Naval Research Laboratory were using MODIS to discern coastal ocean processes, which in part involved utilizing the 250m resolution bands to observe patterns of turbidity. Barney Balch was picking out differences between Particulate Inorganic Carbon (PIC), which was primarily coccolithophorid calcite, and Particulate Organic Carbon (POC), which is the carbon that composes most of the cells of phytoplankton, as well as detrital material.[205]

Paula Bontempi was focused on the upcoming end of the MOBY deployment, scheduled for March 2007. The Ocean science team was asked to consider if there was a need for a new deep-water calibration site, or alternatively, a site that allowed calibration for coastal waters. [206]

Another development for this year was the transfer of the SST data product to the OBPG later that year. SeaDAS was upgraded to include the capability to process data from the MODIS high resolution bands.

Wayne Esaias was now working on the NPP VIIRS instrument team, and there were echoes of previous experience in a brief statement in his report to the MODIS Science Team meeting: "We are worried about how well we can understand the instrument based on pre-launch tests." [207] Given the MODIS-Terra experience, this potential lack of knowledge was ominous.

Later in 2006, funding for the EOS Recompete was cut by $5 million dollars. The OBPG was analyzing the Terra data, hoping that another instrument providing science-quality data could be in orbit while the launch of NPP and eventually the NPOESS satellites slipped further and further into the future.

One of the main topics concerning NASA's ocean research community now was the construction of, and the actual identity of, a "climate data record" for ocean color data. Chlorophyll data had been collected starting in 1978 with the CZCS, and including a decade-long data gap, and starting with SeaWiFS in 1997 and perhaps OCTS several months earlier, chlorophyll data for the global oceans had been collected routinely. But there were serious questions regarding whether or not this data was "climate quality", and whether or not CZCS data could be processed sufficiently well to extend the record back in time. The necessity of MOBY to the construction of a chlorophyll or water-leaving radiance climate

---

[204] NOAA, "MOCE Team Expedition Log (MOBY-L114(M232SB):: Oahu-7)," http://www.star.nesdis.noaa.gov/sod/orad/mot/moce/personnel/moby_l114.html, (accessed 17 August 2009); NOAA, "MOCE Team Expedition Log (MOBY-L115)," http://www.star.nesdis.noaa.gov/sod/orad/mot/moce/personnel/moby_l115.html, (accessed 17 August 2009).

[205] NASA GSFC, "MODIS Science Team Meeting, January 4-6, 2006," http://modis.gsfc.nasa.gov/sci_team/meetings/200601/minutes.pdf, (accessed 17 August 2009).

[206] NASA GSFC, "MODIS Science Team Meeting, January 4-6, 2006," section entitled "Ocean Breakout Session, Day Two".

[207] NASA GSFC, "MODIS Science Team Meeting, January 4-6, 2006," subsection entitled "VIIRS Update".

data record was highlighted, but the current MOBY program that had supported both SeaWiFS and MODIS was scheduled to end in March 2007. [208]

Aqua celebrated its five-year anniversary in orbit on May 4, 2007, with all the instruments still working well.

In 2008, the EOS Recompete resulted in a refreshed MODIS science team, with some familiar members and some new members. Though MODIS-Terra had been in orbit eight years and MODIS-Aqua six years, both instruments were still performing well, despite indications of age-related changes. Once concerned with the acceptance of MODIS science data, Vince Salomonson had seen the total number of publications using MODIS data rise to over 2,000. McClain told the team that Mike Behrenfeld had developed a new fluorescence line height data product, and they were still (emphasis on "still") working to improve the MODIS-Terra ocean color data. The OBPG planned to reprocess the entire suite of data from ocean color sensors in the OBPG archive, and figured that this reprocessing process would be completed by the end of 2008.[209]

Positive Press: The Public's View of MODIS Ocean Color

Despite the difficulties that accompanied MODIS-Terra, which MODIS-Aqua surmounted to provide science-quality ocean color data, the public perception of the missions appeared primarily to be of beneficial and useful science. The fish-finding capabilities of free MODIS ocean color data, data which had been (and still were) for sale from Orbimage from SeaWiFS, were discussed by Mitchell Roffer in a short article in the magazine *Yachting* in 2004, primarily focusing on sea surface temperature.[210] Roffer's own commercial activities using NASA remote sensing data were profiled two years later in the magazine *Salt Water Sportsman*.[211]

Many of the main triggers for the public to notice NASA's ocean remote sensing activities were unusual (and potentially harmful) oceanic phenomena: harmful algal blooms (HABs), water pollution, worrisome effects of hurricanes, and the warming of the ocean's waters and related coral reef bleaching. The "black water" event off of southwest Florida in 2002 that was investigated by the University of South Florida was one such event.[212] The increasing potential to use such data to detect the threat of HABs was described in *Bioscience* in 2003 and *Remote Sensing of Environment* in 2005, the latter paper describing the use of MODIS fluorescence bands to detect the signature of red tides.[213]

---

[208] NASA GSFC, "MODIS Science Team Meeting, October 31-November 2, 2006," Minutes, http://modis.gsfc.nasa.gov/sci_team/meetings/200610/minutes.pdf, (accessed 17 August 2009).
[209] NASA GSFC, "MODIS/VIIRS Joint Science Team Meeting, May 13-16, 2008," http://modis.gsfc.nasa.gov/sci_team/meetings/200805/minutes.pdf, (accessed 17 August 2009).
[210] Mitchell A. Roffer, "NASA Technology," *Yachting*, 196(1), 36, (July 2004).
[211] Gary Caputi, "Fish Eyes in the Sky," *Salt Water Sportsman*, 67(6), 99-107, June 2006.
[212] Noreen Parks, "Black Water Mystery Solved".
[213] Cheryl Lyn Dybas, "Harmful algal blooms: biosensors provide new ways of detecting and monitoring growing threat in coastal waters," *Bioscience* 53(10), 918-923, (October 2003); Chuanmin Hu, Frank E. Muller-Karger, C. Taylor, Kendall L. Carder, C. Kelble, E. Johns, and C. Heil, "Red tide

The 2003 article was on-target: only a few months later, in April 2004, a HAB in Paracas Bay, Peru induced a massive fish kill. Mati Kahru of Scripps described the use of MODIS data (even the 250-meter bands) to investigate this month-long event in the American Geophysical Union weekly *Eos* (Figure 6.16).[214] A potential cause was the dumping of organic matter from anchovy-processing factories located on the shore; the researchers examined patterns of turbidity using a simplified atmospheric correction scheme to observe the intensity of the bloom during the month, using images from both MODIS-Terra and MODIS-Aqua.

Though not considered hazardous, the buoyant aquatic plant *Sargassum* (made famous by the placid and oligotrophic Sargasso Sea in the tropical North Atlantic) can occasionally be pushed ashore by wind onto Gulf of Mexico beaches. Using MERIS and MODIS data, researchers led by Dr. James Gower of the University of British Columbia identified long "slicks" of *Sargassum* in the Gulf of Mexico, the first time this had been accomplished from space.[215]

Another algal species not considered hazardous suddenly became of global interest when thick bright-green algal mats invaded the waters just offshore of China a few weeks before the Olympics – the site of the Olympic sailing regatta. A massive cleanup effort put thousands of people on the beaches and in boats to clear the waters of this extensive bloom.[216] The bloom was observed easily in MODIS imagery.[217]

Florida also entered the remote-sensing news when a wastewater-filled phosphate mining pit at a defunct mining operation threatened to overflow due to heavy rains, which could have released acidic phosphate-laden water into Tampa Bay. Under emergency measures, millions of gallons of the wastewater were removed into a barge and dispersed in the Gulf of Mexico. MODIS data and SeaWiFS data were utilized to monitor the dispersal of the water and its movement in the Gulf.[218] The University of South Florida Institute for Marine Remote Sensing (IMaRS) created a Web site with imagery and interpretive reports, with some images showing very slightly elevated chlorophyll near the discharge site.[219] The

---

detection and tracing using MODIS fluorescence data: a regional example in SW Florida coastal waters." *Remote Sensing of Environment*, 97, 311-321, (2005).

[214] Mati Kahru and B. Greg Mitchell, "MODIS detects a devastating algal bloom in Paracas Bay, Peru," *Eos, Transactions American Geophysical Union*, 85(45), 465 and 472, (9 November 2004).

[215] "Envisat captures first image of Sargassum from space," http://www.spacemart.com/reports/Envisat_Captures_First_Image_Of_Sargassum_From_Space_999.html, (accessed 6 January 2010); Jim Gower, Chuanmin Hu, Gary Borstad, and Stephanie King, "Ocean color satellites show extensive lines of floating *Sargassum* in the Gulf of Mexico," *IEEE Transactions on Geoscience and Remote Sensing*, 44(12), (December 2006).

[216] Jim Yardley, "To Save Olympic sailing races, China fights algae," *New York Times*, 1 July 2008, http://www.nytimes.com/2008/07/01/world/asia/01algae.html, (accessed 6 January 2010).

[217] Michon Scott and Rebecca Lindsey, "Algal bloom along the coast of China," http://earthobservatory.nasa.gov/IOTD/view.php?id=8897, (accessed 6 January 2010).

[218] Florida Department of Environmental Protection, "Cleanup Progress at Piney Point," http://library01.gsfc.nasa.gov/goddardnews/September_1971.pdf, (accessed 6 January 2010).

[219] Chuanmin Hu and Frank Muller-Karger, "Satellite monitoring of the FDEP Gulf dispersal of the Piney Point treated wastewater, Report #12, 14 - 20 October 2003," http://imars.usf.edu/Piney_Point/reports/report12_USF_IMaRS.pdf, (accessed 6 January 2010).

ability to perceive these features was attributed to the extraordinary sensitivity of MODIS. The observed features could not be unambiguously attributed to the wastewater discharge, as they occurred near the Loop Current eddy front, which could also induce the mixing of deepwater nutrients to the surface.

The combined destructive forces of hurricanes Katrina and Rita, and the associated flooding from these two storms, created a public concern that polluted waters discharged into the Gulf of Mexico would flow toward the populated shoreline of Florida. Mitch Roffer and his ROFFS colleagues utilized ocean color data and SST data to observe the movement of the Katrina and Rita discharge in the aftermath of these two storms in September and October 2005. (Figure 6.15)[220]

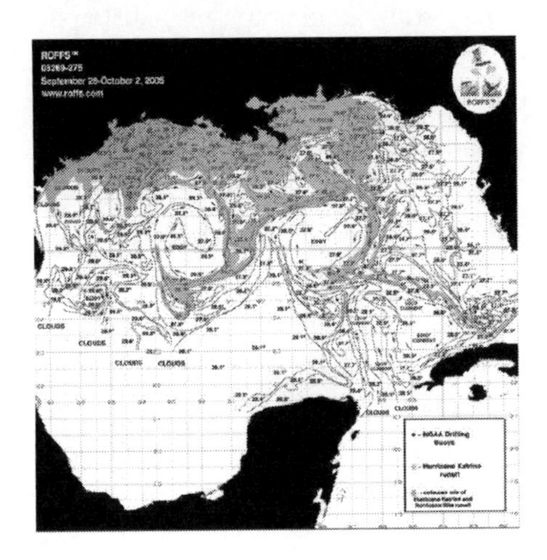

Figure 6.16. ROFFS map of flood waters discharged into the Gulf of Mexico by hurricanes Katrina (green) and Rita (pink), created from SST and ocean color data.

The effects of global warming could be both direct and indirect. In 2006, the activities of Scarla Weeks, who had moved to Australia after researching fisheries and sulphide gas eruptions off the coast of South Africa, described how she utilized MODIS data to rapidly investigate coral reef bleaching events on the Great Barrier Reef. This research was publicized by NASA and was also publicized in the industrial electronic newsletter *Satellite News*.[221]

---

[220] Matt Reed, "Bands of runoff from Katrina spread to Florida," 19 October 2005, http://www.usatoday.com/news/nation/2005-10-18-katrina-runoff_x.htm, (accessed 6 January 2010); Figure 14 image from http://www.roffs.com/katrina.htm, (accessed 6 January 2010).

[221] Goddard Space Flight Center, "NASA helps researchers diagnose recent coral bleaching at Australia's Great Barrier Reef," http://www.nasa.gov/centers/goddard/news/topstory/2006/coral_bleach.html, (5 April 2006, accessed 28 September 2009); "NASA helps researchers diagnose recent coral bleaching at Australia's Great Barrier Reef," *Satellite News*, 29 (14), (10 April 2006).

Summary: The complexity of the MODIS instruments, as well as problems in the fabrication of the MODIS-Terra instrument, vexed the ocean scientists tasked with creating science-quality data products from their observations. The size of EOS and its associated budgets, as well as the top-down management of EOS, also made it difficult to make the necessary programmatic and technical modifications that were ultimately required to make the MODIS-Oceans effort successful. These problems led to considerable evolution of the scientists involved with MODIS, as well as the oceanographic data it produced. For these reasons, the impact of MODIS oceanographic data on oceanography has been a slow process, but as the reliability of the data was realized, as well as its unprecedented global resolution, it has become a vital oceanographic remote sensing instrument. Because MODIS-Aqua is now the sole NASA ocean color instrument in orbit, all efforts will be focused on continuing to produce science-quality data. The lessons learned in the MODIS experience ultimately condense down to the fact that doing science that pushes the envelope with an instrument that does the same requires a large investment of time and money, as well as dedication and persistence. That MODIS has achieved as much as it has now is due to all of those factors.

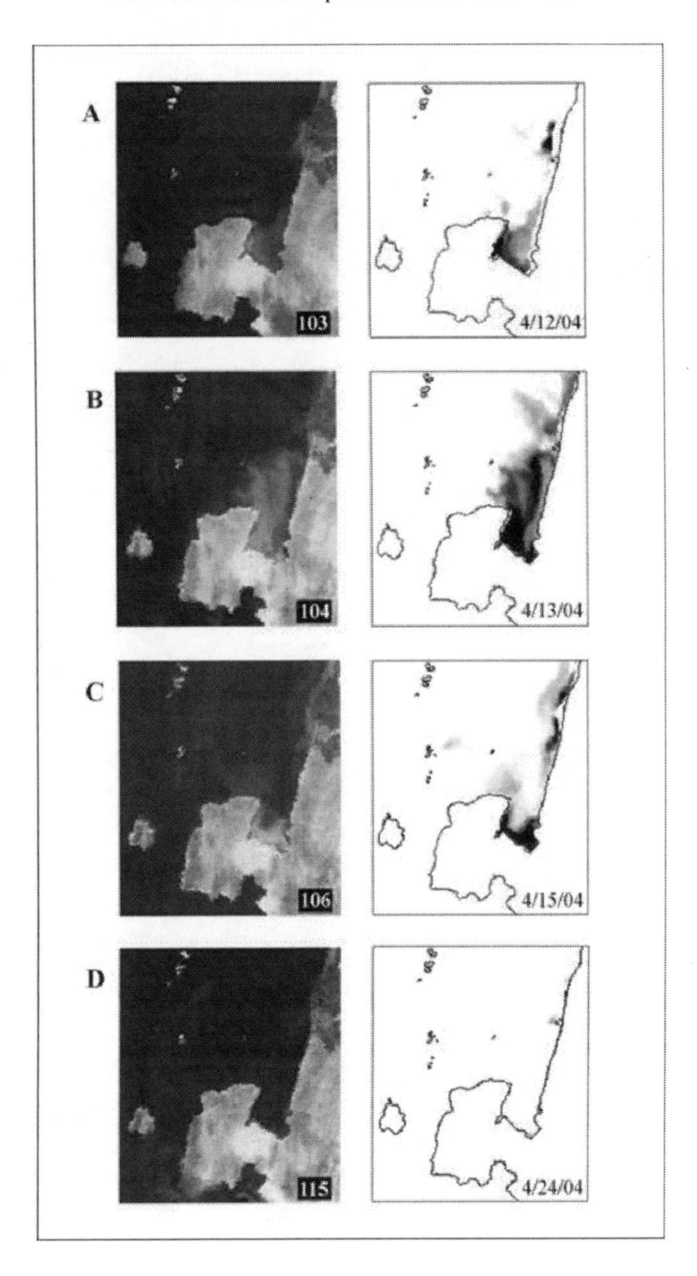

Figure 6.15. MODIS images of the harmful algal bloom in Paracas Bay, Peru, utilizing the 250-meter resolution bands of the instrument.

7

INTERNATIONAL AND INTERAGENCY COLLABORATIONS

Historical perspective

From the beginning of NASA's ocean color remote sensing efforts, the missions included a significant contribution from the international community. The original CZCS NET had two international representatives: Frank Anderson of the National Research Institute for Oceanology in South Africa, and Boris Sturm of the Joint Research Centre, located in Ispra, Italy.[1] As described in Chapter 4, the CZCS NET convened in several international locations, and one CZCS NET meeting was hosted by André Morel in Villefranche-sur-Mer, France.[2] (Gordon and Morel also collaborated on a early treatment of the field, *Remote Assessment of Ocean Color for Interpretation of Satellite Visible Imagery: A Review.*[3] Due in part to the participation of Anderson and Sturm, the Benguela upwelling region on the western coast of South Africa and the Mediterranean Sea received fairly extensive coverage on the limited CZCS observational budget. CZCS NET member Sayed Z. Al-Sayed participated in several cruises in international waters, some off the coast of Egypt.[4]

The necessity of international collaboration became more obvious in the post-CZCS era, as NASA sought a partner for a CZCS follow-on mission. Though the efforts did not result in a joint mission, NASA discussed placing the Ocean Color Imager, a CZCS follow-on instrument, on the French SPOT-3 satellite.[5]

Marlon Lewis of Canada's Dalhousie University was at NASA HQ heading the Ocean Biology program during the final stages of the EOSAT negotiations, and the subsequent events which led to the SeaWiFS mission with Orbimage.[6] Lewis, Trevor Platt, and Shubha Sathyendranath would play key roles in the 1990s in the expansion of international collaboration for ocean color research.

---

[1] James G. Acker, "The Heritage of SeaWiFS: A Retrospective on the CZCS NIMBUS Experiment Team (NET) Program".
[2] James G. Acker, "The Heritage of SeaWiFS: A Retrospective on the CZCS NIMBUS Experiment Team (NET) Program".
[3] Howard R. Gordon and André Y. Morel, *Remote Assessment of Ocean Color for Interpretation of Satellite Visible Imagery: A Review*, Springer-Verlag, New York, 114 pp, (1983).
[4] NASA GES DISC, "CZCS Spatial Coverage", http://disc.sci.gsfc.nasa.gov/guides/GSFC/guide/CZCS_coverage.gd.shtml, (accessed 21 August 2009).
[5] NASA/CNES, "Ocean Color Instrument on SPOT-3: Phase A Study Report, Technical Summary."
[6] Marla Cranston, "Understanding ocean colour – satellite gives scientists a view of ocean changes."

SeaWiFS and JGOFS Set the Stage

The level of international interest and potential collaboration widened rapidly with the formation of the SeaWiFS Science Team. The original SeaWiFS Science Team had 18 members from foreign countries, representing 13 different countries.[7] Due to the activities of the Japanese scientists (Kishino, Fukushima, and Matsumura) with ADEOS and OCTS, Greg Mitchell, who succeeded Lewis at NASA HQ, interacted extensively with the Japanese members of the team. There was particular interest in data sharing between the optical buoy that Kishino was going to deploy off Japan for OCTS data validation and MOBY. [8]

At the same time that the SeaWiFS science team was conducting their international activities, the international JGOFS program was also moving into a higher gear.[9] JGOFS anticipated having data from ocean color missions, and utilized data from other orbiting sensors (notably AVHRR) prior to the launch of OCTS and SeaWiFS.[10] So during the early 1990s there was both a programmatic commitment from NASA for international collaboration, as well as ongoing collaborations between oceanographers in the United States and the international oceanographic community on the extensive JGOFS program that intended to better characterize all aspects of the ocean carbon cycle.

Other international collaborative efforts included the licensing of numerous direct broadcast download stations under the SeaWiFS contract, which would eventually provide partial global coverage at 1km spatial resolution during the mission. The SeaWiFS Project and NASA also worked with German and Russian researchers in cross-calibration activities with the Modular Optoelectronic Scanners (MOS) on the Indian IRS-P3 satellite and the Priroda science module on the Mir space station. There were additional collaborative activities with the European Space Agency on their plans for MERIS, the Medium Resolution Imaging Spectroradiometer that was slated to orbit on Envisat.[11]

International Collaborations
(Official and Unofficial)

Several events in the mid-1990s crystallized the need for international collaborations on

---

[7] Stanford B. Hooker, Wayne E. Esaias, and Lisa A. Rexrode, "Volume 8, Proceedings of the First SeaWiFS Science Team Meeting."

[8] Stanford B. Hooker, Wayne E. Esaias, and Lisa A. Rexrode, "Volume 8, Proceedings of the First SeaWiFS Science Team Meeting," Section 5.19.

[9] Margaret C. Bowles and Hugh D. Livingston, "U.S. Joint Global Ocean Flux Study (JGOFS)."

[10] Wei Shi, John M. Morrison, Emanuele Bohm, and V. Manghnani, "Remotely Sensed Features of the Ocean Circulation in the northern Arabian Sea during the JGOFS Arabian Sea Study", *Deep-Sea Research II*, **46**(8-9), 1551-1575 (1999).

[11] Frank Muller-Karger, email received 18 March 2010.

ocean color research, and created organizational frameworks which enabled such collaborations. The first event was the decision (described in Chapter 6) by Associate Administrator Charles Kennel to cancel the EOS Color mission, which had been planning to fly a SeaWiFS-class sensor in conjunction with the Landsat 7 mission. Because SeaWiFS was extensively delayed, Kennel decided not to fly EOS Color in favor of an international program that would coordinate observations from U.S. and international sensors, and also coordinate the important task of gathering and analyzing calibration and validation "sea truth" data. The first organizational meeting for this international program, convened to draft the project plan, took place in February 1995.[12]

The special meeting of the (take a deep breath before you say this out loud) "Infrared and Visible Optical Sensors (IVOS) sub-group of the Committee on Earth Observations Satellites (CEOS) Working Group on Calibration and Validation (WGCV)" took place in Lanham, Maryland (just down Greenbelt Road from GSFC).[13] The purpose of this meeting was to accomplish what was summarized above: to have many of the international agencies planning to launch ocean color-capable missions share information about those missions, and also set up a framework for the vital calibration and validation activities.

The leader of the meeting was Ian Barton of Australia's CSIRO, who was the chairman of the IVOS group. Howard Gordon and Chuck McClain briefed the group on atmospheric correction and the SeaWiFS Project status, respectively, and Wayne Esaias then delineated Kennel's decision to forego EOS Color in favor of an "international infrastructure" to coordinate at-sea activities. This "international infrastructure" would become the SIMBIOS program. Following Esaias, Masanobu Shimada described the status of the ADEOS mission (launch was expected in about a year). Giuseppe Zibordi from the JRC described European plans to create calibration/validation sites for ocean color missions; one buoy would be deployed in the northern Adriatic Sea in the vicinity of Venice, and another, the *PlyMBody* (Plymouth Marine Optical Buoy), would be anchored south of the United Kingdom. This deployment was part of a larger program to be funded by the European Commission, with plans to have two other sites, one in the Nordic Sea and another in the North Sea.[14]

Robert Frouin of Scripps, who was on rotation as the Ocean Biology project manager at NASA HQ, then described a proposal soon to be presented to CEOS, for the formation of a CEOS standing subgroup for ocean color issues.[15] Following his presentation, discussions of these potential programs converged on agreement that the

[12] Jim Mueller, Chuck McClain, Bob Caffrey, and Gene Feldman, "The NASA SIMBIOS Program", http://oceancolor.gsfc.nasa.gov/DOCS/SIMBIOS/backscat1.html, (accessed 20 August 2009), originally published in *Backscatter*, 29-32, May 1998. Also found at http://www.ioccg.org/reports/simbios/simbios.html, (accessed 20 August 2009).
[13] Ian Barton, "Committee on Earth Observations Satellites (CEOS) Ocean Color Meeting," http://eospso.gsfc.nasa.gov/eos_observ/7_8_95/p50.html, (accessed 20 August 2009). Originally published in *The Earth Observer*, July/August 1995.
[14] Ian Barton, "Committee on Earth Observations Satellites (CEOS) Ocean Color Meeting."
[15] Ian Barton, "Committee on Earth Observations Satellites (CEOS) Ocean Color Meeting."

calibration/validation program, which would involve techniques and procedures, did not come under the aegis of CEOS. So there was, essentially, a tacit endorsement of the "international infrastructure" for calibration and validation Esaias had described, and also a call for an international ocean color group (which could be useful for data management and data sharing issues).

The International Ocean Colour Coordinating Group

Frouin subsequently proceeded with the effort to create an ocean color subgroup under CEOS, assisted by Lisa Shaffer of Scripps and NASA HQ.[16] CEOS officially established the group, named the "International Ocean Colour Coordinating Group" (IOCCG) in March 1996. The first meeting of the IOCCG Committee took place in Toulouse, France in 1996, with Trevor Platt of the Bedford Institute of Oceanography acting as interim Chairman. This meeting was definitely the first, as the main goal of the meeting was to create the group and its charter! The membership, meeting plan, and Project Office plan were established at this meeting. At the end of the meeting, interim Chairman Platt was elected IOCCG Chairman.[17]

The IOCCG set a goal of biannual meetings, a schedule that was generally followed after the first meeting. The committee solicited support from major national space agencies. The IOCCG Project Office was set up at the Bedford Institute of Oceanography in Dartmouth, Nova Scotia (Canada). The second IOCCG meeting in March 1997 in Tokyo, Japan, received briefings on the following instruments: [18]

i. MOS (Modular Optoelectronic Scanner): on the IRS-P3 (India) satellite and the PRIRODA science module on the Russian Mir space station)
ii. POLDER (Polarization and Directionality of the Earth's Reflectances): Midori and Midori-II;
iii. OCTS (Ocean Color and Temperature Scanner): Midori
iv. SeaWiFS (Sea-viewing Wide Field-of-view Sensor): Orbview-2
v. MODIS (Moderate Resolution Imaging Spectroradiometer): EOS Terra and Aqua
vi. MERIS (Medium Resolution Imaging Spectrometer): Envisat
vii. OSMI (Ocean Scanning Multispectral Imager): KOMPSAT (Korea Multi-Purpose Satellite)
viii. GLI (Global Imager): Midori-II
ix. OCI (Ocean Color Imager): ROCSAT-1
x. OCM (Ocean Color Monitor): IRS-P4 (India)

[16] Robert Frouin, email message, 20 October 2008; Robert Frouin, email message, 5 January 2009.
[17] International Oceanographic Commission, "First Session of the International Ocean Color Co-ordination Group (IOCCG)", http://ioc.unesco.org/iocweb/iocpub/iocpdf/i1039.pdf, (accessed 20 August 2009); IOCCG, "First IOCCG Committee Meeting", http://www.ioccg.org/reports/ioccg_meeting1.html (accessed 20 August 2009).
[18] IOCCG, "Second IOCCG Committee Meeting", http://www.ioccg.org/reports/ioccg_meeting2.html (accessed 20 August 2009).

In fact, it was the existence of these ten instruments (with copies, there were actually thirteen) that had motivated NASA to undertake the SIMBIOS program, to coordinate the analysis and calibration/validation activities for this impressive cohort of ocean color instruments.

Since its inception, the IOCCG has published several reports for its community. Report 7, published in 2008, was entitled 'Why Ocean Colour? The Societal Benefits of Ocean- Colour Technology".[19] The IOCCG also commissioned the publication of several articles in *Backscatter* magazine, and created a program of training workshops with international participation, especially in developing countries, a series which began in Olmue, Chile, in 1997.[20] The IOCCG has worked diligently on data sharing issues and advocacy for future missions. In January 2006, James Yoder succeeded Trevor Platt as IOCCG Chairman.[21]

SIMBIOS

The initial organization and activities of the IOCCG paralleled the inception of the SIMBIOS program. Following the organizational meeting in February 1995, NASA sent out an NRA (NASA Research Announcement) in 1996 to gather proposals for SIMBIOS participation.[22] In 1997, the SIMBIOS Project Office was initiated at GSFC, in the same suite of offices in GSFC Building 28 that housed the SeaWiFS Project staff. [23]

CZCS veteran and ocean optics protocols expert Jim Mueller was the interim project manager and the organizer of the first SIMBIOS Science Team meeting, which was held in Solomons, MD in August 1997 – which by happy coincidence ended up taking place just a few days after the successful SeaWiFS launch.[24] SIMBIOS was initially set up to last five years, from 1997 to 2001. The initial SIMBIOS NRA resulted in the selection of 35 investigators. When Mary Cleave moved to NASA Headquarters, Chuck McClain initially

---

[19] Trevor Platt, Nicolas Hoeppfner, Venetia Stuart, and Christopher Brown, editors, *Why Ocean Colour? The Societal Benefits of Ocean-Colour Technology*, Reports of the International Ocean-Colour Coordinating Group, Number 7, 141 pages, (2008).
[20] IOCCG, "Reports of Past IOCCG Training Courses and Workshops", http://www.ioccg.org/training/past.html (accessed 20 August 2009).
[21] IOCCG, "Eleventh IOCCG Committee Meeting", http://www.ioccg.org/reports/ioccg_meeting11.html (accessed 20 August 2009).
[22] Jim Mueller, Chuck McClain, Bob Caffrey, and Gene Feldman, "The NASA SIMBIOS Program."
[23] Jim Mueller, Chuck McClain, Bob Caffrey, and Gene Feldman, "The NASA SIMBIOS Program."
[24] Giulietta Fargion, Bryan Franz, Ewa Kwiatkowska, Christophe Pietras, Sean Bailey, Joel Gales, Gerhard Meister, Kirk Knobelspiesse, Jeremy Werdell, and Charles McClain, "SIMBIOS Program in Support of Ocean Color Missions: 1997-2003," in Robert J. Frouin, Gary D. Gilbert, and Pan Delu, editors, *Ocean Remote Sensing and Imaging II, Proceedings of the SPIE*, 5155, 49-60, (2003).

took over both the SeaWiFS Project Management and the SIMBIOS management.[25] In late 1998, Giulietta Fargion was selected to be the Assistant Project Manager for SIMBIOS and took over most day-to-day management activities. In 2000, she became Project Manager and was thereafter responsible for directing SIMBIOS GSFC staff in the SIMBIOS Program and initiating international collaborations.[26] McClain notes that he meanwhile assumed responsibility for a NASA-wide interdisciplinary carbon research program, which also included NOAA and the National Science Foundation as partners. Mary Cleave had asked GSFC to lead this study, and McClain responded to Vince Salomonson's request for someone to lead the GSFC effort. This carbon research program held three large workshops and several instrument and mission design studies over the course of 6 months.[27]

The 1998 SIMBIOS science team meeting covered aspects of both the in-water bio-optical measurements and atmospheric measurements. One of the topics discussed by the atmospheric measurements group was participation in major atmospheric characterization experiments involving aircraft observations, such as the INDOEX campaign, which was upcoming in spring 1999. SIMBIOS had already accomplished a lot in its first year, including the creation of a shared instrument pool (whereby SIMBIOS researchers could "rent" instruments for field studies that they could not otherwise procure themselves), and the establishment of a satellite overflight prediction service, to tell researchers the exact times that ocean color sensors would be observing their surface location.[28]

During its second year, SIMBIOS was responsible for the fabrication of a second SeaWiFS Transfer Radiometer; creation of software to process data from both MOS and OCTS to Level 3; and the initation of a data bank holding *in situ* calibration and validation data, which received the appellation SeaBASS (SeaWiFS Bio-optical Archive and Storage System).[29] By April 2003, SeaBASS data obtained by researchers from 14 different nations. SeaBASS benefited from the SIMBIOS-funded research programs, as well as participation in other major research campaigns, such as aforementioned INDOEX. The online data archive truly bulged with valuable biological oceanographic data obtained at sea; the April 2003 estimates were that SeaBASS held more than 300,000 measurements of phytoplankton pigments, and over 13,000 valuable surface-to-depth optical property profiles – which was only some of the data that the archive contained. The remarkable success of the SIMBIOS Program was the primary reason for this compendium of oceanographic data, because nearly

---

[25]Giulietta Fargion, Bryan Franz, Ewa Kwiatkowska, Christophe Pietras, Sean Bailey, Joel Gales, Gerhard Meister, Kirk Knobelspiesse, Jeremy Werdell, and Charles McClain, "SIMBIOS Program in Support of Ocean Color Missions: 1997-2003."

[26] Giulietta Fargion and Charles R. McClain, "SIMBIOS Project 2000 Annual Report", *NASA TM-2001-209976*, pages 3-4, (2001).

[27] Charles McClain, email message received 29 December 2009.

[28] NASA SIMBIOS, "Second SIMBIOS Science Team Meeting", http://oceancolor.gsfc.nasa.gov/DOCS/SIMBIOS/science_team98.html (accessed 20 August 2009).

[29] Charles R. McClain and Giulietta S. Fargion, editors, *SIMBIOS Project 1998 Annual Report, NASA Technical Memorandum NASA/TM-1999-208465*, NASA Goddard Space Flight Center, 105 pages, (March 1999); P. Jeremy Werdell, Sean Bailey, Giulietta Fargion, Christophe Pietras, Kirk Knobelspiesse, Gene Feldman, and Charles McClain, "Unique data repository facilitates ocean color satellite validation," *Eos, Transactions of the AGU*, 84 (38), 379, (2003).

90% of the data came from SIMBIOS researchers.[30]

SIMBIOS also initiated analyses of MOS and POLDER and OCTS data, the latter effort involving extensive collaboration with NASDA.[31] Considerable time and care was spent figuring out the complexities of the OCTS instrument and matching up *in situ* data with OCTS observations. Because considerable OCTS data had been acquired at Wallops for the adjacent Atlantic, observations of the Sargasso Sea region were used as a test site for OCTS data validation. SIMBIOS staff also worked on CIMEL sun photometer calibration.[32]

Multiple SIMBIOS-funded projects were initiated, some of them involving very clever exploitation of opportunities to gather data on the ocean waters. Barney Balch started a program that collected measurements from the M/S *Scotia Prince* ferry, which ran from Portland, Maine to Yarmouth, Nova Scotia through 2004.[33] The ferry research cruises were definitely "all-nighters", in the parlance of college students cramming for finals; all of the instruments were set up on the evening cruise from Portland to Yarmouth, and data was collected on the return cruise in the morning from Yarmouth to Portland.[34]

Figure 7.1. M/V *Scotia Prince.*

[30] P. Jeremy Werdell, Sean Bailey, Giulietta Fargion, Christophe Pietras, Kirk Knobelspiesse, Gene Feldman, and Charles McClain, "Unique data repository facilitates ocean color satellite validation."
[31]Menghua Wang and Bryan A.Franz, "Comparing the ocean color measurements between MOS and SeaWiFS: a vicarious intercalibration approach for MOS," *IEEE Transactions on Geoscience and Remote Sensing*, 38(1), 184-197, (2000); Menghua Wang, Alice Isaacman, Bryan Franz, and Charles R. McClain, "Ocean color optical property data derived from the Japanese Ocean Color and Temperature Scanner and the French Polarization and Directionality of the Earth's Reflectance: A comparison study," *Applied Optics*, 41, 974-990, (2002).
[32] Menghua Wang, Bryan Franz, Alice Isaacman, Christophe Pietras, Brian Schieber, Paul Smith, Tom Riley, Gene Feldman, and Sean Bailey,"SIMBIOS Project Activities, 1997-1998," Chapter 4 in Charles R. McClain and Giulietta S. Fargion, editors, *SIMBIOS Project 1998 Annual Report*, *NASA Technical Memorandum NASA/TM-1999-208465*, pp. 11-31.
[33] Ian Boyle, "Stena Olympica – Scotia Prince", http://www.simplonpc.co.uk/StenaOlympica1972.html, (accessed January 8, 2010).
[34] William Balch, "Validation of surface bio-optical properties in the Gulf of Maine as a means for improving satellite primary production estimates," Chapter 5 in Charles R. McClain and Giulietta S. Fargion, editors, *SIMBIOS Project 1998 Annual Report*, *NASA Technical Memorandum NASA/TM-1999-208465*, pp. 32-38.

Even though only 25% of the time the Gulf of Maine is free of clouds, the daily ferry runs provided numerous chances to match SeaWiFS observations with *in situ* data. Thus, during September-October 1998, the period with the characteristically clear autumn days of New England, eleven "cruises" took place on the *Scotia Prince*, and it was only cloudy on two of the eleven trips. Sampling from the ship provided nearly continuous water sampling and many optical measurements. Balch and his crew also participated in other cruises in the Gulf of Maine and on Georges Bank.[35]

Doug Capone, Ajit Subramaniam, and Ed Carpenter focused on the nitrogen-fixing phytoplankter *Trichodesmium*, and in 1998 they searched for *Tricho* in the southwestern Pacific, cruising from New Zealand to Fiji. Despite having to dodge Cyclone Zumann, and despite missing the prime bloom season for *Tricho*, they acquired data at 51 stations. The focus of their SIMBIOS plan was "undersampled" regions, so this data certainly augmented the SeaBASS archive. This particular SIMBIOS program received a boost when Norman Kuring of the SeaWiFS Project spotted a bloom in SeaWiFS imagery exhibiting the characteristic *Trichodesmium* morphology in the Capricorn Channel, which is near the southern end of Australia's Great Barrier Reef (Figure 7.2). This location was already known to be a prime spot for *Tricho*, so the team utilized this perceptive observation to develop their algorithm.[36]

Other SIMBIOS work involved Francisco Chavez of the Monterey Bay Aquarium Research Institute, looking for marine boundaries off of Hawaii; the late reknowned Arctic oceanographer Glenn Cota of Old Dominion University in Arctic waters; David Eslinger of the University of Alaska-Fairbanks doing more Arctic measurements in the Bering Sea; and the **Simbad** radiometer program on research cruises and merchant ships. (Figure 7.3)[37] The Simbad activity was a collaboration between Robert Frouin and David Cutchin of Scripps and Pierre-Yves Deschamps of the Laboratoire d'Optique Atmospherique, Universite des Sciences et Technologies de Lille, France. Simbad is a portable, hand-held, easy-to-operate radiometer that enabled measurements of both water-leaving radiances and atmospheric optical thickness. Deployment of 10 Simbads provided a large number of optical data points; Simbad was used on 21 cruises between September 1997 and December 1999.[38]

---

[35] William Balch, "Validation of surface bio-optical properties in the Gulf of Maine as a means for improving satellite primary production estimates."

[36] Ajit Subramaniam, Raleigh R. Hood, Christopher W. Brown, Edward J. Carpenter, and Douglas G. Capone, "Validation of ocean color data products in under sampled marine areas," Chapter 6 in Giulietta S. Fargion and Charles R. McClain, editors, *SIMBIOS Project 2000 Annual Report, NASA/TM-2001-209976*, pp. 42-50.

[37] Image source: Geochemistry, Phytoplankton, and Color of the Ocean, http://www.locean-ipsl.upmc.fr/gepco/gepco_6.html, (accessed 16 March 2010).

[38] Francisco P. Chavez, "Bio-optical measurements at ocean boundaries in support of SIMBIOS", Chapter 9 in Charles R. McClain and Giulietta S. Fargion, editors, *SIMBIOS Project 1998 Annual Report, NASA Technical Memorandum NASA/TM-1999-208465*, pp. 44-45; Glenn F. Cota, "Remote Sensing of ocean color in the Arctic: algorithm development and comparative validation,",Chapter 10 in Charles R. McClain and Giulietta S. Fargion, editors, *SIMBIOS Project 1998 Annual Report, NASA Technical Memorandum NASA/TM-1999-208465,* pp. 47-49; Dave L. Eslinger, "The high-latitude intercomparison and validation experiment (HIVE)," Chapter 12 in Charles R. McClain and Giulietta

Figure 7.2: SeaWiFS image acquired July 16, 1998 from the Townsville , Australia HRPT station, showing a phytoplankton bloom suspected to be *Trichodesmium* in the Capricorn Channel. The bloom (indicated by the arrow) is south of the blue-green speckles of the Great Barrier Reef. Image provided by Norman Kuring.

In addition to the Simbad radiometers, SIMBIOS researchers collaborated with the Aerosol Robotic Network (AERONET). Chuck McClain said "SIMBIOS provided sunphotometers for 12 coastal and islands sites for the AERONET and worked with [Brent] Holben's group to install them. Christophe Pietras was the SIMBIOS lead and also was responsible for their calibrations on the [Building] 33 roof. When SIMBIOS ended, we turned that equipment, plus all the hand-held Microtops radiometers, over to the AERONET group. The Microtops data collection into the AERONET is being continued by Alex Smirnov."[39]

S. Fargion, editors, *SIMBIOS Project 1998 Annual Report, NASA Technical Memorandum NASA/TM-1999-208465* pp. 52-54; Robert Frouin, David L. Cutchin, and Pierre-Yves Deschamps, "Satellite ocean color validation using merchant ships," Chapter 13 in Charles R. McClain and Giulietta S. Fargion, editors, *SIMBIOS Project 1998 Annual Report, NASA Technical Memorandum NASA/TM-1999-208465*. Chapters with the same titles (not always the same chapter number) also appear in Charles R. McClain and Giulietta S. Fargion, editors, *SIMBIOS Project 1999 Annual Report, NASA/TM-1999-209486*, NASA Goddard Space Flight Center, 128 pages, (December 1999), and Giulietta S. Fargion and Charles R. McClain, editors, *SIMBIOS Project 2000 Annual Report, NASA/TM-2001-209976*. Each chapter provides an annual report from the funded SIMBIOS projects.

[39]Charles McClain, email message received 5 December 2009; Alex Smirnov, Brent N. Holben, I. Slutsker, D. M. Giles, Charles R. McClain, T. F. Eck, S. M. Sakerin, A. Macke, P. Croot, Giuseppe, Zibordi, P. K. Quinn, J. Sciare, S. Kinne, M. Harvey, T. J. Smyth, S. Piketh, T. Zielinski, A. Proshutinsky, J. I. Goes, Norman B. Nelson, Pierre Larouche, V. F. Radionov, P. Goloub, K. Krishna Moorthy, R. Matarrese, E. J. Robertson, and F. Jourdin, "Maritime Aerosol Network as a component of Aerosol Robotic Network," *Journal of Geophysical Research*, 114, D06204, doi:10.1029/2008JD011257, (2009).

Figure 7.3: Researcher using the Simbad radiometer.

The connections between North America and Latin America were evident in two SIMBIOS projects. Ron Zaneveld, Scott Pegau, and Andrew Barnard from Oregon State included their SIMBIOS activities in a larger U.S. and Mexican effort to fully characterize the Gulf of California. Mueller and Trees of CHORS, S. Alvarez Borrego, R. Lara-Lara, G. Gaxiola and H. Maske from the Centro de Investigacion Cientifica y de Educacion Superor de Ensenada (CICESE), Ensenada, Mexico, and Dr. E. Valdez from the University of Sonora were all participants in this 6-year project.[40]

Frank Muller-Karger, working with Ramon Varela (Fundacion La Salle de Ciencias Naturales, Venezuela) acquired data in conjunction with the CARIACO project in the Caribbean Sea off of Venezuela. The Cariaco Basin is an intriguing place to collect data because the waters below 200 meters in the basin are anoxic, so that organic matter and phytoplankton are well-preserved in the basin sediments. The Cariaco Basin thus allows correlation of remote-sensing observations with the data from a naturally-occurring sediment trap, allowing insight into the relationships between terrigenous and oceanic fluxes,

[40] J. Ronald Zaneveld and Andrew H. Barnard, "SIMBIOS data product and algorithm validation with emphasis on the biogeochemical and inherent optical properties," Chapter 21 in Charles R. McClain and Giulietta S. Fargion, editors, *SIMBIOS Project 1998 Annual Report, NASA Technical Memorandum NASA/TM-1999-208465,* pp. 77-79. Annual reports from this group also appear in Charles R. McClain and Giulietta S. Fargion, editors, *SIMBIOS Project 1999 Annual Report, NASA/TM-1999-209486* and Charles R. McClain and Giulietta S. Fargion, editors, *SIMBIOS Project 2000 Annual Report, NASA/TM-2001-209976.*

primary productivity, and tropical climate.[41]

Though Bermuda finished second in the running for the MOBY deployment site, the long-running JGOFS time-series station also allowed the SIMBIOS participation of the jazzy BBOP (Bermuda Bio-Optics Program), headed by Dave Siegel. Siegel worked in conjunction with Norm Nelson, who was at the Bermuda Biological Station for Research at that time. The regular BATS sample collection cruises allowed up to 16 cruise opportunities a year for bio-optical data collection in the oligotrophic Sargasso Sea.[42] A similar campaign was conducted by John Porter and Ricardo Letelier at the HOTS site (Hawaii).[43]

Respected Russian bio-optical scientist Oleg Kopelevich participated in SIMBIOS, with his work focused on validation of the water-leaving radiance products, and conducted cruises in the Black and Aegean seas in 1997, and the Barents Sea in 1998.[44]

[The participation of Kopelevich in SIMBIOS was an extension of a longer period of collaboration between NASA and the Shirshov Institute of Oceanology. Kopelevich and colleagues met with SeaWiFS Project Manager Bob Kirk at GSFC in 1991 and became involved with the development of bio-optical algorithms. Kopelevich's group had actually been working with ocean color data on the MOS instrument orbited on the INTERKOSMOS-20 satellite, which launched in 1979. Subsequently, in 1992 the Shirshov

---

[41] Frank Muller-Karger and Ramon Varela, "Validation of carbon flux and related products for SIMBIOS; the CARIACO continental margin time-series," Chapter 19 in Charles R. McClain and Giulietta S. Fargion, editors, *SIMBIOS Project 1998 Annual Report, NASA Technical Memorandum NASA/TM-1999-208465*, pp. 71-73. An annual report from this group also appears in Charles R. McClain and Giulietta S. Fargion, editors, *SIMBIOS Project 1999 Annual Report, NASA/TM-1999-209486* and Charles R. McClain and Giulietta S. Fargion, editors, *SIMBIOS Project 2000 Annual Report, NASA/TM-2001-209976* ; Robert C. Thunell, Ramon Varela, Martin Lllano, James Collister, Frank Muller-Karger, and Richard Bohrer, "Organic carbon fluxes, degradation, and accumulation in an anoxic basin: Sediment trap results from Cariaco Basin," *Limnology and Oceanography* , 45 (2), 300-308, (2000); "This Week in Science", *Science,* 290 (5498), 1853 (8 December 2000).

[42] David Siegel, "Bermuda Bio-optics program (BBOP)," Chapter 20 in Charles R. McClain and Giulietta S. Fargion, editors, *SIMBIOS Project 1998 Annual Report, NASA Technical Memorandum NASA/TM-1999-208465,* pp. 74-76. An annual report from this group also appears in Charles R. McClain and Giulietta S. Fargion, editors, *SIMBIOS Project 1999 Annual Report, NASA/TM-1999-209486* and Charles R. McClain and Giulietta S. Fargion, editors, *SIMBIOS Project 2000 Annual Report, NASA/TM-2001-209976.*

[43] John N. Porter and Ricardo Letelier, "Measurements of aerosol, ocean and sky properties at the HOT site in the central Pacific," Chapter 23 in Charles R. McClain and Giulietta S. Fargion, Charles R. McClain and Giulietta S. Fargion, editors, *SIMBIOS Project 1998 Annual Report, NASA Technical Memorandum NASA/TM-1999-208465,* pp. 77-79. Annual reports from this group also appear in Charles R. McClain and Giulietta S. Fargion, editors, *SIMBIOS Project 1999 Annual Report, NASA/TM-1999-209486* and Charles R. McClain and Giulietta S. Fargion, editors, *SIMBIOS Project 2000 Annual Report, NASA/TM-2001-209976.*

[44] Oleg V. Kopelevich, "Validation of the water-leaving radiance data product," Chapter 25 in Charles R. McClain and Giulietta S. Fargion, editors, *SIMBIOS Project 1998 Annual Report, NASA Technical Memorandum NASA/TM-1999-208465.* Annual reports from this group also appear in Charles R. McClain and Giulietta S. Fargion, editors, *SIMBIOS Project 1999 Annual Report, NASA/TM-1999-209486* and Charles R. McClain and Giulietta S. Fargion, editors, *SIMBIOS Project 2000 Annual Report, NASA/TM-2001-209976.* In the 2000 Annual Report, Kopelevich listed Vladimir I. Burenkov, Svetlana V. Ershova, Sergey V. Sheberstov, Marina A. Evdoshenko, Genrik S. Karabashev and Constantine A. Pavlov as collaborators.

Institute was visited by B. Greg Mitchell, who was at the time the Ocean Biology and Biogeochemistry Program Manager at NASA HQ. They developed a plan for Russian participation in at-sea calibration/validation activities for SeaWiFS and MOS. Also in 1992, Kopelevich (with V. Burenkov, A. Sudbin, A. Vasilkov, and V. Volynsky) responded to the SeaWiFS NRA with a proposal for algorithm development. Their proposal was accepted, making Kopelevich a member of the SeaWiFS science team. According to Kopelevich, Mitchell's successor at NASA HQ, Frank Muller-Karger, assisted them with the purchase of a SeaDAS-compatible computing system. By participating with SeaWiFS, their program received funding from the Russian Ministry for Sciences, Higher Education, and Technology, during a period that Kopelevich described thusly: "Russian science went through a difficult time then, and nobody could be sure in sufficient funding from our government." Kopelevich and his group subsequently worked on analysis of data from SeaWiFS and the MOS on the Mir space station (in the "Priroda" science module), as well as their SIMBIOS projects.][45]

At the three-year mark, the SIMBIOS program was "renewed" with the selection of a new science team. The new science team was not a complete upheaval; several of the originally selected projects were continued. The renewed team enlarged the international participation aspect, with 12 proposals selected from international organizations. One of the fundamental highlights of this period was an agreement to collaborate with NASDA on a reprocessing of the entire OCTS data set (OCTS was the ocean color and sea surface temperature instrument on the short-lived Midori satellite), to make it into the same format as SeaWiFS data, and thus compatible with SeaDAS processing. Dr. Tasuku Tanaka spent several months at the SeaWiFS Project assisting with this effort. This collaboration resulted in the *NASDA-SIMBIOS-OCTS* data set, which was initially archived at the Goddard DAAC next to the SeaWiFS data archive, and then transferred to the OBPG when the OBPG undertook the archive and distribution responsibilities for NASA ocean color data. [46] This collaborative data set was noted as a "major success" of SIMBIOS, according to Chuck McClain.[47]

The SIMBIOS Project staff working at GSFC continued to analyze the flow of MODIS data and compare it to other data sets, particularly SeaWiFS. They determined that MODIS seasonal data variability was much higher than for SeaWiFS, particularly at polar latitudes. This observation led to the determination that the polarization sensitivity correction had been applied incorrectly, making the effect of polarization worse rather than correcting it.[48]

---

[45] Oleg Kopelevich, "Brief biographic information related to my involvement with NASA's ocean color missions," unpublished manuscript received by email 1 August 2008.

[46] Giulietta Fargion, Bryan Franz, Ewa Kwiatkowska-Ainsworth, Christophe Pietras, Paul Smith, Sean Bailey, Joel Gales, Gerhard Meister, Kirk Knobelspiesse, Jeremy Werdell, Charles McClain and Gene Feldman, "SIMBIOS Project data processing and analysis results," Chapter 3 in Giulietta S. Fargion and Charles R. McClain, editors, *SIMBIOS Project 2001 Annual Report, NASA Technical Memorandum. 2002-210005*, NASA Goddard Space Flight Center, Greenbelt, Maryland, 184 pages, (February 2002).

[47] Charles McClain, email message received 5 December 2009.

[48] Bryan Franz, email message received 4 January 2010; Gerhard Meister, Ewa J. Kwiatkowska, Bryan A. Franz, Frederick S. Patt, Gene C. Feldman, and Charles R. McClain, "Moderate-Resolution

The SIMBIOS staff also published papers regarding intercomparison of ocean color sensor data.[49]

The Simbad ship-of-opportunity project participated in ten cruises in 2001 and 2002, and in 2003 assisted with POLDER-II and GLI calibration/validation on a cruise of the R/V *Ioffe* from Ushuaia (Argentina, Tierra del Fuego) to Kiel, Germany.[50] During this second SeaWiFS phase, Larry Harding and Andrea Magnuson of the University of Maryland's Horn Point Laboratory also performed characterization of the Chesapeake Bay; Harding's familiarity with the Chesapeake Bay extended back through many years of aerial remote-sensing observations, augmented considerably when SeaWiFS was launched.[51] Stan Hooker from NASA, the seemingly tireless organizer of NASA's calibration activities, worked with Giuseppe Zibordi and Jean-Francois Berthon of the JRC and André Morel and David Antoine to improve protocols for measuring apparent optical properties (AOPs); this work included research on the Aqua Alta tower off Venice, and the BENCAL cruise in the Benguela upwelling zone. The latter cruise included an effort to ship ocean water/phytoplankton pigment samples frozen in Dewar flasks filled with liquid nitrogen; the researchers discovered that despite their best efforts to determine how long the liquid nitrogen in the flasks would last, it managed to evaporate considerably faster than expected. The resulting analyses indicated that the amount of chlorophyll in the samples carried in defrosted Dewars was lower than those in the still-frozen samples; such an observation induced important considerations of how to store and analyze samples collected at-sea.[52]

Imaging Spectroradiometer ocean color polarization correction," *Applied Optics*, 44(26), (September 2005).

[49] Menghua Wang, and Bryan A. Franz, "Comparing the ocean color measurements between MOS and SeaWiFS: a vicarious intercalibration approach for MOS," *IEEE Transactions on Geoscience and Remote Sensing*, 38(1), 184-197, (2000); Menghua Wang, Alice Isaacman, Bryan Franz, and Charles R. McClain, "Ocean color optical property data derived from the Japanese Ocean Color and Temperature Scanner and the French Polarization and Directionality of the Earth's Reflectance: A comparison study," *Applied Optics*, 41, 974-990, (2002).

[50] Robert Frouin, David L. Cutchin, and Pierre-Yves Deschamps, "Satellite Ocean-Color Validation Using Ships of Opportunity," Chapter 6 in Giulietta S. Fargion and Charles R. McClain, editors, *SIMBIOS Project 2001 Annual Report, NASA Technical Memorandum. 2002-210005*, 40-49. An annual report from this group also appears in Giulietta S. Fargion and Charles R. McClain, editors, *SIMBIOS Project 2002 Annual Report, NASA/TM-2003-211622*, NASA Goddard Space Flight Center, 157 pages, (February 2003).

[51] Lawrence W. Harding Jr. and Andrea Magnuson, "Bio-Optical and Remote Sensing Observations in Chesapeake Bay," Chapter 8 in Giulietta S. Fargion and Charles R. McClain, editors, *SIMBIOS Project 2001 Annual Report, NASA Technical Memorandum. 2002-210005*, 52-59. An annual report from this group also appears in Giulietta S. Fargion and Charles R. McClain, editors, *SIMBIOS Project 2002 Annual Report, NASA/TM-2003-211622*; Lawrence W. Harding, Jr., Eric C. Itsweire and Wayne E. Esaias, "Determination of phytoplankton chl-a in the Chesapeake Bay with aircraft remote sensing," *Remote Sensing of the Environment*, 40, 79-100, (2002); Lawrence W. Harding Jr., Andrea Magnuson, Michael E. Mallonee, "SeaWiFS retrievals of chlorophyll in Chesapeake Bay and the mid-Atlantic bight," *Estuarine, Coastal and Shelf Science*, 62(1-2), (January 2005).

[52] Stanford B. Hooker, Giuseppe Zibordi,.Jean-François Berthon, André Morel and David Antoine, "Refinement of Protocols for Measuring the Apparent Optical Properties of Seawater," Chapter 9 in Giulietta S. Fargion and Charles R. McClain, editors, *SIMBIOS Project 2001 Annual Report, NASA Technical Memorandum. 2002-210005,* 63-72. An annual report from this group also appears in Giulietta

Ultimately, despite its initial status as a "fallback" position due to launch delays and budget considerations, SIMBIOS became a vital program for the continuing improvement of NASA's ocean color mission data accuracy, and also fostered a remarkable level of international cooperation and collaboration. The heritage of SIMBIOS is still strongly evident, with such aspects as SeaBASS, the NASDA-SIMBIOS-OCTS data set, and the still-ongoing efforts to refine biological oceanographic sampling and ocean optical measurement protocols.[53]

The success of SIMBIOS may have partly contributed to its untimely conclusion. While SIMBIOS was thriving, the MODIS-Terra data calibration effort was foundering, and Ghassem Asrar at NASA HQ wanted more collaboration between the two groups. Because MODIS was not producing the desired environmental data records (EDRs), there was considerable pressure to accelerate the MODIS data release, and this contributed to the "demise" of SIMBIOS. [54] Paula Bontempi said that Asrar's top priority became "getting MODIS fixed", mainly due to external pressure stemming from how well it was known in the remote sensing community.[55] Fargion and McClain even produced a supplementary NASA Technical Memorandum in an attempt to justify the continuation of SIMBIOS.[56] Bontempi poetically characterized the end of SIMBIOS as resulting in the "burning of fifteen bridges with the international community", and that some of the international goodwill lost by the termination of the project was "irrecoverable".[57]

Despite its somewhat premature end, SIMBIOS stands as one of the more remarkable achievements of NASA's ocean color missions. It was a unique amalgamation of at-sea research, instrument development, and algorithm development activities, in addition to community-based cross-calibration, protocol development, and shared satellite data-merging techniques. These activities were greatly benefited by active engagement of project staff with the both the U.S. and international science teams. Furthermore, SIMBIOS set the standard for international cooperation in the data calibration and validation effort for ocean color satellites – a model for the near-future needs of the ocean color community

The Atlantic Meridional Transect Cruises

Ship time is always hard to acquire; ships are expensive to operate and there always

---

S. Fargion and Charles R. McClain, *SIMBIOS Project 2002 Annual Report, NASA/TM-2003-211622.*

[53] Giulietta Fargion, Bryan Franz, Ewa Kwiatkowska, Christophe Pietras, Sean Bailey, Joel Gales, Gerhard Meister, Kirk Knobelspiesse, Jeremy Werdell, and Charles McClain, "SIMBIOS Program in Support of Ocean Color Missions: 1997-2003."

[54] Paula S. Bontempi, comments recorded at the Ocean Color Collaborative Historical Workshop, January 13-14, 2009, St. Petersburg, FL.

[55] Paula Bontempi, interview recorded 27 October 2010.

[56] Giulietta S. Fargion and Charles R. McClain, *MODIS Validation, Data Merger and Other Activities Accomplished by the SIMBIOS Project, 2002-2003, NASA/TM-2003-212249,* NASA Goddard Space Flight Center, 75 pages, (September 2003).

[57] Paula S. Bontempi, comments recorded at the Ocean Color Collaborative Historical Workshop, January 13-14, 2009, St. Petersburg, FL.

seem to be many more oceanographers seeking to be on a ship than there are berths to hold them. So finding a regular berth on a long cruise through multiple oceanic bio-optical regimes is a very valuable opportunity. This opportunity became available to NASA's ocean color missions via participation in the Atlantic Meridional Transect (AMT) series of research cruises.

The AMT program is hosted by the Plymouth Marine Laboratory and the National Oceanography Centre in Southampton.[58] A south-to-north AMT cruise begins in the Southern Hemisphere; most of the cruises left port from the Falkland Islands, and a few of them left from Cape Town, South Africa. The cruise heads north in the southern Atlantic, heading toward England, and can cover over 13,000 km of ocean (Figure 7.4) [59] The north-to-south cruise plan reverses the course, originating in the UK and usually finishing in the Falklands. The various cruise plans, which began in 1995, took the ship (either the RRS *James Clark Ross* or the RRS *Discovery*) through upwelling zones, vast expanses of blue oligotrophic basins, and water temperatures ranging from the frigidity of the sub-polar eddies to the warmth of the balmy horse latitudes. [60] AMT cruises were usually about a month long, give or take a couple of days and depending on how many Pollywogs had to be initiated under the supervision of King Neptune during their first equatorial crossing.

Through the auspices of James Aiken, the AMT cruises, starting with AMT-1, included an ocean color calibration and validation component, which was done in collaboration with Stan Hooker. During each AMT, measurements of classic oceanographic variables (such as salinity, nutrient concentrations) and bio-optical variables (phytoplankton pigments, IOPs) were conducted on underway water samples. Depth profiles of optical properties were obtained with a UOR (*undulating* oceanographic recorder) from depths of 10-80 meters. And surface, above-, and below-water data was collected at station stops occurring as close to noon as possible, to coincide with the SeaWiFS overflight times.[61]

Hooker and Aiken collaborated on an IOCCG-commissioned report about the AMT cruises in 1997; the AMT participation resulted in several publications, and importantly contributed substantially to the SeaWiFS and OBPG data calibration and validation program.[62]

---

[58] "Atlantic Meriodonal Transect," http://www.amt-uk.org/, accessed 24 August 2009.

[59]Cruise track image from "Atlantic Meriodonal Transect," http://www.amt-uk.org/.

[60] "James Clark Ross," http://www.south-pole.com/p0000081.htm, (accessed 24 August 2009). Ross was the discoverer of the Ross Sea and Ross Ice Shelf. "Dreaming on Desolation Island," http://www.channel4.com/history/microsites/E/ends2/island3.html, (accessed 24 August 2009), also notes that Mount Ross on Kerguelen Island is named after him, and that Ross visited Kerguelen in 1840; British Antarctic Survey, "RRS James Clark Ross – Research Ship," http://www.antarctica.ac.uk//living_and_working/research_ships/rrs_james_clark_ross/index.php, (accessed 1 October 2009).

[61] James Aiken and Stanford B. Hooker, "The Atlantic Meriodonal Transect," http://www.ioccg.org/reports/amt1/amt.html, (accessed 24 August 2009).

[62] Stanford B. Hooker and James Aiken, "Calibration evaluation and radiometric testing of field radiometers with the SeaWiFS Quality Monitor (SQM)," *Journal of Atmospheric and Oceanic Technology* 15(4), 995-1007, (1998). AMT Contribution Number 4; Stanford B. Hooker and Stephane Maritorena, "An evaluation of oceanographic radiometers and deployment methodologies," *Journal of Atmospheric and Oceanic Technology* 17(6), 811-830, (2000), AMT Contribution Number 56; Stanford B. Hooker and

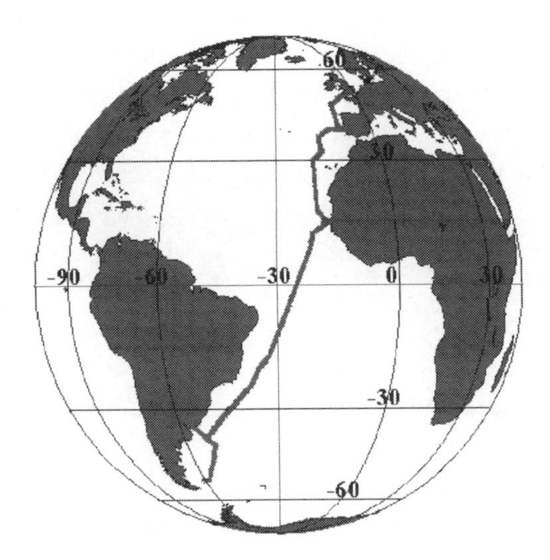

Figure 7.4. Cruise track for Atlantic Meridional Transect cruise 7, September 14 – October 25, 1998, which originated in the UK, included ports-of-call at Lisbon, Portugal, Madeira (Canary Islands); Dakar, Senegal; and Montevideo, Uruguay. The cruise terminated at Stanley, Falkland Islands.

Interagency Collaborations

No less important than the international collaborations entered into by the various partitions of NASA's ocean color activities are collaborations with other agencies (or Administrations or Departments or Surveys or Offices) of the U.S. government. Data from NASA's ocean color sensing missions has been utilized by several different government entities, for purposes ranging from fisheries management to harmful algal bloom predictions to evaluations of water clarity for military operations.

National Institute of Standards and Technology (NIST)

One of the most important interagency collaborations entered into by the NASA ocean color teams was with the National Institute of Standards and Technology (NIST), which back in the CZCS era was called the National Bureau of Standards (NBS). NIST entered into the ocean color arena for two primary reasons; one was to provide calibration support

Charles R. McClain, "The calibration and validation of SeaWiFS data," *Progress in Oceanography* 45(3-4), 427-46, (2000), AMT Contribution Number 22; Stanford B. Hooker, N. Rees, and James Aiken, "An objective methodology for identifying oceanic provinces," *Progress in Oceanography* 45(3-4), 313-338, (2000), AMT Contribution Number 21.

for the MOBYs in Hawaii, and the other was to provide calibration support for researchers on ships-at-sea, performing in-water or above-water radiometric measurements as well as crucial measurements of chlorophyll and pigment concentrations in seawater. At NIST, the main point of contact for the ocean color calibration activities was B. Carol Johnson. Johnson's participation in the SIRREX series spanned the second through the fifth SIRREX, and she was also the lead for the fabrication of the SeaWiFS transfer radiometer.[63] Johnson's group also provided NIST expertise to the calibration of Goddard integrating spheres.[64]

National Oceanic and Atmospheric Administration (NOAA)

Likely the main collaborator with NASA on ocean color data and applications has been the National Oceanic and Atmospheric Administration (NOAA). NOAA has been involved with NASA on the operational (near-real-time) side, the research side, and the applications side. The NASA/NOAA collaboration began with the CZCS; three members of the original team (Clark, Apel, and Baker) were affiliated with NOAA, and early in the 80's, Warren Hovis moved from NASA GSFC to the NOAA National Earth Satellite Service (NESS), which was later renamed to be the NOAA National Environmental Satellite Data and Information System (NESDIS).[65]

During the ocean color interregnum (1986-1996), NOAA continued to participate with NASA; Dennis Clark's work on developing MOBY and his subsequent membership on the MODIS science team (which included his partnership with the SeaWiFS Project) is

---

[63] James L. Mueller, B. Carol Johnson, Christopher L. Cromer, John W. Cooper, James T. McLean, Stanford B. Hooker, and Todd L. Westphal, "Volume 16, The Second SeaWiFS Intercalibration Round-Robin Experiment, SIRREXB, June 1993"; James L. Mueller, B. Carol Johnson, Christopher L. Cromer, Stanford B. Hooker, James T. McLean, and Stuart F. Biggar, "Volume 34, The Third SeaWiFS Intercalibration Round-Robin Experiment.(SIRREX-3), 19-30 September 1994"; B. Carol Johnson, Sally S. Bruce, Edward A. Early, Jeanne M. Houston, Thomas R. O'Brian, Ambler Thompson, Stanford B. Hooker, and James L. Mueller, "Volume 37, The Fourth SeaWiFS Intercalibration Round-Robin Experiment (SIRREX-4), May 1995"; B. Carol Johnson, Howard W. Yoon, Sally S. Bruce, Ping-Shine Shaw, Ambler Thompson, Stanford B. Hooker, Robert E. Eplee, Jr., Robert A. Barnes, Stephane Maritorena, and James L. Mueller, "The Fifth SeaWiFS Intercalibration Round-Robin Experiment (SIRREX-5), July 1996"; B. Carol Johnson, J.B. Fowler, and Christopher L. Cromer, "The SeaWiFS Transfer Radiometer (SXR)."

[64] Fumihiro Sakuma, B. Carol Johnson, Stuart F. Biggar, James J. Butler, J.W. Cooper, Masaru Hiramatsu, and Katsumi Suzuki,"EOS AM-1 preflight radiometric measurement comparison using the advanced spaceborne thermal emission and reflection radiometer (ASTER) visible/near-infrared integrating sphere," in William L. Barnes, editor, "Earth Observing System" *Proceedings SPIE*, 2820, 184-196, (1996); E. A. Early and B. Carol Johnson, "Calibration and characterization of the GSFC sphere," Chapter 1 in Eueng-nan Yeh, Robert A. Barnes, Michael Darzi, Lakshmi Kumar, Early, B. Carol Johnson, James L. Mueller, and Charles C. Trees, "Case Studies for SeaWiFS Calibration and Validation, Part 4," *NASA Technical Memorandum 104566, Volume 41*, NASA Goddard Space Flight Center, 3-17, (1997).

[65] Warren Hovis, "Results of the CZCS Validation Program," *Oceans 1982*, IEEE, Washington DC, September 20-22, 1982, 1273-1276, (1982). Abstract: http://ieeexplore.ieee.org/xpl/freeabs_all.jsp?arnumber=1151915, (accessed 24 August 2009).

perhaps the most scientifically-interactive aspect of this collaboration during this time period, though there were many other important levels of interaction. NOAA was a vital partner in the original SeaWiFS contract negotiations, and their participation enabled 13 real-time licenses for SeaWiFS data download over U.S. waters.[66] As noted earlier, NASA/NOAA collaboration unfortunately did not extend to the procurement of a successful berth for the Ocean Color Imager CZCS follow-on mission on a NOAA polar-orbiting satellite.

Chapter 5 noted that NOAA had contracted for the rights to receive direct data downlinked from OCTS on Midori; and NASA requested that NOAA archive the OCTS data even if they weren't ready to process it (which they weren't when OCTS launched). By keeping the OCTS data, a longer ocean color data record for U.S. coastal waters was acquired.[67] As described earlier in the chapter, this data acquisition was one factor that enabled successful calibration/validation of the NASDA-SIMBIOS-OCTS data set, utilizing OCTS data received for NOAA at the NASA Wallops receiving station.

When the SeaWiFS mission was underway, NOAA incorporated SeaWiFS data into their Coastwatch data system, utilizing the real-time data licenses it had helped insure were in the SeaWiFS contract.[68] Coastwatch featured ocean color data in a common format with other oceanographic data from NOAA satellites, particularly SST from the AVHRR instruments, and Coastwatch also provided ocean wind data from Quikscat and Defense Meteorological Satellite Program (DMSP) satellites. Coastwatch divided the data up regionally into "nodes" (such as the Great Lakes, Caribbean/Gulf of Mexico, or West Coast), and each node acquired SeaWiFS high-resolution data utilizing the real-time licensing agreement for ground stations. The nodes sent the received data to the Central Operations office at NESDIS, which distributed it. Coastwatch expanded its ocean color holdings to include MODIS data when the MODIS instruments were launched, and the "free" data from MODIS allowed Coastwatch to distribute near-real-time ocean color data products. NOAA also purchased the rights to SeaWiFS direct broadcast (DB) data from Orbimage. The Terra and Aqua spacecraft included a DB capability for suitably-equipped receiving stations; processing of the DB data did not include all the data processing steps necessary for the creation of science-quality data products, but with the benefit of being available rapidly, DB data could be used for fish finding, hazardous event monitoring, and other applications requiring rapid data examination and response.[69]

NOAA also created its own satellite ocean color processing system. This effort, which included former Orbimage chief oceanographer Linda Stathoplos, was so successful that it received the Department of Commerce Gold Medal award in 2007. The other team

---

[66] Frank Muller-Karger, email received 18 March 2010.

[67] Mary Cleave and Gene Feldman recorded interview, 21 November 2008.

[68] Frank Muller-Karger, email received 18 March 2010.

[69] "About NOAA Coastwatch," http://coastwatch.noaa.gov/cw_about_offices.html, (accessed 24 August 2009); "NOAA Coastwatch Ocean Color Data Sources," http://coastwatch.noaa.gov/cw_dataprod_color.html, (accessed 24 August 2009); "MODIS chlorophyll products," http://coastwatch.noaa.gov/modis_ocolor_overview.html, (accessed 24 August 2009); Frank Muller-Karger, email received 18 March 2010.

members were Christopher Brown, Mary Culver, Eugene Legg, Kent Hughes, and Richard Stumpf. [70]

When the contracts for continuing SeaWiFS data acquisition by NASA ceased including the high resolution data from HRPT stations, NOAA was able to procure funding to continue to receive HRPT data for U.S. coastal waters. One of the motivations for this continuing data record was that NOAA utilized SeaWiFS data for the Harmful Algal Bloom Forecasting System.[71] The system incorporates remotely-sensed ocean color, wind, and SST along with reports from surface water sampling to determine when blooms of *Karenia brevis* are occurring, and utilizes the data to predict where impacts from the blooms are likely to be felt. The system sends out bulletins directly by email to natural resource managers. (Figure 7.5) [72]

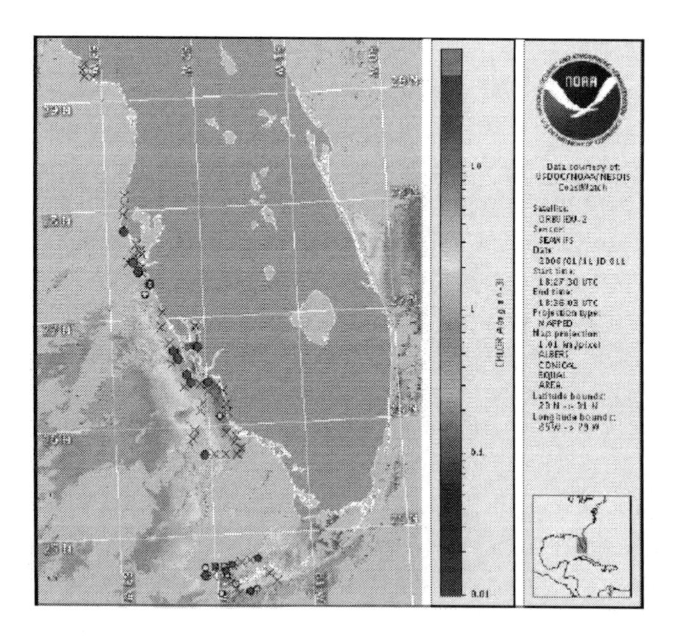

Figure 7.5. Harmful Algal Bloom Bulletin image for January 11, 2006, showing locations of *Karenia brevis* blooms. The caption for this image reads: "Satellite chlorophyll image with possible HAB areas shown by red polygon(s). Cell

[70] Conrad C. Lautenbacher, Jr. "Message From the Under Secretary -- NOAA's 2007 Department of Commerce Gold and Silver Medal Recipients," (16 October 2007), http://www.sp-systems.com/news/awards/2007/KHughs_Gold_Medal_Award.pdf, (accessed 5 January 2010).
[71] "Harmful Algal Bloom Forecasting System", http://tidesandcurrents.noaa.gov/hab/, (accessed 24 August 2009).
[72] NOAA Ocean Service NOAA Satellites and Information Service, "Gulf of Mexico Harmful Algal Bloom Bulletin," 12 January 2006, http://tidesandcurrents.noaa.gov/hab/bulletins/HAB20060112_200604_SFL.pdf, (accessed 1 October 2009).

concentration sampling data from January 2- 6 shown as red squares (high), red triangles (medium), red diamonds (low b), red circles (low a), orange circles (very low b), yellow circles (very low a), green circles (present), and black "X" (not present). This image was copied from an online report.

The SIMBIOS program also included NOAA participation. Christopher Brown (NOAA NESDIS), John Brock (now with the U.S. Geological Survey), Mary Culver of the NOAA Coastal Services Center (CSC) in Charleston, South Carolina, and Ajit Subramaniam conducted several research cruises in the South Atlantic Bight for bio-optical algorithm development and data product validation. This SIMBIOS Project conducted nine cruises in the South Atlantic Bight region, and also visited a few other locations (Lake Erie, New York, and Nantucket) as well.[73] While the original goal was to compare the OCTS and SeaWiFS data products, the loss of OCTS focused the group on SeaWiFS data product validation. Not unexpectedly, the contribution of sediments from rivers and the continental shelf, and organic matter from the lazy rivers of the piedmont Carolinas, caused significant deviations of the actual data from the remotely-sensed chlorophyll concentrations. Alternate algorithms that were better at quantifying and removing the influence of CDOM performed better in this region than the standard algorithms used by NASA.

In the second phase of SIMBIOS, Rick Stumpf led a team to examine Case 2 water algorithms. The team also included Robert Arnone and Rick Gould of the Naval Research Laboratory at the Stennis Space Center, Varis Ransibrahmanukul (a contractor scientist affiliated with NOAA) and Patricia Tester of the NOAA National Ocean Service in Beaufort, North Carolina. This team collected data on high number of cruises (twenty-one) in 2001 and 2002. While 21 cruises is a large number, fortunately the examination of Case 2 waters meant that many of these cruises were short distance and short duration; areas such as North Carolina's Pamlico Sound, the environs of Mobile Bay, Alabama, and coastal New Jersey were the locations of many of these "thick water" investigations. This team's investigations made a significant contribution to SeaWiFS data processing; the near-infrared (NIR) correction developed by the team was incorporated into SeaDAS in July 2002. Among the numerous accomplishments of this group, they also developed an algorithm that utilized data from the SeaWiFS 555 nm band to derive the concentration of total suspended sediments.[74]

---

[73] John C. Brock and Christopher W. Brown, "OCTS and SeaWiFS bio-optical algorithm and product validation and intercomparison in U.S. Coastal Waters," Chapter 5 in Charles R. McClain and Giulietta S. Fargion, editors, *SIMBIOS Project 1998 Annual Report, NASA Technical Memorandum NASA/TM-1999-208465,* pp. 36-40. An annual report from this group also appears in Charles R. McClain and Giulietta S. Fargion, *SIMBIOS Project 1999 Annual Report, NASA/TM-1999-209486* and Charles R. McClain and Giulietta S. Fargion, editors, *SIMBIOS Project 2000 Annual Report, NASA/TM-2001-209976.* Subramaniam and Culver were added as co-investigators in the 1999 and 2000 reports.

[74] Richard P. Stumpf, Robert A. Arnone and Richard W. Gould, Jr., Varis Ransibrahmanakul, and Patricia A. Tester, "Algorithms for Processing and Analysis of Ocean Color Satellite Data for Coastal Case 2 Waters," Chapter 17 in Giulietta S. Fargion and Charles R. McClain, editors, *SIMBIOS Project*

Coral Reefs

NASA's ocean color missions have been a partner in a program with both international and interagency participation to map the global extent of coral reefs as an aid to monitoring their status in the face of escalating environmental pressure. NASA utilized SeaWiFS data to create a global shallow bathymetry map, and also created a tool that allowed comparison of the SeaWiFS bathymetry to the United Nations Environmental Programme (UNEP) / World Conservation Monitoring Centre (WCMC) coral reef map. Data from both satellite (Landsat, Ikonos, MOS) and airborne (AVIRIS) instruments were also utilized for this mapping program. Participants in the program included Gene Feldman, Norman Kuring, and Bryan Franz of the OBPG; Rick Stumpf of NOAA; Julie Robinson of the Earth Observations Laboratory at the Johnson Space Center, which allowed inclusion of astronaut photographs; Edmund Green of UNEP/WCMC; and Marco Nordeloos of Reefbase and the World Fish Center.[75] A presentation based on this effort won the "Best in Plenary" award at the Sixth International Conference on Remote Sensing for Marine and Coastal Environments, which convened in Charleston, South Carolina in May 2000.

The Department of Defense/ Navy

The history of NASA ocean color data utilization would be very incomplete if the collaborative contributions of the Office of Naval Research, and the Ocean Optics Section of the Ocean Sciences Branch of the Naval Research Laboratory were not included. The Ocean Optics Section – led for many years by Robert Arnone, and including the participation of other Navy scientists (Rick Gould, John Kindle, and Alan Weidemann) and several dedicated contractor scientific programmers (Sherwin Ladner, Paul Martinolich) – has taken the foundational data from SeaWiFS and MODIS and created an extensive number of additional data products. This prolific group has produced a large number of publications, and has studied ocean color data in many different water bodies around the world. Arnone's research with CZCS data included research on the Alboran Sea area of the western Mediterranean; he also performed extensive analysis of the Arabian Sea monsoon system and the seasonal upwelling driven by the monsoon wind patterns.[76] Gould's

---

2001 Annual Report, NASA Technical Memorandum. 2002-210005, pp. 112-117. An annual report from this group also appears in Giulietta S. Fargion and Charles R. McClain, editors, *SIMBIOS Project 2002 Annual Report, NASA/TM-2003-211622.*

[75] Julie A. Robinson, Gene C. Feldman, Norman Kuring, Bryan Franz, Ed Green, Marco Nordeloos, and Richard P. Stump, "Data fusion in coral reef mapping: working at multiple scales with SeaWiFS and astronaut photography," *Proceedings Sixth International Conference on Remote Sensing for Marine and Coastal Environments*, Charleston, South Carolina May 1-3, 2000, ERIM, Ann Arbor, Michigan, http://eol.jsc.nasa.gov/newsletter/DataFusionInCoralReefMapping/, (accessed 25 August 2009).

[76] Robert A. Arnone, Denis Wiesenburg, and K. Saunders, "The origin and characteristics of the Algerian current"; Robert A. Arnone, "The temporal and spatial variability of chlorophyll in the

research included estimation of total suspended sediments, classification of water masses based on optical properties, and the ocean color and SST variability in waters near Japan.[77] Kindle, an ocean modeler, has worked with the use of ocean color data in models intended to provide forecasts of coastal ocean conditions, including temperature and optical properties.[78]

Arnone's early work involved the creation of a Secchi depth atlas, created about the time of the CZCS launch, in 1978. He said that the diffuse attenuation coefficient (K) algorithms created by Ros Austin revolutionized the way the Navy was going to do optics – most of their work had been image analysis that was not radiometrically accurate, and he said "I've been trying to convince people of the importance of calibration over 30+ years." Especially critical to Navy operations and research was real-time data and analysis of changing conditions. Arnone said that he requested that the operators of the Hyperion instrument on the Earth Observer-1 satellite (EO-1) turned on to observe a research campaign, and he was told they would get to it "in the next six months". Because he had a ship at sea, he repeated his request with some urgency, and was told that it cost $3,000 to turn on the system. Arnone pointed out to them that ship operations cost around $14,000 a day. [79]

Arnone also described a real-time use of MODIS data during the Iraqi Freedom campaign – Navy divers needed to inspect the hulls of tankers coming out of the Persian

---

western Mediterranean: coastal and estuarine studies"; Robert A. Arnone, Sherwin Ladner, Paul E. La Violette, John C. Brock, and Peter A. Rochford, "Seasonal and interannual variability of surface photosynthetically available radiation in the Arabian Sea," *Journal of Geophysical Research,* 103, 7735 – 7748, (1998); Sherwin Ladner, Robert A. Arnone, Richard W. Gould, Jr. and Paul M. Martinolich, "Evaluation of SeaWiFS optical products in coastal regions," *Sea Technology*, 43(10), 29-35, (October 2002).

[77] Richard W. Gould, Jr. and Robert A. Arnone, "Remote sensing estimates of inherent optical properties in a coastal environment," *Remote Sensing of the Environment*, 61, 290-301, (1997); Richard W. Gould, Jr., Robert A. Arnone, and Michael Sydor, "Absorption, scattering, and particle size relationships in coastal waters: Testing a new reflectance algorithm," *Journal of Coastal Research*, 17, 328-341, (2001); Richard W. Gould Jr., Robert A. Arnone, R. Smith, Sherwin D. Ladner, and Paul M. Martinolich, "Coastal transport of organic and inorganic matter from ocean color remote sensing," in *Proceedings: Oceanography Society Annual Meeting,* New Orleans, Louisiana, The Oceanography Society, Rockville, MD, (2003); Richard W. Gould, Jr. and Robert A. Arnone, "Optical water mass classification for ocean color imagery," in I. Levin and G. Gilbert, editors, *Proceedings: Second International Conference, Current Problems in Optics of Natural Waters*, St. Petersburg, Russia, September 9-12, 2003; Grace C. Chang and Richard W. Gould, "Comparisons of optical properties of the coastal ocean derived from satellite ocean color and in situ measurements," *Optic Express,* 14, 10149-10163, (2006).

[78] John Kindle and Jason K. Jolliff, "Integrating ocean color observations and nowcast/forecast of bio-optical properties into the Naval Research Laboratory coastal ocean model (NCOM)," http://www.onr.navy.mil/sci_tech/32/reports/docs/05/obkindle.pdf, (accessed 25 August 2009); John C. Kindle, "On the road toward real-time, coupled bio-physical, coastal forecast systems," *Abstract, Center for Coastal Physical Oceanography, Old Dominion University, 2005 Spring Seminar Series,* http://www.ccpo.odu.edu/Misc/kindle32105.html, (accessed 25 August 2009).

[79] Robert Arnone, comments recorded at the Ocean Color Collaborative Historical Workshop, January 13-14, 2009, St. Petersburg, FL; "Hyperion Instrument", http://eo1.gsfc.nasa.gov/Technology/Hyperion.html, (accessed 11 January 2010).

Gulf for mines, and this inspection would be accomplished most rapidly in clear water. Arnone's group used the MODIS 250-meter bands to identify the best areas to "park the ships" to do the inspections expeditiously.[80] One of the Navy group's other areas of interest is using ocean color data in combination with models, to forecast ocean optical conditions – Arnone noted that the concept of ocean color "weather" is really new, though meteorologists do it all the time for the atmosphere.[81]

One of the main areas of effort from this group at the Naval Research Laboratory has been the utilization of inherent optical properties (IOPs) derived from the ocean color data. IOPs, particularly backscattering and absorption coefficients for various constituents in the water, allow better estimation of suspended sediments and organic matter concentrations. Working close to the murkily complex waters of the northern Gulf of Mexico, the Stennis group has thus been involved with development of Case 2 water algorithms, such as their SIMBIOS-sponsored research.[82]

Office of Naval Research

After the Scripps VisLab disbanded in 1987, Eric Hartwig of the Office of Naval Research (ONR) realized that the topic of ocean optics, including ocean color remote sensing, was too important to the Navy and the nation to be without strong leadership. In that same year he therefore established the Ocean Optics program and recruited Richard Spinrad as the first program officer. The program grew rapidly within the first several years and changed leadership several times; Curtis Mobley and Gary Gilbert took their turns as program manager until 1992. In 1993, Steven Ackleson took up the management reins and was joined in 1996 by Joan Cleveland.[83]

The primary role of ONR in ocean color remote sensing research was to continue the defense-related work of the VisLab – developing new and novel means of accurately measuring *in situ* inherent optical properties guided by more accurate and faster running radiative transfer models and investigating ocean processes of optical consequence. *In situ* sensors for measuring the inherent and apparent optical properties of natural waters, often developed with contracts to small businesses such as Western Environmental Laboratories (WET Labs); Satlantic; Sequoia Scientific; Hydro-Optics, Biology, and Instrumentation Laboratories (HOBI Labs); and Biospherical have become standard oceanographic infrastructure equipment in research and monitoring programs, both domestically and overseas, and have become integral components of the U.S. Integrated Ocean Observation System. Ron Zaneveld and Scott Pegau were instrumental in bringing accurate *in situ*

---

[80] Robert Arnone, comments recorded at the Ocean Color Collaborative Historical Workshop, January 13-14, 2009, St. Petersburg, FL.
[81] Robert Arnone, comments recorded at the Ocean Color Collaborative Historical Workshop, January 13-14, 2009, St. Petersburg, FL.
[82] Richard P. Stumpf, Robert A. Arnone and Richard W. Gould, Jr., Varis Ransibrahmanakul, and Patricia A. Tester, "Algorithms for Processing and Analysis of Ocean Color Satellite Data for Coastal Case 2 Waters."
[83] Manuscript provided by Steven Ackleson and Joan Cleveland, January 7, 2011.

absorption and attenuation sensors to market and available to researchers outside the discipline of ocean optics.[84] Likewise, Robert Maffione developed an array of light scattering sensors that, together with absorption, form the basic ground truth information set for validating ocean color water -leaving radiances.[85] Marlon Lewis has led the development of *in situ* hyperspectral radiance and irradiance measurements. Complicated radiative transfer simulations that required main frame computers and hours of CPU time two decades ago can now be solved in minutes on standard laptop computers. The radiative transfer model, Hydrolight, developed by Curtis Mobley, has become the gold standard for basic and applied research as well as operations.[86]

A hallmark of ONR investments is the multidisciplinary, multiple investigator research initiative focused on specific oceanographic processes. Focused field investigations such as Bioluminescence and Optical Variability of the Sea (Biowatt; 1987–1992), Marine Light, Mixed Layer (MLML, 1989-1994), Coastal Mixing and Optics (CM&O; 1990-1995), Hyperspectral Coastal Ocean Dynamics Experiment (HyCODE; 1999-2003), and Radiance in a Dynamic Ocean (RaDyO; 2005-2010) helped define the field of optical oceanography and generated new knowledge about subsurface light fields in response to near surface ocean physical and biological forcing.[87] These short duration research initiatives have added greatly to our knowledge of ocean processes critical to interpreting remotely sensed ocean color.[88]

One common thread running through each of these research initiatives was the development of new approaches to measuring bio-optical and physical properties with high spatial and temporal fidelity. With *in situ* instruments developed by Biospherical, Ray Smith deployed the first bio-optical profiling system.[89] Tommy Dickey, for example, pioneered the development of instrumented moorings capable of collecting interdisciplinary ocean time series measurements over the course of months to years.[90] ONR facilitated the development

---

[84] W. Scott Pegau, Joan S. Cleveland, W. Doss, C. D. Kennedy, Robert A. Maffione, James L. Mueller, R. Stone, Charles C. Trees, Alan D. Weidemann, W. H. Wells, and J. Ronald V. Zaneveld, "A comparison of methods for the measurement of the absorption coefficient in natural waters," *Journal of Geophysical Research* 100 (C7), 13,201-13,220 (1997).
[85] Robert A. Maffione and David R. Dana. "Instruments and methods for measuring the backward-scattering coefficient of ocean waters," *Applied Optics* 36, 6057-6067 (1997).
[86] Curtis D. Mobley, *Light and water: radiative transfer in natural waters.* Academic Press, San Diego, 592 pages (1994).
[87] Thomas Dickey, T. Granata, John Marra, C. Langdon, Jerome Wiggert, Z. Chai-Jochner, M. Hamilton, J. Vazquez, M. Stramska, Robert Bidigare and David Siegel, "Seasonal variability of bio-optical and physical properties in the Sargasso Sea," *Journal of Geophysical Research* 98(C1), 865-898 (1993); Grace C. Chang, Thomas D. Dickey, Oscar M. Schofield, Alan D. Weidemann, Emmanuel Boss, W. Scott Pegau, Mark A. Moline, and Scott M. Glenn, "Nearshore physical processes and bio-optical properties in the New York Bight," *Journal of Geophysical Research* 107, 3133, doi:10.1029/2001JC001018 (2002); Grace C. Chang, Thomas D. Dickey, Curtis D. Mobley, Emmanuel Boss, and W. Scott Pegau, "Toward closure of upwelling radiance in coastal waters," *Applied Optics* 42, 1574-1582 (2003).
[88] Manuscript provided by Steven Ackleson and Joan Cleveland, 7 January 2011.
[89] Ray C. Smith, C.R. Booth and J.L. Star, "Oceanographic biooptical profiling system," *Applied Optics* 23, 2791-2797 (1984).
[90] Thomas D. Dickey, "The emergence of concurrent high-resolution physical and bio-optical measurements in the upper ocean and their applications," *Reviews of Geophysics*, 29, 383-413 (1991).

of autonomous, mobile sampling platforms equipped with bio-optical and physical sensors. Oscar Schofield and Scott Glenn played leading roles in the development of this capability, providing unprecedented views of how optically-important properties are distributed within open ocean and coastal environments, even when conditions prohibit the safe operation of oceanographic vessels.[91]

The Ocean Optics Program has in many respects played a leading role in developing predictive ocean process models that realistically couple ocean physical, biological, and optical processes — models that can be used to predict ocean color and that will in the future be able to efficiently assimilate remotely sensed water leaving radiance. Curt Mobley developed a simplified version of his Hydrolight model for use in predictive ocean process and ecosystems models. Paul Bissett participated in HyCODE and worked to link phytoplankton ecosystem models within the Regional Ocean Modeling and Simulation framework (ROMS) and investigated approaches to assimilating ocean color data.[92] Fei Chai linked his phytoplankton ecosystem model, CoSINE, with ROMS to investigate temporal and spatial changes in phytoplankton abundance and ocean color response in Monterey Bay.[93] Many feel that future advances in ocean color interpretation rest not in stand-alone algorithms, but in the assimilation of spectral water-leaving radiance into accurate, interdisciplinary models.[94]

Another area of ONR investment directly related to ocean color analysis, especially in coastal waters, is the source, transport, transformation and optical properties of colored dissolved organic matter (CDOM). Knowledge of how CDOM contributes to the ocean color signature is key to mapping chlorophyll distributions within coastal waters, Paula Coble performed ground-breaking research on the sources and spectral fluorescent signatures of CDOM along the west coast of Florida and within the Mississippi river plume.[95] Bob Chen reported on CDOM sources and transport within estuarine and coastal systems, including the Gulf of Mexico, the South Atlantic Bight, New York Harbor and adjacent North Atlantic Bight, and the Yellow Sea.[96] Bill Miller has been a leader in the study of photochemical properties of CDOM found in both coastal and open ocean waters and development of algorithms for estimating CDOM absorption from satellite ocean color

---

[91] Oscar Schofield , J. Kohut, and Scott M. Glenn, "Evolution of coastal observing networks," *Sea Technology*, 49, 31-36 (2008).

[92] Paul W. Bissett, Robert Arnone, S. DeBra, D. Dye, G. Kirkpatrick, Curtis D. Mobley, and Oscar M. Schofield, "The integration of ocean color remote sensing with coastal nowcast/forecast simulations of harmful algal blooms (HABS)," in Marcel Babin, Colin Roesler, and John J. Cullen, editors, *Real Time Coastal Observing Systems for Ecosystem Dynamics and Harmful Algal Blooms*, UN Education, Science, and Cultural Organization, Paris, pp. 85–108 (2008).

[93] Fei Chai and Emmanel Boss, "Physical-Biological-Optics Model Development and Simulation for the Monterey Bay, California," *Technical Report, University of Maine, ADA516870* (2008).

[94] Manuscript provided by Steven Ackleson and Joan Cleveland, 7 January 2011.

[95] Carlos E. Del Castillo, Fernando Gilbes, Paula G. Coble, and Frank E. Muller-Karger, "On the dispersal of riverine colored dissolved organic matter over the West Florida Shelf," *Limnology and Oceanography*, 45(6), 1425-1432 (2000).

[96] Robert F. Chen, Paul Bissett, Paula Coble, R. Conmy, G.B. Gardner, M.A. Moran, X.C. Wang, M. Well, P. Whelan, and R.G. "Chromophoric dissolved organic matter (CDOM) source characterization in the Louisiana Bight," *Marine Chemistry* 89, 257-272 (2004).

data.[97]

The pursuit of new knowledge of ocean optics and ocean color remote sensing requires a high degree of interaction across disciplinary boundaries, especially physics and biology. Realizing the need to sensitize students to this mode of research, ONR supported an ocean optics summer course series, conducted roughly every two years. The goal of these courses was to provide graduate students with intensive theoretical, practical, and hands-on experiences in ocean optics and ocean color remote sensing through lectures, modeling workshops, field observations, and independent research projects The first course was offered in 1985 at Friday Harbor Laboratories, University of Washington, and taught by Mary Jane Perry and Ken Carder. In 1987 the course was enhanced with invited lectures by leading experts from around the world, including John Kirk, André Morel, Ken Carder, Howard Gordon, Niels Højerslev, Dale Kiefer, Charlie Yentsch, and Motoaki Kishino. The lectures were eventually compiled and published in the book *Ocean Optics*, with lead author Richard Spinrad.[98] Later, after Perry moved to the University of Maine, courses were convened at the Darling Center and aided by Collin Roesler, Emmanuel Boss, and Curt Mobley. These courses created an opportunity for graduate students from diverse disciplines to interact with leading researchers in biological and optical oceanography and played an important role in educating a new generation of young researchers in modern approaches to measuring and modeling ocean optical processes and formulating new ocean color algorithms.[99]

The Navy initiated the now famous Ocean Optics conference series in the early 1960s. The conference has grown in the past two decades to be a major international gathering of ocean optics and ocean color remote sensing experts and students, and is jointly supported by ONR and NASA. The 20th conference in this series convened in Anchorage, Alaska and attracted over 300 participants.

In 1998 ONR, NASA and the Oceanography Society agreed to jointly support an award for outstanding achievements in ocean optics and ocean color remote sensing research, education, and community service. The award was named after Nils Jerlov, an internationally recognized leader in the field of ocean optics. The first award was presented to André Morel in 2000 and since has been awarded to Ray Smith, Howard Gordon, Ron Zaneveld, and Charles Yentsch, all having contributed significantly to our knowledge of ocean color remote sensing.[100]

After 24 years of dedicated ocean optics funding and untold advances in sensor and observational capabilities, theoretical modeling of underwater light fields, contributions to

---

[97] William L. Miller and M. A. Moran,"Interaction of photochemical and microbial processes in the degradation of refractory dissolved organic matter from a coastal marine environment," *Limnology and Oceanography* 42(6), 1317-1324 (1997); Cedric G. Fichot, Shubha Sathyendranath, and William L. Miller, "SeaUV and SeaUVC: Algorithms for the retrieval of UV/Visible diffuse attenuation coefficients from ocean color," *Remote Sensing of Environment*, 112, 1584–1602, doi:10.1016/j.rse.2007.08.009 (2008).
[98] Spinrad, RT. W., K. L. Carder and M. J. Perry, *Ocean Optics*. Oxford University Press, New York, 304 pages, (1993).
[99] Manuscript provided by Steven Ackleson and Joan Cleveland, 7 January 2011.
[100] Manuscript provided by Steven Ackleson and Joan Cleveland, 7 January 2011.

ocean color and LIDAR remote sensing, underwater imaging, and fundamental understanding of ocean processes of optical consequence, the ONR Ocean Optics Program was discontinued as of 30 September, 2010.[101]

Summary: This chapter covers several of the more obvious international and interagency collaborations that NASA's ocean color missions have participated in. What is more difficult to express is the level of usage of NASA ocean color data around the world. Because of the necessity of registering as Authorized Users to obtain SeaWiFS data, it was possible, however, to know the locations of many scientists around the world who acquired and used the data. A count of the countries represented on this list exceeded 80 in early 2004.[102] During the early days of the SeaWiFS mission, it was not uncommon to receive plaintive, almost desperate requests for the data, and several times unusual methods had to be employed to deliver the data on tapes or even CD-ROMs to such places as Ghana and southern China. The increased effectiveness of the Internet, allowing electronic data transfer, has made it considerably easier to get the data, and thus the international impact of the data is less easy to evaluate, particularly in the case of MODIS. It is, however, relatively safe to say that NASA's ocean color data has enabled research in nearly every region of the global oceans and coastal seas. It has been utilized by oceanographic agencies in the United States, as well as by oceanographic agencies and scientists in many different countries.

[101] Manuscript provided by Steven Ackleson and Joan Cleveland, 7 January 2011.
[102] David A Siegel, Andrew C. Thomas, and John Marra, "Views of ocean processes from the Sea-viewing Wide Field-of-view Sensor mission: introduction to the first special issue," *Deep Sea Research Part II: Topical Studies in Oceanography: Views of Ocean Processes from the Sea-viewing Wide Field-of-view Sensor (SeaWiFS) Mission*: 51, pp. 1-3, (January-February 2004).

8

LESSONS FROM THE PAST – UNCERTAINTY FOR THE FUTURE

One of the main lessons that NASA learned from its experience with the progression of missions from CZCS to SeaWiFS to MODIS was to attempt to plan ahead in order to eliminate gaps in the observational data record, such as the 11-year gap between CZCS and SeaWiFS. Such advance planning would, *it was thought*, insure that there would be successor sensors in space to overlap the observations being made by active sensors in space, before the active sensors reached the limit of their observational lifetimes and ceased their observations.

It has turned out (as this chapter is written in the latter part of 2010) that advance planning can only go so far. At this point in time, where history converges with current events, there is considerable uncertainty as to whether the exemplary, unprecedented, and nearly continuous record of ocean color observations from space can continue without a significant gap, either in actual observational data or (perhaps more troubling) in the quality of oceanographic observational data. Also in question is whether NASA can continue to maintain its leadership role in this branch of remote-sensing science, or become relegated, at least for a few years, to a partnership role utilizing and analyzing data from sensors launched and operated by foreign space agencies.

It is also useful here to evaluate what was learned from the missions that brought the scientific community to this point; particularly the difference in the level of success achieved by the SeaWiFS and MODIS ocean color projects. Thus, first we look back at what has occurred thus far, and then we gaze into a murky crystal ball to evaluate the prospects for the future of NASA in the realm of ocean color remote sensing.

Lessons Learned: SeaWiFS and MODIS

It is clear from the histories of the CZCS, SeaWiFS, and MODIS missions that each team – the CZCS NET, the SeaWiFS Project, and MODIS Oceans – contended with different challenges and operated in a different manner. The CZCS NET was in itself an experiment; NASA had never before had science teams interacting with spacecraft engineers on the previous Nimbus missions; Nimbus 7 was the first time that science took a somewhat equal footing with just getting data and keeping the instrument on the spacecraft operating[1] Previously, as Al Fleig noted, most of the work was done by "data techs and programmers", while the scientists were elsewhere leading the research and doing university stuff."[2] The CZCS NET also conducted its activities during a period of evolution and revolution in data processing capabilities; as an example, the speed of computation limited Howard Gordon's ability to test his atmospheric correction algorithm before launch, and Warren Hovis marked two minute scenes with a red pen on quick-look CZCS photographic film.[3] Late in the program, CZCS data in the basement of Goddard Building 314 was analyzed to try and find all the non-cloudy scenes.[4] Even years after the mission ended, the global reprocessing effort had to conduct night raids on the drawers in the office of Warren Hovis in Suitland, Maryland, to find and digitize CZCS photographic film strips on a light table and add them to the data that was being transferred, using a laserfax, to 9-track tapes.[5]

Chuck McClain noted several lessons that were engendered from the fabrication and mission of the CZCS. The instrument itself provided a few of the lessons; one was the obvious next-generation design consideration that near-IR bands would be required for atmospheric correction. Another instrumental lesson was the need for additional blue-range bands to separate the chlorophyll signal from associated phaeopigments.[6] The abrupt calibration shifts experienced by CZCS predicated the need for a way to track the stability of the instrument, which motivated the monthly moon views carried out by SeaWiFS. [7]

The experiences of the CZCS NET members also provided lessons for the subsequent missions. The first-year data calibration and validation cruises, particularly the post-launch cruises, were not enough – in order for the data to continue to be valid, a mission-long Cal/Val program would be necessary. Also, in order to evaluate the accuracy of the data, essentially immediate availability of the data to the broader research community – beyond a small group of investigators, like the CZCS NET, was required. Furthermore, keeping the community in the dark was not advisable; providing software for the community to process and analyze data, as well as detailed documentation of the decisions made by the project's engineers and scientists, was definitely better than having oceanographic researchers guess at (or second-guess) why NASA engineers and scientists decided to do what they did with the instrument and its data.[8]

---

[1] "Nimbus Program History," NASA GSFC, page 10.

[2] Al Fleig, interview recorded 21 April 2011.

[3] Howard Gordon, comments recorded at the Ocean Color Collaborative Historical Workshop, January 13-14, 2009, St. Petersburg, Florida.

[4] Gene Feldman, comments recorded at the Ocean Color Collaborative Historical Workshop, January 13-14, 2009, St. Petersburg, Florida.

[5] Howard Gordon, comments recorded at the Ocean Color Collaborative Historical Workshop, January 13-14, 2009, St. Petersburg, Florida.

[6] Even before the CZCS was launched, the desirable characteristics of the next-generation sensor were discussed at a meeting; the IUCRM Colloquium on "Passive Radiometry of the Ocean" at the Institute of Ocean Sciences, Patricia Bay (near Victoria, British Columbia, Canada from June 14 to 21, 1978. A report on this meeting is found in André Morel and Howard R. Gordon, "Report of the Working Group on Water Color," *Boundary-Layer Meteorology*, 18, 343–355, (1980).

[7] Charles McClain, email message on ocean color lessons learned, received 20 April 2009.

[8] Charles McClain, email message on ocean color lessons learned, received 20 April 2009.

The reason that the CZCS data experience directly influenced the SeaWiFS Project was that it emphasized the need to have a robust, flexible data processing capability in place before the launch of the sensor – which perhaps seems eminently logical, but is not always achievable. Computer technology was still catching up to the instrument and data at the beginning of the CZCS mission. The four-year SeaWiFS launch delay allowed the Project to fully develop and fully test the data processing system and data distribution protocols and pathways before launch. This advantage of time thus enabled the apparent remarkable ease of data acquisition, processing, and distribution that occurred when SeaWiFS initiated its observational mission.

The SeaWiFS Project many times ultimately benefited from what may have initially seemed detrimental aspects of its project structure. The SeaWiFS Project never had what was (even by most NASA mission standards) an excess of funding; thus, the team had to be relatively small and focused on necessary activities, with an "agile" staff utilizing persons with specific expertise. As Cleave and Feldman pointed out, everyone was critical and there was very little redundancy.[9] The size of the Project and its relatively autonomous operation allowed the managers to identify needs and make decisions and implement them.

By contrast, the MODIS team's requirements were more "rigid and bureaucratic", imparting a "here's what you have to do" flavor from upper management levels.[10] Bob Evans reiterated the importance of flexibility and quick data turnarounds as having "as much a contributor to success as the instrument calibration".[11] Evans noted that they a lot was learned setting up the CZCS data processing and carrying it through to SeaWiFS – they had scoped their system with the goal of being able to process the data at least 10 times faster than it was received, which would allow them to play "a lot of 'what-if' games". This capability was a "stark contrast with EOSDIS", where the projected processing speed was only about two times greater than the rate that the data was received. This constrained capacity meant there were limited "opportunities for enhancement of algorithms and correction of mistakes".[12]

Wayne Esaias noted that for much of the mission, the Oceans team was on their own, because program managers expressed to him that they had several higher priorities than MODIS. He also noted that the MODIS science team had eight members, each responsible by contract for the accurate performance of their designated algorithms, while the SeaWiFS science team had more than 50 members, and operated by consensus (or at least by attempts to *achieve* a consensus). These differences seemed to create a much greater sense of ownership in the ocean color scientific community for SeaWiFS compared to

[9] Mary Cleave and Gene Feldman recorded interview, 21 November 2008.

[10] Charles McClain, comments recorded at the Ocean Color Collaborative Historical Workshop, January 13-14, 2009, St. Petersburg, Florida.

[11] Robert Evans, comments recorded at the Ocean Color Collaborative Historical Workshop, January 13-14, 2009, St. Petersburg, Florida.

[12] Robert Evans, comments recorded at the Ocean Color Collaborative Historical Workshop, January 13-14, 2009, St. Petersburg, Florida.

MODIS.[13]

The MODIS experience provides a few more lessons, not all of which have been transferred to the VIIRS instrument, which will be discussed shortly. While MODIS is a remarkable instrument, it also demonstrates that in terms of ocean color data quality, it is not optimal to have visible bands and thermal IR bands in the same instrument.[14] Furthermore, the combined experience of SeaWiFS and MODIS demonstrates that achieving accurate vicarious calibration results is a goal that may take several mission-years to attain, and there are several factors that must be combined to make it possible: orbit maintenance, specification of detailed field and laboratory protocols, and participation of outside experts in calibration activities.[15] Gene Feldman indicated that one perceived problem with VIIRS is that there is no organization given the task of reprocessing VIIRS data.[16] Al Fleig emphasized the need for this, pointing out that a long-term dataset doesn't get constructed until 2-3 years after launch, and that the data has to be "reprocessed almost constantly", taking into account instrument and calibration changes.[17] MODIS also indicated what it would be nice to have in future ocean color sensors: better-refined shortwave IR bands for aerosol correction, and even UV bands (despite the difficulties of atmospheric absorption) to enable better separation of the signals from CDOM and chlorophyll.[18]

Obviously, it would be desirable if the lessons learned and the expertise gained via the experience of successful missions would be transferred to the next generation of instruments, in any field of remote sensing. In the case of ocean color, however, the evolution of how the United States elected to continue remote sensing observations of the Earth significantly impacted the design and operation of instruments still waiting to be launched. The travails of the program go hand-in-hand with the travails of the sensor which is designated to continue the record of ocean observations initiated by CZCS, SeaWiFS, and MODIS.

The Woes of NPOESS

As this history is primarily about the NASA heritage of successful ocean color missions; as such, it will not provide a detailed history of the NPOESS program. The NPOESS

---

[13] Wayne Esaias, comments recorded at the Ocean Color Collaborative Historical Workshop, January 13-14, 2009, St. Petersburg, Florida.

[14] Charles McClain, email message on ocean color lessons learned, received 20 April 2009.

[15] Charles McClain, email message on ocean color lessons learned, received 20 April 2009; "Preface," in James L. Mueller, Giulietta S. Fargion, and Charles R. McClain, editors, *Ocean Optics Protocols For Satellite Ocean Color Sensor Validation, Revision 4, Volume VI: Special Topics in Ocean Optics Protocols and Appendices*, Mueller, *NASA Technical Memorandum 2003-211621, Revision 4, Volume 6*, NASA Goddard Space Flight Center, 141 pages, (April 2003).

[16] Gene Feldman, comments recorded at the Ocean Color Collaborative Historical Workshop, January 13-14, 2009, St. Petersburg, Florida.

[17] Al Fleig, interview recorded 21 April 2011.

[18] Charles McClain, email message on ocean color lessons learned, received 20 April 2009.

program was envisioned to save money and reduce redundancy by combining the remote-sensing activities of NASA, NOAA and the Department of Defense (DoD). All three agencies have operated polar-orbiting environmental observation satellites; NASA had the Nimbus heritage carried on by SeaWiFS and the EOS satellites; NOAA had its series of Polar-Orbiting Environmental Satellites (POES), which notably carried the AVHRR used for sea surface temperature observations; and the DoD operated the Defense Military Satellite Program (DMSP) system, which featured the Special Sensor Microwave Imager (SSM/I) of oceanographic interest, particularly for sea ice and winds. The initial idea for this amalgamation came from a 1992 National Space Council Report, which was followed by joint NASA/DoD studies.[19] In May 1994, a plan was submitted to the U.S. Congress, which was then affirmed by Presidential Decision Directive NSTC-2. NOAA and DMSP were directed to join forces and converge their programs, with the participation of NASA to provide "new remote sensing and spacecraft technologies that could potentially improve the capabilities of the operational system".[20] In the directive, it is clearly stated that the new program had a goal of "reducing the cost of acquiring and operating polar-orbiting environmental satellite systems", partly by "incorporating appropriate aspects of NASA's Earth Observing System." In conception, combining the polar-orbiting observational activities of the three different agencies into one single polar-orbiting system would likely save money. NPOESS would consist of a series of the same satellite with the same instruments, similar to how the POES and DMSP systems worked, with new satellites launching as the expected lifetime of the predecessors neared its end, allowing for instrument cross-calibration and a continuous operational observational record.

NOAA created the Integrated Program Office (IPO), which would manage the NPOESS program. With the directive and the creation of the IPO essentially initiating the NPOESS program, development of instruments commenced. In October 1996, Chris Scolese reported to the MODIS Science Team meeting that some members were participating in NPOESS studies.[21] The sensor which would inherit the task of ocean color remote sensing was going to be VIIRS (Visible Infrared Imaging Radiometer Suite), and the contractor specified to build VIIRS was Raytheon, which now managed the Santa Barbara Research Center (SBRC) that had built SeaWiFS and MODIS. The center was now called Santa Barbara Remote Sensing.[22]

The NPOESS program initiated a "Risk Reduction" strategy, one aspect of which was the NPOESS Preparatory Project (NPP), a precursor satellite that would demonstrate the operability of the some of the NPOESS instruments, and also overlap with the ongoing NASA Terra and Aqua satellite missions. The NPP spacecraft would be built by Ball Aerospace constructed under the auspices and direction of NASA. In 2002, it was expected

---

[19] NOAA, "The National Polar-orbiting Operational Environmental Satellite System (NPOESS)," http://www.publicaffairs.noaa.gov/grounders/npoess.html, (accessed 14 September 2009).
[20] "Presidential Decision Directive NSTC-2," http://www.ipo.noaa.gov/index.php, (accessed 14 September 2009).
[21] NASA GSFC, "MODIS Science Team Meeting Minutes, October 10-11, 1996."
[22] NOAA, "The National Polar-orbiting Operational Environmental Satellite System (NPOESS)."

that NPP would launch in 2006, carrying VIIRS, the Cross-Track Infrared Sounder (CrIS), and NASA's Advanced Technology Microwave Sounder (ATMS). The first NPOESS satellite was slated to launch in 2009; the eventual satellite "constellation" was planned to consist of three satellites all operating at the same time, following the same orbit, so that the Earth could be observed at different times of the day: early morning, mid-morning, and early afternoon.[23]

Briefly summarized, the development of the NPOESS program did not accrue the expected cost-saving benefits. Program costs exceeded expectations, and the program fell further and further behind schedule. The program came under the oversight of laws passed by Congress, specifically the Nunn-McCurdy Amendment. The Nunn-McCurdy amendment was intended to rein in runaway costs on programs that were dramatically exceeding estimates, specifically by 25% or more. Nunn-McCurdy was initially aimed at expensive DoD weapons programs.[24]

Because NPOESS was partly a DoD program, the cost overruns on NPOESS came under Nunn-McCurdy scrutiny. There were two thresholds in the Nunn-McCurdy amendment that initiated a review process, overruns of 15% and 25%. NPOESS went over the 15% boundary in September 2005 and broke the 25% barrier in January 2006. Due to these events, a process was initiated requiring the Nunn-McCurdy certification process to allow the program to continue. The review process resulted in a restructuring of NPOESS; one of the three simultaneous satellites was deleted, the instrument manifest was revised, with some instruments taken off (and later replaced), some instruments redesigned and other instruments replaced with instruments having a working record of success, such as CERES.[25]

When the NPOESS situation became so critical, there was concern the DoD would abandon its collaborative commitment, though it did not. The situation did, however, lead to calls from Congressional Democrats for the removal of the NOAA Administrator, Vice Admiral Conrad Lautenbacher, and also the removal of Department of Commerce Deputy Undersecretary for Oceans and Atmosphere General John J. Kelly, Jr., on grounds of mismanagement.[26] At this time, the estimated costs for NPOESS had risen an initial $6.8 billion dollars to $13.8 billion, significantly higher than NOAA's entire annual operating budget. Despite the controversy, Lautenbacher served in his position until near the end of the George W. Bush presidential administration.[27]

[In 2010, the NPOESS program was terminated, and replaced by the Joint Polar Satellite System Program (JPSS), a partnership of only NASA and NOAA. One of the

---

[23] NOAA, "The National Polar-orbiting Operational Environmental Satellite System (NPOESS)."
[24] Center for Defense Information, "The Nunn-McCurdy Amendment," http://www.cdi.org/missile-defense/nunn-mccurdy.cfm, (accessed 14 September 2009).
[25] NOAA, "About NPOESS," http://www.ipo.noaa.gov/index.php, (accessed 14 September 2009).
[26] U.S. House of Representatives Committee on Science and Technology, "Democrats call for NOAA Leadership's removal," http://science.house.gov/press/PRArticle.aspx?NewsID=1115, (accessed 14 September 2009).
[27] Scientific American, "NOAA chief Conrad Lautenbacher resigns," http://www.scientificamerican.com/blog/60-second-science/post.cfm?id=noaa-chief-conrad-lautenbacher-resi-2008-09-23, (accessed 14 September 2009).

reasons for this was that the requirements for a climate and weather mission were different that for a military mission. So the DoD was separated from NASA and NOAA to build a morning satellite mission, when clouds are not as prevalent, and NASA and NOAA would collaborate on an afternoon mission.][28]

The Woes of VIIRS

Through all of this, the VIIRS instrument was maintained, and viewed as a vital, central NPOESS instrument. But VIIRS managed to develop its own set of problems – and the major impact of those problems fell on the category of ocean color data.

As noted above, the contractor selected to build VIIRS was Raytheon, and the laboratory that would build VIIRS was the Santa Barbara Remote Sensing Group, which had experience with both SeaWiFS and MODIS. The design selected for VIIRS returned to the successful experience with SeaWiFS; a rotating telescope rather than a rotating mirror. The actual name for the telescope is a "3-mirror anastigmat all reflective rotating telescope". The critical design review for VIIRS was completed in spring 2002; soon after, the management and contract situation became a little complicated, as TRW (which became Northrop Grumman Space Technologies, NGST) inherited VIIRS when they received a contract to manage the NPOESS program and build the NPOESS satellites. SBRS was then solely responsible for building VIIRS in accordance with the instrument specifications.[29]

In 2003, it was expected that the first flight model for VIIRS would be completed in 2005 and ready for launch on NPP in 2006, while the Engineering Development Unit (EDU) would be completed in 2004.[30] That schedule turned out to be overly ambitious, in terms of both NPP and VIIRS, and the NPOESS program in general, partly because of the cost overruns and the Nunn-McCurdy review process described earlier. All of these circumstances led to delays, but by April 2007, the EDU had undergone Thermal/Vacuum testing, and the flight unit was in development. (Figure 8.1)[31]

---

[28] National Environmental Satellite Data and Information Service, "Joint Polar Satellite System," http://www.nesdis.noaa.gov/pdf/jpss.pdf, acquired 22 April 2011.

[29] Carl Schueler , J. Ed Clement, Russ Ravella, Jeffery J. Puschell, Lane Darnton, Frank DeLuccia, Tanya Scalione, Hal Bloom, and Hilmer Swenson, "VIIRS sensor performance," in *Proceedings International TOVS Study Conference XII*, Sainte Adèle, Canada, 29 October 2003 - 4 November 2003, http://cimss.ssec.wisc.edu/itwg/itsc/itsc13/proceedings/session10/10_2_schueler.pdf, (accessed 15 September 2009).

[30] Carl Schueler , J. Ed Clement, Russ Ravella, Jeffery J. Puschell, Lane Darnton, Frank DeLuccia, Tanya Scalione, Hal Bloom, and Hilmer Swenson, "VIIRS sensor performance."

[31] Jim Gleason, "NPOESS Preparatory Project (NPP) Status," presented at the Ocean Color Research Team meeting, Seattle, Washington, April 11-13, 2007, http://oceancolor.gsfc.nasa.gov/DOCS/ScienceTeam/OCRT_Apr2007/Gleason_NPP_Overview_OCRT2007.pdf, (accessed 15 September 2009); Raytheon, http://www.raytheon.com/capabilities/rtnwcm/groups/sas/documents/image/rtn_p_sas_viirs_4hi-res.jpg, (accessed 8 January 2010).

Figure 8.1. VIIRS being prepared for Thermal/Vacuum testing by Raytheon technicians.

By this time in 2007, the most significant problems with VIIRS had already been recognized, but at the Ocean Color Research Team (OCRT) meeting in April, a large sector of the interested scientific community likely became fully cognizant of them. It had been determined in testing that VIIRS had "optical crosstalk" between the most vital visible bands used for ocean color algorithms, a problem identified as coming from the filter assembly.[32] "Optical crosstalk" basically means that the light which was intended to be measured for one band was contaminated with light from other bands, rendering the data for each band somewhat suspect – or worse. The problem was particularly acute for a situation akin to the "ringing" observed in CZCS data, where the scan transitioned from a bright to a dark target. The assessment at the OCRT meeting was that the modifications to the hardware had not fully fixed the problem, and the problem was going to be "significant" with respect to ocean color. The alternative was fixing the problem via software. But there was another problem with that; because of the contract transfers, SBRS was only responsible for building and testing the instrument to see it if met specifications; if it didn't, and if the fixes were going to be required in the software (which initially required characterizing the Relative

[32] Jim Gleason, "NPOESS Preparatory Project (NPP) Status."

Spectral Response (RSR) of VIIRS), SBRS wasn't responsible for doing that.[33] Despite this, in June 2007 testimony before the House of Representatives Subcommittee on Energy and Environment, Brigadier General Susan K. Mashiko recognized the problem explicitly and said that "One major technical issue, optical cross talk, remains and we are pursuing several potential solutions", then noting that "The VIIRS product most at risk is ocean color." [34]

The level of those risks seemed unacceptable to the ocean color community in the United States. By October, a letter from the community was sent jointly to NOAA Administrator Lautenbacher and NASA Administrator Michael Griffin, expressing the marked concerns of the ocean color community regarding the situation with VIIRS and its impact on ocean color data quality – the quality that the oceanographic scientists had come to expect over the previous decade from the record of SeaWiFS and MODIS-Aqua. The letter was co-authored by David Siegel of the University of California – Santa Barbara and James Yoder, now the chairman of the IOCCG as well as a senior scientist at Woods Hole, having moved there from the University of Rhode Island. The letter had 56 supporting signatures, including Ralph Keeling, the carbon-dioxide measuring guru responsible for the famous "Keeling Curve" of rising atmospheric CO_2 concentrations.[35] In the letter, it was suggested that there were three possibilities to address the situation. One was to pursue a dramatically increased level of effort on characterizing, modifying, and fixing the existing VIIRS. The second was to essentially abandon NPP VIIRS for ocean color, concentrate on getting the VIIRS for the first NPOESS satellite right, and launch a "quick" global ocean color mission, in the tradition of QuikSCAT (which made it to orbit and is still performing admirably in place of the lost scatterometer on Midori), and QuikTOMS (which was not nearly so successful). The third alternative was to be stuck with the deficient VIIRS on NPP. The letter recognized that NASA's other plans for instruments capable of acquiring acceptable ocean color data were on the far side of the 2013-2016 timeframe, and sharing data with other countries and agencies with working sensors (notably MERIS on Envisat) wasn't a very good alternative for many researchers, either.[36] Paula Bontempi indicated that NASA has had somewhere between 5-7,000 users that had obtained data from the OBPG, while MERIS has had around 600.[37] The other clear consideration was the age of the sensors in orbit; SeaWiFS had just passed ten years in space, and MODIS-Aqua five. The warranty on these sensors had been exceeded, and just about everyone remembered the 11-year data gap between CZCS and SeaWiFS

NASA and NOAA didn't eagerly jump on the idea of a gap-filling ocean color mission

---

[33] Jim Gleason, "NPOESS Preparatory Project (NPP) Status."
[34] Susan K. Mashiko, "Written Statement of Brigadier General Susan K. Mashiko, Program Executive Officer for Environmental Satellites," http://www.ogc.doc.gov/ogc/legreg/testimon/110f/Mashiko0607.pdf, (accessed 15 September 2009).
[35] David Siegel and James Yoder, "Community Letter to NASA and NOAA Regarding Concerns Over NPOESS Preparatory Project VIIRS Sensor," http://www.spaceref.com/news/viewsr.html?pid=25593, (accessed 15 September 2009).
[36] David Siegel and James Yoder, "Community Letter to NASA and NOAA Regarding Concerns Over NPOESS Preparatory Project VIIRS Sensor."
[37] Paula Bontempi, interview recorded 27 October 2010.

to compensate for the inadequacies of VIIRS; instead, the program intended to address the problems via testing and software fixes. NOAA's Deputy Undersecretary for Oceans and Atmosphere, Mary Glackin, sent a letter to Siegel and Yoder in response to their concerns about the NPP VIIRS instrument; the letter indicated that while it would be impossible to fix the NPP VIIRS instrument, NOAA would make concerted efforts to fix the VIIRS instrument on the first NPOESS satellite. NOAA would also investigate expediting the acquisition of MERIS data for U.S. researchers, as well as acquiring data from India's Oceansat-2 mission (more on this below). Finally, although cost was obviously a major consideration, NOAA would discuss an ocean color free-flyer with NASA.[38] Furthermore, the National Research Council was requested to convene a study project entitled "Assessing Requirements for Sustained Ocean Color Research and Operations", a project which would begin in February 2010, chaired by Jim Yoder.[39]

Meanwhile, the schedule was pushed back even further; while a 2009 launch of NPP and a 2011 launch of the first NPOESS satellite had been planned in 2007, by 2009 the NPP launch had been moved to 2011, and the first NPOESS launch was now scheduled for 2014. In the IOCCG News for September 2009 (where history and current events intersect in this chapter), it was also noted that NPP had been elevated from its testing and risk-reduction role to a more vital operational status – ultimately meaning that it had been promoted to be the first NPOESS satellite.[40] This short article also indicated that the Obama Administration was evaluating the alternatives to the NASA-NOAA-DoD management strategy – which by this stage in the NPOESS saga, obviously hadn't functioned very well.

Still, the program had been trying to address the VIIRS problems; two presentations at the 2009 American Meteorological Society Annual Meeting indicated what was being done. The first presentation indicated that the pesky problem with measuring the RSRs of VIIRS had been accomplished; the influence of the optical crosstalk problem on the Environmental Data Records (EDRs) from this VIIRS had been evaluated. According to the presentation, there would be minimal effects on the VIIRS Cloud Mask, Aerosol, and Land data products. The ocean color problems would be mitigated by the RSRs used in the atmospheric correction algorithm.[41] The other presentation indicated that the ocean color products would likely be lower quality than specification, but that NGST was making algorithm corrections, and – *maybe* – these corrections would get the data closer to the

---

[38] Mary Glackin, email message, forwarded to the author on 2 December 2009 by Stan Wilson.
[39] The National Academies, "Project Information: Assessing Requirements for Sustained Ocean Color Research and Operations," http://www8.nationalacademies.org/cp/projectview.aspx?key=49185, (accessed 11 January 2010).
[40] "NPOESS Program", IOCCG News September 2009, http://www.ioccg.org/news/Sept2009/news.html, (accessed 15 September 2009).
[41] David Lewis, Steve Mills, Jim McCarthy, Justin Diehl, Mau Song Chou, Gary Grimm, Jackie Jaron, Rush Patel, and Brendan Robinson, "VIIRS sensor radiometric performance overview," *Paper 151332*, January 1, 2009, http://npoess.noaa.gov/ams/2009/posters/AMS_09_&AGUVIIRSRadiometric_PerformancePoster.Rev4.12.05.pdf, (accessed 15 September 2009).

specifications, particularly at a good distance from bright targets like clouds.[42] (The idea of removing cloud cover over the oceans as a mitigation strategy was apparently not considered.)

Thus, more than 30 years after the launch of CZCS, the NPP VIIRS instrument is still bedeviled by the difficulty of sensing the dark ocean next to the bright boundary, echoing the sensor ringing problem of the CZCS and the bright target recovery tail of SeaWiFS, and even the stray light problems of MODIS. The travails of VIIRS emphasize that the color of the atmosphere with the ocean below is still the most difficult remote-sensing target to examine from space. It's unfortunate that someone can't make another contemplative stroll across a cow pasture and come up with a clever solution to this particular problem.

If not VIIRS… ?

There are potential alternatives to VIIRS; some of these alternatives belong to NASA, but are several years from fruition; some belong to other countries; and some would likely be feasible for a few hundred million dollars or so.

The primary alternative currently in space to MODIS-Aqua and SeaWiFS is the ESA's MERIS.[43] One problem with MERIS is the same as the problem with MODIS-Aqua; they were both launched in 2002, and so both are seven years into their operating lifetime in 2009. So MERIS faces the same uncertainty as MODIS-Aqua in terms of how long it will continue to operate in space. The community letter to the NOAA and NASA administrators noted that MERIS has a much narrower scanning swath than SeaWiFS, and that is difficult to acquire the base level of data that would allow examination of calibration issues and their impact on the algorithms for geophysical products. Furthermore, the letter noted that MERIS does not have a vicarious calibration program akin to MOBY. [44] Thus, while MERIS has demonstrated comparable data quality to MODIS-Aqua, relying on MERIS as a solution to loss of data from SeaWiFS or MODIS-Aqua is not what the scientists were looking for.

The European Space Agency utilized data from all three of the world's finest ocean color sensors to construct the GlobColour data set, which was a merger of 55 Terabytes of data.[45] Globcolour was expected to continue as part of the Global Monitoring and Security (GMES) program, which will also construct and launch ESA's next ocean color satellite.

---

[42] Bonnie Reed, B. Guenther, C. Hoffman, H. Kilcoyne, G. Mineart, and K. St. Germain, "Early performance predictions of ocean environmental data records for the NPOESS Preparatory Project VIIRS sensor," *Paper 151260*, http://npoess.noaa.gov/ams/2009/posters/AMS_09_OCEANS_EDRs_NPP_VIIRS-Reed_R5.pdf, (accessed 15 September 2009).
[43] ESA Earthnet, "The Medium Resolution Imaging Spectrometer Instrument."
[44] David Siegel and James Yoder, "Community Letter to NASA and NOAA Regarding Concerns Over NPOESS Preparatory Project VIIRS Sensor."
[45] European Space Agency, "Ocean carbon cycle research gets boost from satellite data," Science Daily, (7 May 2008), http://www.sciencedaily.com/releases/2008/05/080505094125.htm, (accessed 30 October 2009).

The community letter also highlighted that while there are several other ocean color scanners in space, they all have similar concerns: limited data acquisition capabilities compared to SeaWiFS and MODIS-Aqua, and more pressingly, they are not nearly as well understood in terms of calibration and data quality as the NASA mission instruments.[46] Indeed, it is likely difficult to emulate either the dedication of the NASA personnel or the funding that NASA, NOAA, and NIST were able to devote to calibration and validation of the sensors to insure their preeminent data quality.

So the next potential consideration is future missions. One mission stands right at the convergence of history and headline news: the Indian Oceansat-2 mission. Oceansat-2 is a sequel to India's successful Oceansat-1 mission, which launched on May 26, 1999.[47] Oceansat-1 carried the Ocean Colour Monitor (OCM), an instrument with a similar complement of bands to SeaWiFS. The similarities end there, however. SeaWiFS, MODIS-Aqua, and other remote-sensing instruments that use scanning mirrors (or telescopes) are classified generically as "cross-track" scanners: the scan of the instrument sweeps across the ground track of the satellite as it moves over the Earth.[48] The design element of the scanning mirror (or telescope) requires technological sophistication; the advantage of cross-track scanners is that they only require a small number of detector elements. In contrast, OCM is a "pushbroom", or "along-track" scanner: this type of sensor doesn't need a rotating scanner, and is thus simpler in terms of the machinery required to perform its remote-sensing observations. A pushbroom scanner collects all of the data for one band across the width of the scan area. As the satellite moves over the scan area, data is collected successively for each band; the speed of an orbiting satellite makes this essentially a simultaneous data acquisition.[49] The downside to a pushbroom scanner is that it requires many more detectors to be able to acquire all the data "at once" over the scan area; and it is difficult (ranging to impossible) to fully characterize the calibration for all the detectors. OCM, in fact, had 3,740 detectors; the problems with detector calibration cause OCM data to be strongly striped (and not in the same way for each band), requiring sophisticated algorithms for de-striping so that useful data can be derived.[50]

Despite the potential deficiencies of such a design, India had been planning to launch Oceansat-2 with OCM-II by 2002; but as seems to happen in the game of orbital satellites, they have experienced delays, such that as of September 2009, the launch of Oceansat-2 was scheduled to occur in late September.[51] In fact, it did: Oceansat-2 launched successfully on

---

[46] European Space Agency, "Ocean carbon cycle research gets boost from satellite data."
[47] T. Srinivasa Kumar, "Update on OCM on Oceansat-1 & scheduled OCM on Oceansat-2," http://www.ioccg.org/sensors/ocm/Kumar%20-OCM.pdf, (accessed16 September 2009).
[48] Nicholas Short, "Sensor Technology: Types of Resolution," http://rst.gsfc.nasa.gov/Intro/Part2_5a.html, (accessed 16 March 2010).
[49] Nicholas Short, "Sensor Technology: Types of Resolution."
[50] Paul E. Lyon, "An automated de-striping algorithm for Ocean Colour Monitor imagery," *International Journal of Remote Sensing*, 30 (5-6), 1493-1502, (March 2009).
[51] V Jayaraman, V. S. Hegde, Mukund Rao, and H Honne Gowda, "Future earth observation missions for oceanographic applications: Indian perspectives,", *Acta Astronautica* 44(7-12), 667-674, April-June 1999; "India to launch Oceansat-2 on September 23: ISRO,"

September 23, 2009 (Figure 8.2).[52] If it continues to function successfully in space, OCM-II could be a potential new source of ocean color data that could cover the gap if SeaWiFS and MODIS-Aqua data acquisition falters. The potential of using OCM-II data has not gone unnoticed by NASA; in January 2007, Ocean Biology and Biogeochemistry Program Manager Paula Bontempi told the annual IOCCG meeting that NASA was discussing the possibility of getting Oceansat-2 OCM-II data and making it possible to process with SeaDAS; at the time, Bontempi indicated that there was a chance they might even be able to get the data free of charge.[53] As of May 2008, OCM-II was still in the mix, and Bontempi noted to the Ocean Color Research Team (meeting jointly at NASA'S Carbon Cycle and Ecosystems meeting) that its design provided more potential for global climate research than some of the alternatives.[54]

In November 2009, during a meeting of the Group on Earth Observations in Washington, D.C., NASA, NOAA, and ISRO signed an agreement (a "Letter of Intent") which allow U.S. agencies to use data from Oceansat-2 for "research, education and activities of public good." Bontempi noted that it took 2 years of complex negotiations to reach that stage.[55] The described activities included data calibration and validation, and algorithm development. So Oceansat-2 data could potentially provide a data bridge to the middle of the 21st century's teen-age years.[56] But Bontempi also noted that the vital first step of actually sharing data with ISRO has yet to be realized.[57]

One of the other alternatives is ESA's Sentinel-3 satellite. The Sentinel series is envisioned as follow-on missions to Envisat and the previous ERS-1 and ERS-2 missions, which carried synthetic aperture radar (SAR) instruments. Thus, Sentinel-1 will be a SAR satellite, and Sentinel-2 will be primarily devoted to land surface imaging, following the heritage of the SPOT satellite series.[58] Sentinel-3 will carry both the Ocean Land Colour Instrument (OLCI) and the Sea Land Surface Temperature Radiometer (SLSTR) – apparently the European science community understood the lesson about putting visible and

---

http://www.ptinews.com/news/284743_India-to-launch-Oceansat-2-on-Sep-23--ISRO, (accessed 16 September 2009).

[52] "Oceansat-2 launch is a success," *The Gulf Times*, September 23, 2009, http://www.gulf-times.com/site/topics/article.asp?cu_no=2&item_no=316272&version=1&template_id=40&parent_id=22, (accessed 1 October 2009).

[53] "Twelvth IOCGG Committee Meeting," http://www.ioccg.org/reports/ioccg_meeting12.html, (accessed 16 September 2009).

[54] Paula Bontempi, "State of the Program: NASA Ocean Biology and Biogeochemistry," http://cce.nasa.gov/meeting_2008/oceans_pres/Bontempi_OCRT2008_NASA%20Vision.ppt, (accessed 16 September 2009).

[55] Paula Bontempi, interview recorded 27 October 2010.

[56] IOCCG, "NASA signs agreement with ISRO for Oceansat-2 data," *December 2009 News*, IOCCG, http://www.ioccg.org/news/Dec2009/news.html, (accessed 28 December 2009).

[57] Paula Bontempi, interview recorded 27 October 2010.

[58] ESA, "Contract signed for building of GMES Sentinel-1 satellite," (18 June 2007), http://www.esa.int/esaEO/SEMBRT7OY2F_index_0.html, (accessed 16 September 2009); ESA, "Sentinel-2," November 17, 2008, http://www.esa.int/esaLP/SEMM4T4KXMF_LPgmes_0.html, (accessed16 September 2009).

Figure 8.2. Launch of Oceansat-2. Six small European "nanosatellites" were also carried into space by this launch.

thermal bands on different instruments.[59] Sentinel-3 is a component of the GMES program, a joint European program for earth observation remote sensing. Sentinel-3 will also carry a radar altimeter and accompanying microwave radiometer.

The Sentinel-3 instrument of highest interest to the ocean color community will be OLCI, which is a five-camera, 21-band, pushbroom, *sun-glint free* instrument, capable of 300 meter resolution, but it will only collect data over the opean ocean at 1.2 km resolution, saving the higher resolution data acquisition (and its correspondingly higher data volume) for coastal areas.[60] The lack of sun glint is accomplished by aligning the cameras at a permanent 12.2° tilt. As Paula Bontempi had noted, one of the main drawbacks of the OLCI is a narrow swath width.[61] The revisit specification for OLCI is less than 2 days, but like the paired MODIS system envisioned for the original EOS, accomplishing this would require an

---

[59] Miguel Aguirre, Bruno Berruti, Jean-Loup Bezy, Mark Drinkwater, Florence Heliere, Ulf Klein, Constantinos Mavrocordatos & Pierluigi Silvestrin, "Sentinel-3: The Ocean and and Medium Resolution Land Mission for GMES Operational Services," *ESA Bulletin*, 131, 25-29, (August 2007), http://www.esa.int/esapub/bulletin/bulletin131/bul131c_aguirre.pdf, (accessed 17 September 2009); Samantha Lavender, "Sentinel 3 – Ocean Observations," *GEO-CAPE: Geostationary Coastal and Air Pollution Events Science Definition Planning Workshop*, August 18-20, 2008, http://geo-cape.larc.nasa.gov/docs/workshop2008/Lavender.ppt, (accessed 17 September 2009).

[60] Samantha Lavender, "Sentinel 3 – Ocean Observations," *GEO-CAPE: Geostationary Coastal and Air Pollution Events Science Definition Planning Workshop*, August 18-20, 2008, http://geo-cape.larc.nasa.gov/docs/workshop2008/Lavender.ppt, (accessed 17 September 2009).

[61] Paula Bontempi, "State of the Program: NASA Ocean Biology and Biogeochemistry."

OLCI on more than one satellite. However, currently only the first Sentinel-3 is being built.[62] While in 2007 a 2011/2012 launch was planned, by 2008 the launch year had quietly slipped to 2013.[63] Sentinel-3 is envisioned as a series of satellites, and the ESA has committed €830 million the second phase of GMES, which plans the launch of Sentinel-3B for 2017.[64] There will therefore not be much of a mission overlap, and this indicates that a single OLCI in orbit will not provide global ocean color data with high temporal resolution.

The extended time horizon includes other missions, including missions from NASA. Japan is planning to try ocean observations again with their Global Change Observation Mission for Climate research (GCOM-C) mission, expecting a launch in January 2014. The GCOM-C manifest includes, importantly, the Second-generation Global Imager (SGLI), and the current instrument characteristics feature 250-meter resolution in coastal areas and 1 km for open ocean, similar to OLCI, and a 1400 km swath width – both of which would be very desirable to the ocean color science field.[65]

NASA has two missions with specific ocean observational capabilities under consideration: the Aerosol/Clouds/Ecosystem (ACE) and the Geostationary Coastal and Air Pollution Events (GEO-CAPE) missions. ACE was mentioned in the letter from the ocean science community to the NOAA and NASA administrators; the letter stated that "recommends a 2013-2016 launch date for the Aerosol/Cloud/Ecosystems (ACE) mission", and then adds, "Our understanding, however, is that even a 2016 launch for ACE is optimistic and would require funding for mission planning to begin no later than 2009." [66] This quote references the "Decadal Survey", which was a National Research Council effort sponsored by NASA, NOAA and the U.S. Geological Survey to determine a ten-year strategy for the deployment of earth-observing instruments to optimize the data available to examine global environmental change.[67] The Decadal Survey was formulated to guide the decision-making process for NASA by prioritizing the data needed to better characterize the global environment.[68] The Decadal Survey ultimately placed the highest priority on missions to characterize solar and earth radiation, ice sheet dynamics, and soil moisture, such that

[62] GMES, "SWIFT E-News No. 3," February 2, 2009, http://www.gmes.info/pages-principales/library/newsletter/newsletters-2009/?no_cache=1&download=SWIFT_E-News03.pdf&did=81, (accessed 17 September 2009).

[63] Samantha Lavender, "Sentinel 3 – Ocean Observations."

[64] Miguel Aguirre, Bruno Berruti, Jean-Loup Bezy, Mark Drinkwater, Florence Heliere, Ulf Klein, Constantinos Mavrocordatos & Pierluigi Silvestrin, "Sentinel-3: The Ocean and and Medium Resolution Land Mission for GMES Operational Services"; GMES, "SWIFT E-News No. 3."

[65] Kazuhiro Tanaka, Yoshihiko Okamura, Takahiro Amano, Masaru Hiramatsu, and Koichi Shiratama, "Development status of the Second-generation Global Imager (SGLI) on GCOM-C1," http://suzaku.eorc.jaxa.jp/GCOM_C/presen/SGLI_at_SPIE_Berlin_2009a.pdf, (accessed 17 September 2009).

[66] David Siegel and James Yoder, "Community Letter to NASA and NOAA Regarding Concerns Over NPOESS Preparatory Project VIIRS Sensor."

[67] Space Studies Board, National Research Council of the National Academies, *Earth Science and Applications from Space: National Imperatives for the Next Decade and Beyond*, The National Academies Press, 428 pages, (2007).

[68] Space Studies Board, "Executive Summary," in *Earth Science and Applications from Space: National Imperatives for the Next Decade and Beyond*, The National Academies Press, 1-16, (2007).

missions addressing those elements were recommended for launch in 2010-2013, so that the ACE and GEO-CAPE missions were recommended for the 2013-2016 period. The report expressed concerns about the NPOESS costs and delays, the deficiencies of VIIRS, and also expressed the need to continue the ocean color data record from CZCS, SeaWiFS, and MODIS.[69]

The mission profile for ACE would improve ocean color measurements by placing three instruments intended to measure atmospheric aerosol, cloud and precipitation properties – a lidar, microwave radar, and polarimeter – side-by-side with an ocean color radiometer. The characterization of aerosols by these instruments addresses the recognized uncertainties in climate change due to aerosol radiative effects.[70] ACE was considered to be a "Large" mission, with an estimated cost of $800 million. The ocean color radiometer considered for ACE is tentatively called GOCEP (Global Ocean Carbon, Ecosystems, and Coastal Processes), and would be a 19-band instrument covering the UV, visible, and near-IR range. GOCEP is projected to be able to meet the science requirements that VIIRS could fall short of. As stated, "It was noted that new observational requirements for advancing research in ocean biology and biogeochemistry would not be met by the VIIRS sensor, or any simple expansion of the SeaWiFS or MODIS designs. GOCECP [sic] was the path for meeting the science needs of today and into the future." The drawback – ACE is not slated for launch until 2015 – at the earliest.[71]

[Because of this delay, and the expectation that a 2015 launch date isn't too likely (see below), NASA is also planning a mission simply called "PACE", which stands for "Pre-ACE". The mission would be funded under a new climate initiative, to maintain the continuity of ocean color, climate, and carbon data. The PACE mission will include an ocean color radiometer.[72] PACE is currently scheduled for a 2018 launch, because ACE has been pushed back to at least 2020. PACE is being planned as a joint NASA-CNES mission.[73]]

GEO-CAPE is a different kind of mission altogether; rather than being deployed on a conventional polar-orbiting satellite, GEO-CAPE would be placed in geostationary orbit, to perform repeated measurements of the same hemisphere of the Earth with very high temporal resolution. The emphasis of GEO-CAPE is the human impact on the oceanic coast, and the goals of the mission would be to characterize air pollution impacts and the

---

[69] Space Studies Board, "Chapter 7: Land-Use Change, Ecosystems, and Biodiversity," in *Earth Science and Applications from Space: National Imperatives for the Next Decade and Beyond*, pages 190-216, (2007).
[70] Space Studies Board, "Chapter 4: Summaries of Recommended Missions," in *Earth Science and Applications from Space: National Imperatives for the Next Decade and Beyond*, pages 79-140, (2007).
[71] International Ocean Colour Coordinating Group, "Minutes, 13th IOCCG Committee Meeting Paris, France, 12-14 February 2008," http://www.ioccg.org/reports/Minutes-IOCCG-13.pdf, (accessed 24 December 2009).
[72] International Ocean Colour Coordinating Group, "March 2011 News," http://www.ioccg.org/news/March2011/news.html, accessed 28 April 2011.
[73] Charles McClain, "Overview & Status of The Decadal Survey Aerosol-Cloud- Ecology (ACE) Mission," 4 May 2009, http://oceancolor.gsfc.nasa.gov/DOCS/ScienceTeam/OCRT_May2009/ACE_Overview_OCRT_McClain_May09.pdf, accessed 2 May 2011.

dynamics of coastal ecosystems, utilizing the high repeat capability of observations from a geostationary satellite. For coastal ecosystems, these observations would allow assessment of storm impact, tidal forcing, short-term transient events like floods, and hazards such as oil spills and toxic algal blooms.[74] GEO-CAPE was classified as a Medium-size mission, with a price tag around $550 million dollars.

Neither ACE nor GEO-CAPE is sufficiently advanced in development to have a named ocean color instrument on its manifest. Thus, the Decadal Survey emphasized the need to continue and improve ocean color observations in two different ways, but by placing higher priorities on other measurements, the survey pushed any potential gap-filling missions to the outer boundaries of expected mission duration for MODIS-Aqua. This outcome puts greater pressure on VIIRS to somehow provide better data than it is currently expected to provide for the purposes of research (particularly in the United States), or it requires innovative strategies, such as SIMBIOS-like international partnerships with countries possessing working ocean color instruments.

NASA has one other mission being considered that offers intriguing possibilities for ocean data, though it is primarily being considered to observe land surfaces. This satellite is called HyspIRI (Hyperspectral Infrared Imager), and is potentially scheduled to launch in the 2013-2016 timeframe. The HyspIRI concept includes two instruments, one a visible and shortwave IR instrument with bands between 400 and 2500 nm, and a thermal IR instrument with bands between 8 and 12 μm. It will be possible to make atmospheric corrections on HyspIRI data to allow hyperspectral scanning of the ocean surface and the acquisition of ocean surface reflectances. The white paper on HyspIRI published in May 2009 delineates a large set of capabilities for marine studies: "… data from HyspIRI will allow for better separation of phytoplankton pigments and phytoplankton FGs [functional groups] such as carbon exporters (diatoms), nitrogen fixers (*Trichodesmium* sp.), calcium carbonate producers (coccolithophores), and the microbial loop organisms (*Prochlorococcus* sp.). More spectral information helps improve the accuracy and diversity in retrievals of absorption and backscattering coefficients as well as other environmental properties through inversion algorithms, including effective discrimination of biogeochemical constituents of the water and seafloor (e.g., colored dissolved organic matter [CDOM], phytoplankton concentration and composition, suspended sediments, bottom type) and physical properties (e.g., temperature, bathymetry, light attenuation)…" The price tag for HyspIRI is attractive, estimated at $300 million dollars.[75]

---

[74] Charles McClain, "Overview & Status of The Decadal Survey Aerosol-Cloud- Ecology (ACE) Mission."

[75] National Academy of Sciences, "Summaries of Recommended Missions: Hyperspectral Infrared Imager (HyspIRI)," http://cce.nasa.gov/pdfs/HYSPIRI.pdf, accessed 2 April 2010; Bo-Cai Gao, "Derivation of Level 2 Products: Terrestrial and Aquatic Surface Reflectances," *Second HyspIRI NASA Decadal Survey Mission Science Workshop*, http://hyspiri.jpl.nasa.gov/downloads/public/2009_Workshop/day1/day1_PMA_8._pres_Gao_HyspIRI_L2_Refl_2009_V1rog.pdf, accessed 2 April 2010; HyspIRI Group, "NASA 2008 HyspIRI Whitepaper and Workshop Report," *JPL Publication 09-19*, http://hyspiri.jpl.nasa.gov/downloads/public/2008%20HyspIRI%20Whitepaper%20and%20Science%20Workshop%20Report-r2.pdf, accessed 2 April 2010.

There was, however, one other possible alternative path directly to ocean color data, with its own set of technical and scientific hurdles. The ocean color community letter also described this alternative: "A gap-filling mission in the spirit of SeaWiFS can be implemented and flown. SeaWiFS is a success story of the "better, faster, cheaper" version of NASA. A dedicated, single-instrument ocean mission can be flown easily and cost-effectively."[76] The twist on this concept was that one of the most important components of such a mission was already in existence: a second SeaWiFS, sitting on the shelf (or more likely in a clean room).[77] The prospects for such a resurrection were quite daunting, and SeaWiFS engineer Alan Holmes did not think it was feasible.[78] The electronics boards for the second SeaWiFS have not been built, and this would likely present significant challenges. Further, SeaWiFS demonstrated than an instrument with a few more bands that it possessed would make much greater contributions to 21st century science. So the second SeaWiFS still sits.

Why Ocean Colour?

It is somewhat ironic that at this crucial point in the history of ocean color remote sensing, the IOCCG would have recently published a report entitled "Why Ocean Colour? The Societal Benefits of Ocean Colour Technology".[79] The report extensively discusses the many scientific aspects of oceanography that have been influenced by ocean color observations from space: ocean physics; biogeochemical cycling (particularly the cycling of carbon); marine ecosystems and the interaction of marine trophic levels – ultimately influencing fisheries; water quality and hazards, both natural and anthropogenic; and climate change. The introductory chapter also discussed how ocean color observations are related to the public interest; ocean color can allow insight into concerns such as pollution, the state of marine mammal populations, and how the biological effects of an El Niño thousands of miles away in the central Pacific Ocean can be felt and observed along the inhabited coastline.[80]

The IOCCG report was written as SeaWiFS was nearing its unprecedented completion of a decade of nearly continuous ocean color remote sensing, and thus it also nearly coincided with the 30-year anniversary of the commencement of ocean color

---

[76] David Siegel and James Yoder, "Community Letter to NASA and NOAA Regarding Concerns Over NPOESS Preparatory Project VIIRS Sensor."

[77] M. Gregory Hammann, and Jeffrey J. Puschell, "SeaWiFS-2: an ocean color data continuity mission to address climate change," in Philip E. Ardanuy and Jeffrey J. Puschell, *Proceedings SPIE*, 7458(745804), "Remote Sensing System Engineering II", doi: 10.1117/12.828949, (2009).

[78] Alan Holmes, telephone interview recorded on 14 April 2009.

[79] IOCCG, *Why Ocean Colour? The Societal Benefits of Ocean-Colour Technology*, Trevor Platt, Nicolas Hoeppfner, Venetia Stuart, and Christopher Brown, Editors, IOCCG Report Number 7, International Ocean Color Coordinating Group, Dartmouth, Nova Scotia, ISSN 1098-6030, 139 pages, (2008).

[80] James Acker, "Ocean-Colour Radiometry and the Public," in *Why Ocean Colour? The Societal Benefits of Ocean-Colour Technology*, edited by Trevor Platt, Nicolas Hoeppfner, Venetia Stuart, and Christopher Brown, pages 3-12, (2008).

observations from space on Nimbus 7. Because the final sentences of the introductory chapter summarize the state of the science and its relationship to the public (and because the authorship of that chapter and this history is the same), it seems fitting to close this assessment of NASA's leading role in the remarkably difficult field of ocean color observations from satellites in the same way: "At the end of the first decade of routine ocean colour observations, the increased use of ocean-colour data in applications that are directly related to the public interest is becoming more evident. It is certain that future ocean-colour radiometry missions will provide data addressing an increasing variety of topics where the interests of the scientific research community and the public would derive mutual benefit - a significant reason to invest in the continuation of accurate ocean-colour remote sensing from space."

It all comes down to *accuracy*: it isn't difficult to observe the Earth from space, but successfully acquiring the slivers of light that escape the ocean and the atmosphere, and deriving scientifically useful information from them, is as difficult (or more) than capturing the light from distant nebula and galaxies on the mirrors and detectors of the Hubble Space Telescope. While the HST has been renowned for both its billion-dollar repairs and the wonders of space that it has shown to human eyes, the wonders of the variegated colors of the ocean have been quietly captured by NASA instruments as well, and they have shown our eyes similar breathtaking views of the oceanic realm that could only be captured from the vantage point of space.

Addendum: The period covered by this history essentially concludes in December 2010 with the termination of the SeaWiFS mission. Some of the prospective launch dates in the previous sections are already outdated! However, three significant events which took place prior to publication of this history have had an important bearing on the topics of this chapter.

The first event of note was the end of the tri-agency NPOESS program and the creation of a partnership program between NASA and NOAA. This change, instituted on February 1, 2010 by the Obama administration, resulted in a new name for the program, the Joint Polar-orbiting Satellite System (JPSS). The Department of Defense would operate its own weather satellites, under the direction of the Air Force.[81] (This change was the result of the administration's evaluation of the NPOESS program, which is mentioned in the body of the chapter.)

The second event was the launch of NPP with VIIRS. The satellite, renamed Suomi (after the "father of satellite meteorology," Verner Suomi, with the mission retitled the Suomi National Polar-orbiting Partnership, to preserve the NPP acronym) launched into

[81] Turner Brinton, "White House dissolves NPOESS partnership in blow to Northrop," Space News, February 2, 2010, http://www.spacenews.com/article/white-house-dissolves-npoess-partnership-blow-northrop/ , accessed July 31, 2013.

Vandenberg's dawn sky on October 28, 2011.[82] There were still lingering scientific concerns about how NOAA would process the ocean color data products from VIIRS, along with the previously described worries about the optical crosstalk in the instrument. As the instrument commissioning phase began, however, VIIRS immediately began providing Earth observation imagery of remarkably high quality, with additional features including a high-sensitivity day-night observational band that returned beautiful images of nighttime lights on the Earth's surface, and even clouds illuminated by light from the full Moon.[83]

Unfortunately, however, as the commissioning phase continued, another optical concern arose – and again, similar to MODIS-Terra, this concern would impact ocean color data product quality more than any other data type. Calibration data acquired during the commissioning phase indicated that several bands, most notably the near-IR bands that provided the data for atmospheric correction of ocean color data, were degrading much faster than expected.[84] The commissioning phase for all of the Suomi instruments was halted to investigate this degradation.

The investigation concluded that the vitally important main mirrors of the instrument had been contaminated with tungsten oxide during the coating process, and the exposure of the mirror surfaces contaminated with tungsten oxide to UV radiation resulted in a "darkening" of the surface.[85] There was a relatively happy outcome to the investigation, as the degradation curves indicated that the initial rapid rate of change would stabilize into a much slower rate that could be monitored and corrected for with ongoing calibration efforts. The result for the ocean color community was that they had to wait for almost a year, until early 2013, for the instrument to begin gathering data of sufficient accuracy that ocean color data products could be generated. Thus, during 2013, NASA was evaluating VIIRS data to determine if it could be processed to meet its science quality requirements.[86]

---

[82] http://www.nasa.gov/mission_pages/NPP/news/suomi.html, accessed 29 July 2013; http://earthobservatory.nasa.gov/Features/Suomi/, accessed 29 July 2013; http://www.nasa.gov/mission_pages/NPP/launch/, accessed 29 July 2013.

[83] Quanhua (Mark) Liu, Changyong Cao, and Fuzhong Weng, "NPP VIIRS emissive band radiance calibration," in James J. Butler, Xiaoxiong Xiong, and Xingfa Gu, editors, *Proceedings SPIE 8510, Earth Observing Systems XVII*,. doi: 10.1117/12.930201.

[84] Robert E. Eplee, Jr., Kevin R. Turpie, Gwyn F. Fireman, Gerhard Meister, Thomas C. Stone, Frederick S. Patt, Bryan A. Franz, Sean W. Bailey, Wayne D. Robinson, and Charles R. McClain, "VIIRS On-Orbit Calibration for Ocean Color Data Processing," in James J. Butler, Xiaoxiong Xiong, and Xingfa Gu, editors, *Proceedings SPIE 8510, Earth Observing Systems XVII*, doi:10.1117/12.930483; F.J. De Luccia, D.I. Moyer, E.H. Johnson, K.W. Rausch, N. Lei, K. Chiang, X. Xiong, E.M. Haas, J. Fulbright, and G. Iona, "Discover and characterization of on-orbit degradation of the VIIRS rotating telescope assembly," in James J. Butler, Xiaoxiong Xiong, and Xingfa Gu, editors, *Proceedings SPIE 8510, Earth Observing Systems XVII*, doi:10.1117/12.930544.

[85] Quanhua (Mark) Liu, Changyong Cao, and Fuzhong Weng, "NPP VIIRS emissive band radiance calibration"; Slawomir Blonski and Changyong Cao, "Monitoring and predicting rate of VIIRS sensitivity degradation from telescope contamination by tungsten oxide," in Khanh D. Pham, Joseph L. Cox, Richard T. Howard, Genshe Chen, editors, *Proceedings SPIE 8739, Sensors and Systems for Space Applications VI*, doi: 10.1117/12.2016008.

[86] Kevin R. Turpie, Barney Balch, Bruce Bowler, Bryan A. Franz, Robert Frouin, Watson Gregg, Charles R. McClain, Cecile Rousseaux, David Siegel, and Menghua Wang, "NASA Science Team Assessment of S-NPP VIIRS Ocean Color Products," *International Ocean Color Science Meeting*, May 8,

The third event was the sudden end of the ENVISAT mission.[87] ENVISAT was important for the ocean color community, particularly the European community, because it carried MERIS, which had become an established source of reliable ocean color data, and was a reliable alternative to MODIS-Aqua. As noted in the body of this chapter, the on-orbit age of ENVISAT and MERIS was a concern of the scientists planning future ocean color missions.

In a situation eerily similar to the events that shortly followed the 10th anniversary of the SeaWiFS mission, ENVISAT abruptly stopped communicating with the ground systems on April 8, 2010, soon after celebrating its 10-year orbital anniversary. There were virtually no indications that the satellite was experiencing problems prior to this loss of communications. Efforts were made over the next several weeks to reestablish communications, but they were unsuccessful, so ESA declared the end of the mission on May 9, 2012.[88] At that time, this left MODIS-Aqua as the only reliable and accurate ocean color sensor in orbit, due to the discovery of the VIIRS mirror contamination and its impact on the atmospheric correction of VIIRS ocean color data.

And a final footnote: As this history was being readied for publishing, NASA announced that the PACE mission, described earlier in the chapter, was approved, with its launch year now slated for 2022, and a line in the budget worth $805 million dollars. According to the press release, "Goddard will build PACE's ocean color instrument. This PACE sensor will allow scientists to see the colors of the ocean, from the ultraviolet to near infrared, and obtain more accurate measurements of biological and chemical ocean properties, such as phytoplankton biomass and the composition of phytoplankton communities." So the scientific and technical heritage that began with CZCS and continued with SeaWiFS and MODIS will now be carried forward.[89]

Good luck and godspeed, PACE.

2013, http://iocs.ioccg.org/wp-content/uploads/1420-kevin-turpie-nasa-science-team-assessment-of-s-npp-viirs-ocean-color-products.pdf, accessed 31 July 2013.

[87] European Space Agency, "ESA declares end of mission for ENVISAT," *European Space Agency PR 15 2012*, May 9, 2012, http://www.esa.int/Our_Activities/Observing_the_Earth/Envisat/ESA_declares_end_of_mission_for_Envisat (accessed 30 July 2013).

[88] European Space Agency, "ESA declares end of mission for ENVISAT."

[89] NASA, "New NASA mission to study ocean color, airborne particles, and clouds," NASA Press Release 15-037, March 13, 2015, http://www.nasa.gov/press/2015/march/new-nasa-mission-to-study-ocean-color-airborne-particles-and-clouds (accessed April 27, 2015).

9

CONCLUSION

The Influence of Satellite Ocean Color Observations on Oceanography

Until the existence of aircraft, and subsequently satellites, oceanography was somewhat of a paradox as a science. While it was quite obvious that the oceans were vast and covered wide areas of the Earth's surface, the view afforded to oceanographers from the shore or from ships was therefore limited, discrete, and restricted in both space and time. Thus, oceanographers could really only study very small slices and sections of the vast oceans. Extrapolations from a limited and discrete data set were made to characterize the much larger, largely unseen, and relatively unknown extent of the global oceans.

There isn't any definitive historical event that marks the first scientific observations of the oceans from a high altitude; certainly someone on a balloon or an airplane must have noticed that you could see a much larger area from altitude than while on the deck of a ship or (even worse) on land. Aerial photography dates back to the early 1900s, and was conducted from kites, balloons, airplanes, and eventually homemade rockets, particularly ones built by Robert H. Goddard.[1]

Though there must have been aerial photography of oceanic targets, there is scant mention of such an activity. However, the term "remote sensing" is attributed to Evelyn Pruitt, a geographer and oceanographer from the Office of Naval Research.[2] Until there was systematic coverage of the oceans from satellites, the view from high altitude seems to have little impact on the field of oceanography, even though the idea clearly intrigued Gifford Ewing, as described in Chapter 1. The potential impact of all aspects of remote sensing on the ocean sciences, which apparently included high-altitude aircraft like the ERS-2 that could fly near the edge of space, was the main topic at the "Oceanography from Space" conference held in 1964.[3] By that time, photographs taken by astronauts from space, including views of the ocean, had already been seen by the public. For example, the Pacific Ocean, Gulf of California, and Gulf of Mexico made a dramatic backdrop for Edward White's spacewalk

---

[1] Paul R. Bauman, "History of Remote Sensing, Aerial Photography,",http://www.oneonta.edu/faculty/baumanpr/geosat2/RS%20History%20I/RS-History-Part-1.htm, accessed November 10, 2011.

[2] Nicholas Short, "Introduction: Technical and Historical Perspectives on Remote Sensing," in *Remote Sensing Tutorial* Part 2, Page 1, http://rst.gsfc.nasa.gov/Intro/Part2_1.html, accessed November 10, 2011.

[3] *Oceanography from Space; Proceedings of the Conference on the Feasibility of Conducting Oceanographic Explorations from Aircraft, Manned Orbital and Lunar Laboratories.* Edited by Gifford C. Ewing. GC2.C748 1964 c.4. Reference Number 65-10, Woods Hole Oceanographic Institution, Woods Hole, MA USA, 1965.

from Gemini 4. (Figure 9.1) [4]

Figure 9-1: Edward White photographed from the Gemini-4 capsule over El Paso, Texas. The body of water behind him is probably the Gulf of California.

Thus, since scientists tend to be pragmatic about new sources of data, it does not appear that there was particular anticipation that data and imagery from satellites would be important to the field of oceanography. However, as Richard Barber is quoted in Chapter 3, the first images from the Coastal Zone Color Scanner profoundly changed his oceanic worldview, and likely changed many other perspectives in the field. There was no doubt that the view from satellites significantly changed the previous basic understanding of the connectivity and discontinuities of oceanic processes. It was also immediately clear that satellite remote sensing allowed observations of oceanic regions far removed from easy access by research vessels. One other aspect of these observations deserves mention: satellites could observe phenomena occurring in the open ocean that were unsuspected and undetected by researchers on land. A prime example was the observation of the phytoplankton bloom in the Alboran Sea by Lohrenz et al. using CZCS data.[5] Were it not for the CZCS, this bloom would have been completely unnoticed by the oceanographic community. When the SeaWiFS mission commenced and began to provide virtually uninterrupted observations of the global oceans, the complexity of oceanic processes and the variability of the oceans in space and time, which the CZCS observations had only sampled, became even more apparent. One particular event early in the mission, the observation of the rapid El Niño – La Niña transition in the spring of 1998, demonstrated the remarkable usefulness of continuous observations to allow full characterization of large-scale oceanographic processes.[6]

[4] LIFE Magazine, Cover, June 18, 1965.

[5] Steven E. Lohrenz, Robert A. Arnone, Denis A. Wiesenburg, and Irene P. DePalma, "Satellite detection of transient enhanced primary production in the western Mediterranean Sea."

[6] F. P. Chavez, P. G. Strutton, G. E. Friederich, R. A. Feely, G. C. Feldman, D. G. Foley, and M. J. McPhaden, "Biological and Chemical Response of the Equatorial Pacific Ocean to the 1997-98 El Niño."

Even though it was clear from the advent of the first CZCS images that the view from space provided an altered (and improved) perspective on the oceans, it would take somewhat longer – and require some colored pencils at the beginning – to establish that the data the images from space depicted were accurate enough to be valuable for improved quantification of oceanic processes. This was true not only of ocean color, but sea surface temperature, winds, and other quantities that were now available from satellite measurements. Studies had to be performed that established that the data were good enough to be used for scientific research, and weren't just pretty pictures.[7]

Now that ocean color remote sensing has been firmly established as a research tool and a research expectation, the concerns of scientists about the NPOESS program – particularly the uncertainty of the ocean color data quality from the VIIRS instrument on the NPP satellite – have provided a good demonstration of how entrenched ocean color remote sensing has become to the oceanographic community.[8] Scientists from numerous disciplines now expect to be able to consult ocean color remote sensing data (and other types) as part of oceanographic and marine biological research in order to characterize environments and influential processes, even if the research topic itself is largely unrelated to what the instruments observe. The simplified data access system developed by the Goddard Earth Sciences Data and Information Services Center (GES DISC), named "Giovanni", has been used widely for diverse oceanographic research subjects that are much different than the quantification of primary productivity and surface biological variability which ocean color data has been traditionally used for.[9]

So to place ocean color observations in their proper position with regard to the advance of science, the breakthroughs that these data have provided for oceanography can be compared to other types of imaging advances in science and medicine. A classic example of seeing something with a new technology that could not previously be seen is X-rays. The discoverer of X-rays, Wilhelm Röntgen, perceived very quickly that this form of penetrating radiation would likely have an impact on medicine; the first picture taken with X-rays was of the hand of Röntgen's wife, showing her finger bones and wedding ring very clearly.[10] Within only a few year of the discovery of X-rays, radiology was established as a medical practice, with X-rays providing internal views of the living human body that were previously impossible. The use of X-rays in medicine was the beginning of a new era in medicine in which many other technologies to view the internal workings of the body were invented.[11]

As described in Chapter 1, the preliminary research that set the stage for ocean color observations from space was part of the rapid advances in science and technology that

---

[7] Wayne Esaias, comments recorded at the Ocean Color Collaborative Historical Workshop, January 13-14, 2009, St. Petersburg, Florida.

[8] David Siegel and James Yoder, "Community Letter to NASA and NOAA Regarding Concerns Over NPOESS Preparatory Project VIIRS Sensor."

[9] James G. Acker and Gregory Leptoukh, "Online analysis enhances use of NASA Earth science data," *Eos, Transactions of the American Geophysical Union*, 88(2), 14, doi:10.1029/2007EO020003.

[10] Otha W. Linton, "Medical Applications of X-rays," *Beamline*, Stanford Linear Accelerator Laboratory, Summer 1995, 25-34.

[11] Otha W. Linton, "Medical Applications of X-rays."

occurred in the 20th century. In particular, the need for improved data on light in the oceans, likely to help predict bottom visibility for landing craft, required the invention of underwater radiometers. This technical advance provided data that were utilized by Nils Jerlov in his paper on the optical properties of the global ocean.[12] One important aspect of this paper and subsequent research papers by Jerlov was that oceanic water masses possessed sufficient uniformity in optical properties over large areas to allow for their characterization from altitude, which was explicitly described by Charlie Yentsch in 1964.[13]

Therefore, just as it would be unimaginable at this point in time to conduct the practice of medicine without the capability to acquire images of the internal structures and processes of the human body, it is similarly unimaginable to conduct global and regional oceanographic studies without the capability to acquire images of the ocean from space. From the 1960s to the present day, ocean color observations from space evolved from a proof-of-concept novelty, to a useful tool, to their current status as a vital and fundamental data type for oceanographic research. Our knowledge of the way the oceans work is now indelibly colored by the numerous images that have been provided by ocean color instruments in space.

Because of these observations, we can no longer look at the oceans as simply the blue-green abyss that is perceived by our eyes. Rather, the technical achievement of ocean color that NASA pioneered and fostered allows us to see the marbled surface of the ocean with great clarity, and to understand the interactions of wind and water, light and organisms, and currents and continents that give rise to its ever-changing panoply of pattern. As humanity faces concerns about increasingly polluted waters, ocean changes related to climate change, declining and overexploited fisheries, and loss of biodiversity in degraded marine and coastal biomes, this source of continual ocean observations remains an important tool for both scientists and decision-makers to help understand, preserve, and restore the oceanic environment.

The Past and Current State of the Science

At this point in time, ocean color observations from space could be said to be at a crossroads, hackneyed though that cliché may be. NASA, which is dedicated to research, is not planning to launch a new and advanced ocean color sensor for several years. VIIRS data may require, and benefit from, extended efforts to refine them to scientific quality, but NOAA is not currently set up to conduct the calibration and validation and reprocessing of the data that has been required for this purpose in the past. Meanwhile, other nations are launching ocean color sensors, but it would not be unexpected that they will encounter similar difficulties to those experienced by NASA with SeaWiFS and MODIS.

One of the overriding lessons learned from the three NASA missions is the need for "agility" in the acquisition and processing of the data. The SeaWiFS mission in particular

[12] Jerlov, N.G., "Optical studies of ocean water."
[13] Charles Yentsch, "Distribution of chlorophyll and phaeophytin in the open ocean."

benefited from its four-year launch delay, which provided ample time for the data processing system to be ready at launch, which has rarely been the case for many other missions. Even so, the advances in computational power that occurred while the mission was active enabled faster and faster reprocessing of the data, to the point that the entire mission data set could be reprocessed in a day. This level of computational speed enables a full characterization of the impact of changes to data product algorithms – essentially, experiments can be run on the entire data set, not just a single scene or short period of time. Hearkening back to the CZCS, when processing a single scene could take a day (or more), this ability to test algorithms and the ever-changing calibration of the instrument (a fact of life for satellite sensors) makes the data more reliable and accurate, making the goal of creating a "climate data record" (CDR) realistic. For significantly larger data volumes like those produced by MODIS, the capability to perform such vital data reprocessing is difficult to achieve, but just as important to qualify the data for scientific research. The potential that there will be insufficient funding and computational capability to perform similar data evaluation for VIIRS is a concern of the oceanographic community.

As this history has shown, the scientists and technical staff associated with these missions surmounted numerous obstacles that could have significantly limited the success that they eventually achieved. The CZCS succeeded far beyond its modest goals, partly because it was quickly recognized that its data were unique, useful, and unprecedented – thus, resources were identified that provided the data to the scientific community and broadened its utilization, in parallel with increasing computational power both where the data resided and in the individual laboratories of oceanographers. It is also clear that independence and flexibility of the small and dedicated staff of the SeaWiFS mission, as well as the sole focus of the sensor on ocean color observations, aided its remarkable success compared to the more unwieldy early years of MODIS – and substantial improvements in MODIS ocean data were ultimately achieved by following the SeaWiFS model, both in management style and in a substantial reduction of the programmatic goals. The remarkable MODIS instruments were governed by a triumvirate of scientific interests (land, ocean, and atmosphere), and management decisions for the instrument had to be made with the overall mission in mind, not just the optimization of ocean color data.

So now, as the oceanographic community waits for more data and hopes that MODIS will continue to operate for a few more years, we can look back on these missions knowing that each of them in their own manner was a clear advance in the state of the science. While the process was not always linear, smooth, and without perturbations, NASA stayed the course with each of these missions and allowed them to establish an important niche in the broad field of oceanography. Although NASA has deployed many other types of remote sensing instruments in space, and successfully provided data from them to scientists, there are few other fields of remote sensing for which it can be said that had it not been for NASA's leadership, the entire branch of science would not have been established and successfully matured. For ocean color observations from space, that statement can be made.

APPENDICES

Appendix 1: Introduction to Ocean Optics – the Straight and Wayward Paths of Light

Without sufficient light, life on Earth (if there were any) would be far, far different.

Yes, there are communities of deep-sea organisms clustered around superheated deep-sea hot springs that do not depend on light from the Sun to produce the essential carbon that is the foundation of their simple food chain. It is therefore possible, on a planet dimly illuminated by a faint star, that some forms of life could exist in the pitch-black depths of a primordial ocean kept liquid mainly by the geothermal energy of the planet – forms of life that are hard to imagine. From what we know of the deep-sea vent communities, and the bacterial carpets that are found in the saline lakes of Antarctica and around seeps of salt on the sea floor, it would be difficult to envision the evolution of remarkably complex life forms in such an unforgiving and strange environment.

That is not the nature of the planet on which we live, however. The Earth is brightly illuminated by the Sun, heated by the Sun, and subject to the variability of climate systems that are powered by the energy of the Sun. The basic elemental building blocks of life, the carbon that composes all known forms of life, are produced by the biological activity of organisms that photosynthesize organic carbon using the energy of sunlight (with the exception of those unusual deep-sea hot spring organisms).

So if we are to investigate how oceanographers utilize light emanating from the oceans to better understand how the oceans work, using sensitive observing instruments carried on satellites, we must begin with light – the properties of light, and particularly the ways that light interacts with both living organisms and inorganic substances.

The Nature of Light

Sir Isaac Newton may be most noted for his description of gravity, but Newton also performed early investigative experiments into the nature of light, most notably by placing a glass prism in a beam of sunlight and noting how the prism separated the sunlight into the visible color spectrum.[1] He then placed a second prism oriented so that it collected the separated spectrum and reconverted it back into a beam of sunlight, showing that the all the colors were contained in the "white" light from the Sun.

This simple experiment, which anyone can perform, demonstrates the existence of a small but very important segment of the electromagnetic spectrum – the visible light range. While it was not suspected in Newton's time, science has determined that the universe is pervaded by electromagnetic energy. The prism *refracted* the light from the Sun because light is composed of energy with different *wavelengths* – and for many centuries after Newton, light was perceived as a phenomenon having many of the properties of waves (Figure A1-1).[2]

The next major milestone in the examination of the nature of light was the insight of British physicist James Clerk Maxwell, who synthesized many of the experiments of Faraday and others into a unified theory described by what have become to be known as "Maxwell's Equations," determined that light was an electromagnetic phenomenon.[3] Waves are characterized by a wavelength [the distance between two adjacent crests of the wave, and usually denoted by the Greek letter lambda (λ)] and a frequency [the time required for the wave to travel a distance λ, and denoted by the Greek letter nu (υ)]. The product of the wavelength and the frequency of the waves is just the speed of the wave, so for an electromagnetic wave $\lambda \times \upsilon = c$, the speed of light. In an electromagnetic wave in Maxwell's theory, the quantities that are "waving" or oscillating are the electric and magnetic fields. These fields are "vectors" meaning that they have a size (magnitude) and they point in a particular direction. The direction in which the electric field points is called the direction of *polarization* of the wave. The energy associated with an electromagnetic wave is proportional to the square of its electric field (its *amplitude*). (Figure A1-1)

We now know that the electromagnetic spectrum extends in both directions from the visible spectrum, ranging from X-rays and gamma rays (short wavelengths and high frequencies) to radio waves and microwaves (long wavelengths and low frequencies). The visible range lies somewhat in the middle, with the red end of the visible spectrum having longer wavelengths and lower frequencies than the blue end. Despite the fact that we can't see beyond the red end and the blue end, the lower frequencies and longer wavelengths of infrared (IR) radiation, and the higher frequencies and shorter wavelengths of ultraviolet (UV) radiation, are very important elements of the segment of the electromagnetic spectrum occupied by light. (Note that IR and UV radiation was carefully not referred to as "light",

---

[1] Gale E. Christianson, *Isaac Newton*, Oxford, pages 31-33, 2005.

[2] Image of electromagnetic spectrum acquired from Web page for the Laboratory for Computer Science and Engineering, University of Minnesota, http://www.lcse.umn.edu/specs/labs/glossary_items/em_spectrum.html, accessed 6 February 2009; Image of an electromagnetic wave acquired from Department of Physics and Astronomy, University of Rochester, "Astro 105: The Milky Way – Lecture 5: The Light", http://www.pas.rochester.edu/~afrank/A105/LectureV/LectureV.html, accessed 11 June 2009

[3] Isaac M. McPhee, "The Speed of Light: How the Study of Electromagnetism Affected Science", http://physics.suite101.com/article.cfm/special_relativity_pt_25 (accessed 6 February 2009)

but we have infrared lights – also called heat lamps – and ultraviolet lights – which can induce colorful *fluorescence* – so clearly IR and UV are closely related to visible light.).

Unfortunately, there was an experiment that could not be explained by the wave nature of light: the *photoelectric effect.* In the photoelectric effect, light illuminates a metal sheet placed in a vacuum and electrons are ejected. Investigators measured the influence of the frequency of the light, the brightness of the light (the square of the amplitude in the wave theory) and the time delay between when the light was first turned on and the first electron was ejected. Maxwell's wave theory could not explain *any* of the measurements. It took Albert Einstein's insight to unify these perplexing observations; Einstein described light as made up of *photons*, which can be envisioned as discrete particles of light, each particle defined by its own particular ν (or equivalently λ).[4] The energy of a photon is proportional to its ν and has nothing to do with its amplitude, i.e., the energy is *quantized* – divided into very little parts, like a Japanese steak house chef turning a big slab of steak into bite-size pieces with swift cuts of a sharp knife. Einstein's theory explained *all* of the experimental observations in regard to the photoelectric effect. It is for this achievement, *not* the theory of relativity, that he was awarded the Nobel prize.[5]

How do we reconcile Maxwell's theory – light consisting of electromagnetic waves – with Einstein's explanation of the photolectric effect – light consisting of massless particles? To make things more complicated, the *quantum theory* demands that what we think of as particles (electrons, protons, and yes, even golf balls) can all display wave-like characteristics – matter waves. In the case of light, the reconciliation was accomplished by R.P. Feynman's quantum electrodynamics – the union of Maxwell's theory and the quantum theory. The easiest way to visualize the wave-particle nature of light is to think of light as a stream of photons of a given frequency. Each photon has an energy proportional to its frequency. Longer wavelengths and lower frequencies are less energetic than higher frequencies and shorter wavelengths. (That's the reason that if you hear about extraordinarily high-energy radiation being detected from distant stars by X-ray or gamma ray telescopes, they are talking about the high-energy end of the electromagnetic spectrum). The total number of photons provides the energy in a beam of light.

The energy of a single photon of visible light is very small, only about 4×10^{-19} Joules.[6] Incident solar radiation, e.g., light from the sun, contains an enormous number of photons, so the energy transfer in such a beam can be quite large.

The preceding discussion summarized quite a lot of important physics in just a few paragraphs. The process of determining the true nature of light is a fascinating part of the history of physics, which is only hinted at above. But at this critical juncture, we now know that light sometimes behaves like a wave, and other times light behaves like a particle. Both

[4] Paul Heckert, "The Photoelectric Effect: Einstein's 1905 Explanation Earned the Nobel Prize", http://particle-physics.suite101.com/article.cfm/the_photoelectric_effect, (accessed 6 February 2009)
[5] Paul Heckert, "The Photoelectric Effect: Einstein's 1905 Explanation Earned the Nobel Prize".
[6] A Joule is the SI unit for energy. Power is the rate at which energy is transferred: power = energy/time. The SI unit of power is the Watt: 1 Watt = 1 Joule/sec.

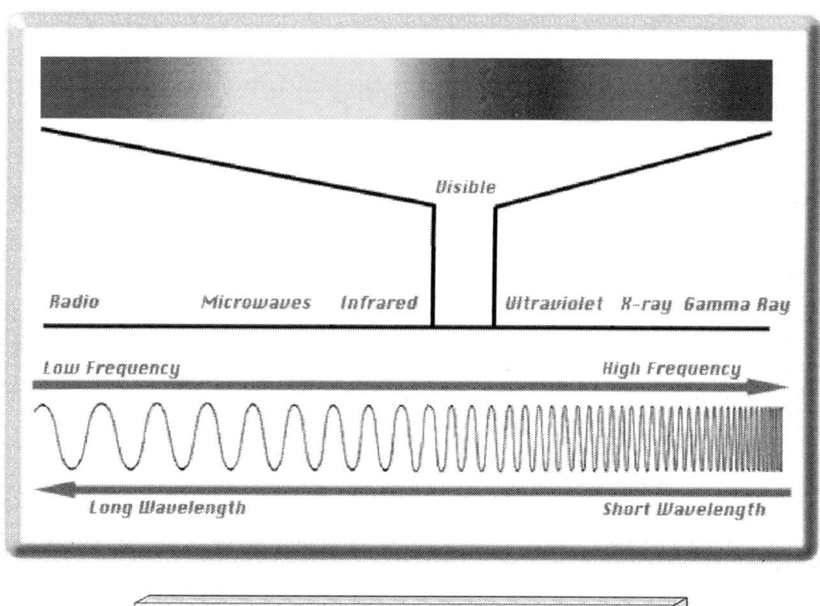

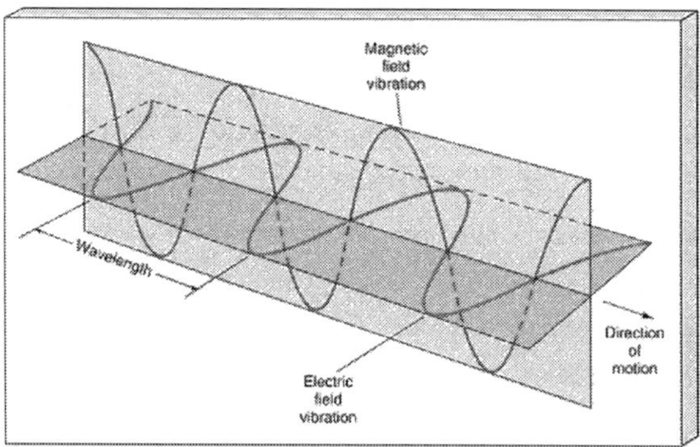

Figure A1-1. (top) Electromagnetic spectrum. (bottom) Illustration of an electromagnetic wave.

of these behaviors are important to understand how light influences and interacts with the various components of the Earth: the atmosphere, land, and oceans, and in particular on this peculiar planet, its life.

The Role of Light (Solar Radiation) on Earth

Before we get started with the Earth – which is going to be complicated – let's start with the Sun, whence cometh our light. Solar physics is another branch of science, but in order to understand the light from the Sun, it's sufficient to say that the Sun is a fusion reactor, producing its self-sustaining energy by the process of hydrogen fusion, which creates helium and a great deal of energy in the process. This energy is released by the Sun in the form of virtually every kind of photon in the electromagnetic spectrum, as well as a few bizarre neutrinos, and all these photons head out into space in every direction.

How do we characterize the "strength" of the sunlight reaching the Earth? The simplest way is to place a surface perpendicular to the beam and count the energy of each photon that crosses the surface in a given interval of time. The "irradiance" of the beam is then defined to be the total energy crossing the area, divided by the product of the area of the surface and the time interval. The units of irradiance are then Joules /(meter^2 × second) = Watts/ meter^2. (If instead we only count the energy associated with photons with wavelengths within a certain wavelength range ($\Delta\lambda = \lambda_2 - \lambda_1$), then the "spectral irradiance" is defined as the energy counted divided by the product of the area of the surface, the time interval, and $\Delta\lambda$. The unit for the spectral irradiance has an additional length, the wavelength interval, in the denominator. The wavelength in the middle of the visible spectrum is about 550 nanometers (10^{-9} meters) or 0.55 micrometers (10^{-6} meters) and the unit for spectral irradiance is then Watts/ (meter^2 nanometer) or Watts/ (meter^2 micrometer). The total solar irradiance from the sun reaching the "top of the atmosphere" (TOA, a fictitious altitude above which there is no atmosphere) is called the *solar constant*, E_s; although it's not quite constant. There is variability in the amount of radiation received by the Earth from the Sun, as much as ±50 W m^{-2}, due to the varying distance of the Earth from the Sun as the Earth makes a full orbit around the Sun. The value of the solar constant is E_s = 1367 W m^{-2}. If all of the solar irradiance falling on one square meter at the TOA could be turned into electricity, it could light 13 one-hundred Watt light bulbs plus one 60 Watt light bulb, continuously. The solar irradiance provides the energy that makes the Earth a habitable place.

Regarding the influence of solar radiation on the Earth, we're most concerned with photons that are in the mid-range of the electromagnetic spectrum: IR, visible, and UV. (The spectral irradiance received from the sun (usually indicated by $E_s(\lambda)$), has a maximum near the center of the visible spectrum.) Specifically, we want to understand how the light energy from the Sun makes the Earth a habitable place. The Earth is located at a critical distance from the Sun: the temperature of the Earth varies, of course, with extreme low temperatures of perhaps around -60 degrees Celsius, and extreme high temperatures a little under +60 degrees Celsius.[7] Most of us, fortunately, don't live where it gets anywhere close to that cold or that hot. More important is the fact that the temperature range of the Earth allows water to exist in all three phases: ice, liquid water, and water vapor. It's also important that the Earth is not so close to the Sun that fierce solar radiation does not strip the planet of its vital atmosphere; the atmosphere is necessary for the maintenance of the habitable range of temperatures that are found on the Earth. (The size of the Earth is also important; Earth has sufficient gravity to keep gaseous atmospheric molecules from flying

---

[7] The actual maximum temperature range is -88°C to +58° C, according to NASA, "Earth: Facts and Figures", (http://solarsystem.nasa.gov/planets/profile.cfm?Object=Earth&Display=Facts&System=Metric, accessed 10 June 2009).

off into space. Our solitary moon was too small, and thus had too little gravity, to retain an atmosphere.)

So Earth has its own set of crucial elements: an atmosphere, and specific components of the atmosphere, notably water vapor. Water vapor is the most important atmospheric component that causes Earth's *greenhouse effect*, which is something that scientists know quite well. The reason that water vapor is so important is due to the fact that it absorbs radiation.

However, water vapor doesn't absorb just any kind of radiation; it absorbs IR radiation, which is also known familiarly as *heat*. At this point, we start to get into the interactions of solar radiation with the components of the Earth system.

All of the light from the Sun that enters and penetrates the Earth's atmosphere is called *shortwave radiation*: light with wavelengths in the range of 0.4 to 4 microns (μm) or 400 to 4000 nanometers (nm).[8] The visible light range is about 390 (deep violet) to 780 (fiery red) nanometers. While a small part of this light interacts with the atmosphere – more on that later in the chapter – much of it makes it to the Earth surface, where it is absorbed by all of the various elements of the Earth's surface. This absorption of energy is manifest as a "heating" of the surface (soil, water, etc.) Some of the energy is utilized or stored, but most of it is re-emitted, as *longwave radiation*.[9] All objects emit radiation that is characteristic of their temperature. This is called "black body radiation." This radiation could not be explained by the wave theory of light, and Max Planck in 1900 proposed that energy of radiation in a closed cavity is quantized, and provided a relationship between the frequency of radiation and the energy of the quanta required to explain this radiation.[10] This relationship was identical to the one used by Einstein to explain the photoelectric effect, and gave Einstein confidence in his explanation. The Sun emits such radiation (what we see as sunlight) that is roughly characteristic of a black body at approximately 6000 degrees Kelvin. For the Earth's surface, which is at about 300 degrees Kelvin, the radiation has wavelengths longer than 4 μm, hence the term longwave radiation, and the maximum irradiance is near 10 μm.

The longwave radiation radiates outward and upward into the atmosphere. In the atmosphere, molecules of water vapor (and other famous greenhouse gases, but remember, water vapor is the most important here) absorb the longwave radiation. This keeps it from escaping back into space. Then the molecules re-radiate the longwave radiation, allowing other molecules to absorb it, and in the process raising and maintaining the temperature of the Earth.[11] If all the longwave radiation escaped to space, Earth would be a lot colder;

---

[8] Arctic Climate and Meteorology Glossary, "Shortwave Radiation", National Snow and Ice Data Center, http://nsidc.org/arcticmet/glossary/short_wave_radiation.html (accessed 6 February 2009)

[9] Arctic Climate and Meteorology Glossary, "Longwave Radiation", National Snow and Ice Data Center, http://nsidc.org/arcticmet/glossary/long_wave_radiation.html (accessed 6 February 2009)

[10] Helge Kragh, "Max Planck: the reluctant revolutionary", http://physicsworld.com/cws/article/print/373, accessed 10 June 2009.

[11] Michael E. Ritter, The Physical Environment: an Introduction to Physical Geography, "Radiation and Energy Balance of the Earth System", 2006, http://www.uwsp.edu/geo/faculty/ritter/geog101/textbook/energy/radiation_balance.html (accessed 6 February 2009).

on average, about 33° C colder.[12]

Thus, one of the noteworthy results of the greenhouse effect on Earth is to make Earth a suitable environment for the existence of living things. And the existence of life on Earth has a very important relationship with the light from the Sun. Even though the origin of life is not yet well understood (and may never be) the salient fact remains that life began on Earth, and over the course of evolution, some of the primitive life forms began to utilize the light energy from the Sun to enhance their metabolism, giving rise to the Kingdom Plantae, otherwise known as plants. Plants have a very distinct difference from the members of Kingdom Animalia; they are *autotrophs*, which means they utilize light energy to create carbon compounds, in contrast to *heterotrophs*, which consume carbon compounds and derive energy from them.

The process that plants use to create carbon compounds is *photosynthesis*.[13] Photosynthesis relies on a key chemical compound, *chlorophyll*, to absorb the light from the Sun and convert it into energy, which can then be used to synthesize carbon inside the cells of the plant.[14] Photosynthesis is obviously one of the most important processes that takes place in organisms living on the surface of the Earth; without it, life on Earth (if it existed at all) would be far different. It is not in the scope of this history, however, to provide a detailed explanation of how photosynthesis works. Rather, for the sake of ocean optics, the most important aspect is the absorption of light by the chlorophyll molecules inside plants; specifically for the oceans, the chlorophyll molecules inside of phytoplankton and other varieties of plant life which live in the ocean.

There's one more term that needs to be mentioned before we delve further into the subject of light and optics. That term is *primary production*; it means the amount of carbon that is produced by plants by the process of photosynthesis.[15] One of the main goals of oceanographers utilizing ocean color data from satellite instruments is to estimate global primary productivity in the oceans, which is not a particularly easy goal to accomplish. But without ocean color data from satellites, it would be nearly impossible to accomplish with any degree of reasonable accuracy.

The Downward Path: From the Sun to the Ocean Surface

Now we will briefly return to the surface of the Sun. In the following presentation of the basics of ocean optics, we will follow the paths of photons as they travel through space

[12] http://www.ldeo.columbia.edu/edu/dees/V1003/readings/Kump.Chapter.3.pdf
[13] Michael J. Farabee, "Photosynthesis", chapter in the *On-Line Biology Book*, http://www.emc.maricopa.edu/faculty/farabee/BIOBK/BioBookPS.html, (accessed 9 February 2009)
[14] Michael J. Farabee, "Photosynthesis".
[15] Encyclopedia Brittanica online, "Marine ecosystem: biological productivity", http://www.britannica.com/EBchecked/topic/365256/marine-ecosystem/70730/Biological-productivity#ref=ref588556, accessed 10 June 2009.

to the Earth, as they interact with the Earth system, and as they emerge, somewhat altered, by these interactions and return toward space. The diagram in Figure A1-2 will help us follow along the various paths of light.[16]

The first path that light follows is a direct one, from the Sun to the Earth. Given the distance to the Earth and the invariant speed of light in a vacuum, photons escaping from the Sun's photosphere take about eight minutes to travel to the Earth. Not much gets in the way or interferes with the quick journey of the photons to the top of Earth's atmosphere. So, essentially, the sunlight that reaches the top of the atmosphere is "undiluted" and full strength – with irradiance $E_s(\lambda)$.

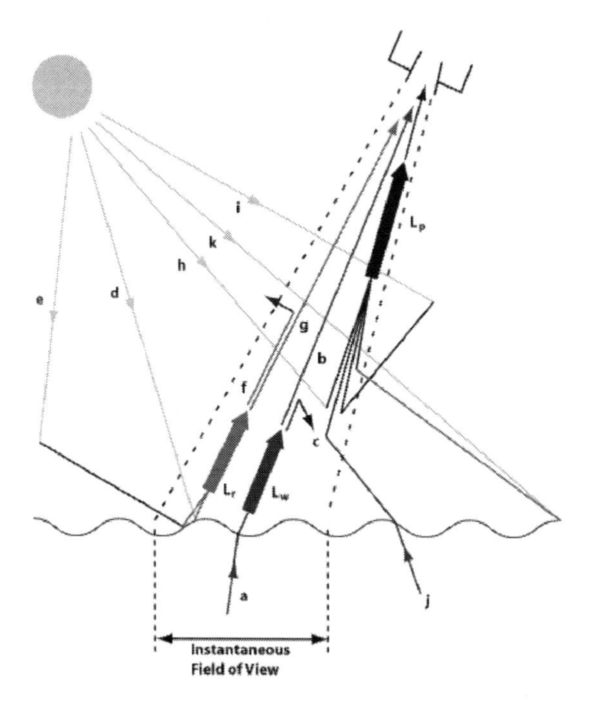

Figure A1-2. Various paths of light from the Sun as they interact with Earth's atmosphere and ocean.

Once the sunlight enters the atmosphere, however, significant alterations start to happen. There are three main things that can happen to the photons as they travel through the atmosphere: *absorption*, *scattering*, and *diffuse reflection* and *transmission*. These terms will be encountered repeatedly as we follow the paths of the photons. The process of absorption is shown in Figure A1-3, in which seven photons (all traveling parallel to one another) are incident on a thin layer of a material from the left. Only 4 of the seven photons come out

[16] GES DISC, "Chapter 11: On the Level from Radiation to Scientific Imagery", http://daac.gsfc.nasa.gov/oceancolor/scifocus/classic_scenes/11_classics_radiation.shtml, accessed 10 June 2009.

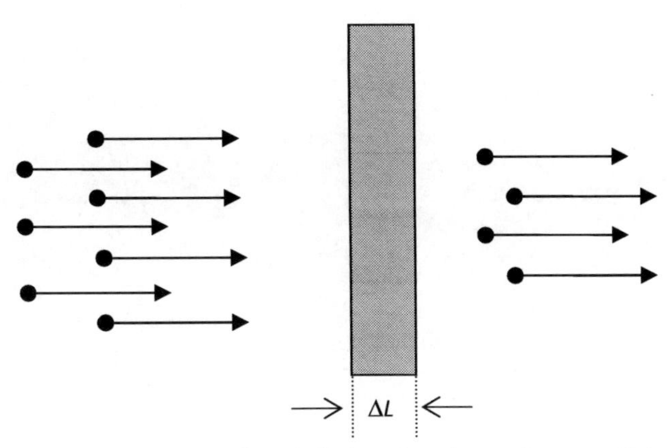

Figure A1-3. The geometry employed in defining the absorption coefficient *a*.

the other side and they are all going in the original direction. The rest are lost. They are said to be absorbed. Their energy will heat the material, or in the case of phytoplankton, may contribute to photosynthesis. The absorption coefficient *a* is defined to be the fraction of photons lost or absorbed $(N_{left} - N_{right})/N_{left}$, or in this case 3/7, divided by the thickness of the layer ΔL. That is

$$a = \frac{\Delta N_{absorbed}}{\Delta L \times N_{incident}} = \frac{\Delta P_{absorbed}}{\Delta L \times P_{incident}},$$

where $N_{incident}$ is the number of incident photons and $\Delta N_{absorbed}$ is the number lost. Since the radiant power in a beam of photons is proportional to the number of photons crossing an area in a given unit of time, we have also written *a* in terms of the power lost ($\Delta P_{absorbed}$) from a beam incident on the slab with power $P_{incident}$. The absorption coefficient has units of 1/meters, usually written m^{-1}.

In a manner that is similar to absorption, *scattering*, is determined by what happens when a beam of photons traverses a thin slab of material. Figure A1-4 depicts scattering. In this case a beam of 9 photons are incident on the slab from the left. Four are transmitted in the same direction; however, the remaining 5 (indicated by dashed arrows) are going in directions that are *different* from that of the 9 incident photons. These 5 photons are said to have been *scattered*— their direction of travel has been changed. As in the case of absorption, we can define the scattering coefficient (*b*) as the fraction scattered (5/9 in the figure) divided by ΔL, i.e.,

$$b = \frac{\Delta N_{scattered}}{\Delta L \times N_{incident}} = \frac{\Delta P_{scattered}}{\Delta L \times P_{incident}},$$

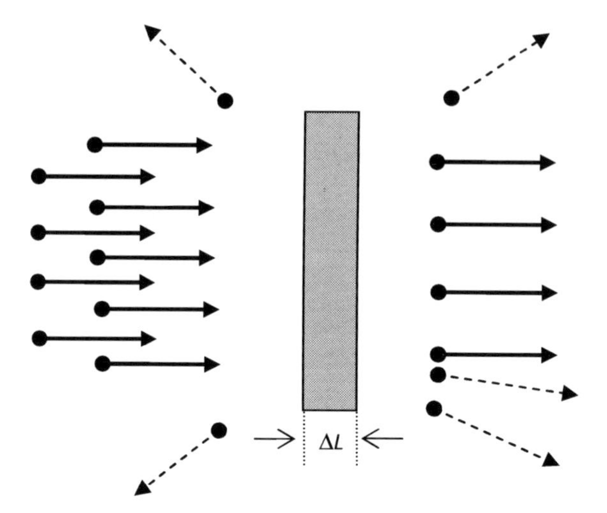

Figure A1-4. The geometry employed in defining the scattering coefficient *b*, and the back- and forward scattering coefficients b_b and b_f.

The two photons (at the top and bottom to the left of the slab) that are traveling away from the slab toward the left are said to have been *backscattered* by the slab, while the three on the other side have been *forward scattered*. The fraction backscattered is called the backscattering coefficient b_b, while the fraction scattered to the right is called the forward scattering coefficient b_f. That is,

$$b_b = \frac{\Delta N_{\text{scattered to left}}}{\Delta L \times N_{\text{incident}}} = \frac{\Delta P_{\text{scattered to left}}}{\Delta L \times P_{\text{incident}}}, \text{ and } b_f = \frac{\Delta N_{\text{scattered to right}}}{\Delta L \times N_{\text{incident}}} = \frac{\Delta P_{\text{scattered to right}}}{\Delta L \times P_{\text{incident}}}.$$

It remains to describe the photons that are scattered at certain angles from the incident beam, rather than just to the left or to the right in Figure A1-4. The volume scattering function $\beta(\theta)$, where θ is the angle between the direction of the incident beam and the direction of scattering, describes the power that is scattered in various specific directions in Figure A1-4. For this we need the notion of "*solid angle*." Take a sphere or radius r and draw a curve enclosing a small area ΔA on its surface. The solid angle $\Delta\Omega$ associated with the area is defined to be

$$\Delta\Omega = \frac{\Delta A}{r^2}.$$

If the sample from which the light is scattered is place at the center of a sphere and the number of photons scattered into ΔA is $\Delta N_{\text{into } \Delta A}$, then the volume scattering function is defined according to

$$\beta(\theta) = \frac{\Delta N_{\text{into } \Delta A}}{\Delta L \times N_{\text{incident}} \times \Delta\Omega} = \frac{\Delta P_{\text{into } \Delta A}}{\Delta L \times P_{\text{incident}} \times \Delta\Omega}.$$

Scattering occurs when a photon strikes something, e.g., an air or water molecule, a microscopic particle of dust in the air, phytoplankton cells in the water, etc. Usually the particles causing scattering are too small to be seen with the naked eye. (Technically of course, any objects that we see, we see because they scattered light.)

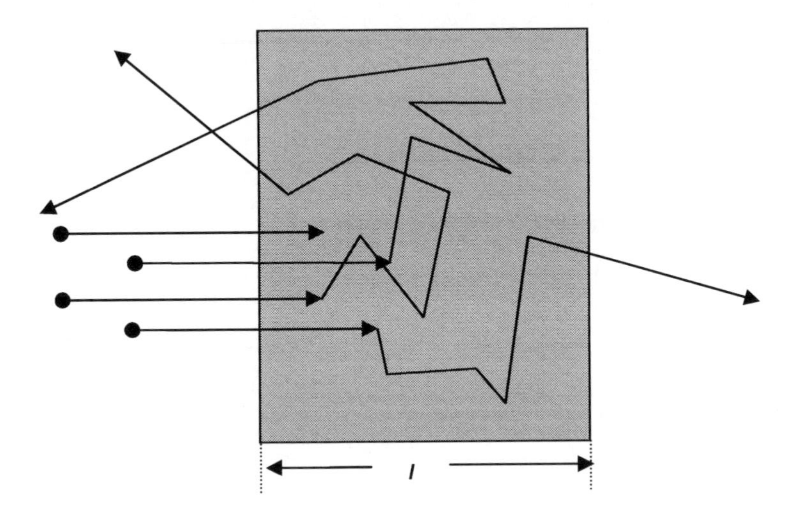

Figure A1-5. Diffuse reflection of light from a thick slab.

Implicit in the definitions above is that the slab is so thin that photons only scatter once within it, that is, a photon's direction is never changed more than once. When this condition is satisfied, the slab is said to be *optically thin*. When the slab is so thick that photons scatter many times before exiting, as depicted in Figure A1-5, the radiation that emerges on the left is said to be *diffusely reflected*, while the radiation propagating away from the slab on the right is *diffusely transmitted*. Clouds are a good example of diffuse transmission — a cloud thick enough to obscure the solar orb still appears bright (few photons are absorbed in the cloud), but there is no hint of the exact position of the sun. In contrast, when we simply look at the sun in the cloud-free atmosphere, there is no difficulty in locating its exact position. There are no simple formulas that specify the diffuse transmission and reflection; their determination is a complex process referred to as *radiative transfer*.

At high altitudes in the atmosphere, in the stratosphere, the first thing that will take place is absorption – ultraviolet photons will be absorbed by ozone (O_3). Denizens of the Earth surface should be grateful to the ozone molecules in the stratosphere, because they significantly reduce the amount of UV light that reaches the surface. UV is the light that causes sunburn, and UV also affects plants and animals in the ocean.[17] The importance of ozone for UV absorption has been emphasized in the past few decades when it was discovered that chlorofluorocarbons (CFCs) released chlorine compounds in the stratosphere when they were broken down by "cosmic rays" (actually high-energy particles: protons, alpha particles, and electrons), and this chlorine, through a number of different reactions, reacted with ozone and reduced its stratospheric concentration.[18] In particular, these reactions contribute to the seasonal reduction of ozone over Antarctica called the "ozone hole".[19]

Also high in the stratosphere are ice clouds called *cirrus*, which will be the first kind of cloud encountered by the solar photons. Cirrus will reflect some of the photons back into space before they get very far in the atmosphere. As the photons travel deeper into the atmosphere, they will encounter more clouds, which will also reflect the light back into space. Reflection from clouds thus reduces the total amount of sunlight which reaches the surface of the Earth.

The third process which affects the solar photons as they travel downward through the atmosphere is scattering. Electromagnetic scattering can be described by two different mathematical approximations, both of which can happen in the atmosphere. The two approximations are called *Rayleigh* scattering and *Mie* scattering. Rayleigh scattering accurately describes the type of scattering when the particle size is much smaller than the wavelength, the type that occurs when light encounters very small particles, such as air or water molecules.[20] Rayleigh scattering describes the cause of the light blue color of the sky, or the deeper blue color of water in a lake or ocean. This is due to the fact that Rayleigh scattering is wavelength-dependent, and it therefore affects light in the blue range the most. Rayleigh scattering will scatter light in all directions just about equally.

Mie scattering describes scattering for larger particles, which in the atmosphere usually means *aerosols*: small particles, such as dust, ash from volcanic eruptions, and gas molecules like sulfur dioxide (SO_2) which can react with water vapor to form droplets of sulfuric acid.[21] Mie scattering is not wavelength dependent, but it is direction dependent – it will scatter more strongly in the forward direction (the direction that the photons were originally traveling) than at right angles or in reverse directions. On a hazy, humid summer day, Mie scattering is the cause of the bright grayness, rather than the clear blue, of the sky. Mie scattering is also the cause of the lighted "glow" of fog. [Literally, the Mie scattering model gives exact results for homogenous spheres of any size.]

---

[17] Anne Renaud, "Solar erythemal ultraviolet radiation", http://www.iac.ethz.ch/en/research/chemie/tpeter/www_uv.html, accessed 10 June 2009; Daniel K. Lubin, Kevin R. Arrigo, and Gert L. van Dijken, "Increased exposure of Southern Ocean phytoplankton to ultraviolet radiation", *Geophysical Research Letters*, 31, L09304, doi:10.1029/2004GL019633.

[18] Mario J. Molina & F. S. Rowland, "Stratospheric sink for chlorofluoromethanes: chlorine atom catalysed destruction of ozone", *Nature* 249, 810 – 812, (28 June 1974), doi:10.1038/249810a0

[19] Ozone Hole Watch, http://ozonewatch.gsfc.nasa.gov/, (accessed 10 February 2009)

[20] C.R. Nave, "Blue Sky", from the Hyperphysics Web site, Georgia State University, http://hyperphysics.phy-astr.gsu.edu/hbase/atmos/blusky.html (accessed 10 February 2009)

[21] C.R. Nave, "Blue Sky".

The result of scattering in the atmosphere is a further reduction in the strength of the sunlight, the *solar irradiance*, reaching the surface of the Earth and the surface of the ocean. While reflected light is completely lost, some scattered light will still continue to travel downward. This type of scattered light is called the *diffuse irradiance*, in contrast to the *direct* (or beam) *irradiance*, essentially an unaltered path straight downward. In Figure A1-2, the downward paths (d) and (e) show these types of irradiance.

Another important factor, which will be discussed in detail later, is the angle of the Sun. Simply put here, if the Sun is directly overhead (which is also called at *zenith*), the photons will travel through less atmosphere than if the Sun is lower in the sky. That means that when the Sun is lower in the sky, less sunlight reaches the Earth's surface. Elementary, perhaps – but very important as more photons are lost as they encounter the Earth's various surfaces.

That's what happens next. After traveling through the atmosphere by their various paths, the photons that weren't reflected or scattered sideways and backwards finally reach the Earth's surface: and here many of them will be diffusely reflected and/or absorbed. Optically, the terrestrial surface essentially acts like an infinitely thick slab (as in the last figure with $L = \infty$). If the photons encounter snow or ice (for which the absorption coefficient is very small in the visible), these cold white substances will also reflect them back toward the sky; and it should be noted here that clouds, snow, and ice are significant factor in the Earth's *planetary albedo*, which is the ratio of the amount of sunlight reflected back to space to the amount of incident sunlight. Lightly-colored rock and sand have small absorption coefficients and so will also reflect much of the light that falls on them.

If they are not reflected, the other fate of the photons on paths that lead to the terrestrial surfaces of the Earth is absorption. Dark materials, such as soil or rock, have very large absorption coefficients and absorb much of the light that falls on them, and re-radiate as heat – earlier the conversion of incoming shortwave radiation to longwave radiation was discussed in relation to the greenhouse effect. Alternatively, the light photons can fall on vegetation, where most of the light is absorbed and utilized by chlorophyll for photosynthesis, and just a few particular wavelengths, particularly in the green region of the visible spectrum, are reflected. Chlorophyll is a green pigment because it reflects green light and absorbs the rest of it, in particular utilizing the blue wavelengths for energy. There is one other optical effect that happens in chlorophyll – *fluorescence* – which will be discussed in more detail shortly.

If the photons don't fall on the terrestrial surfaces that make up about 30% of the Earth's surface, the alternative is the aquatic surface; the surface of the ocean (or lakes or ponds or rivers, but mostly ocean). A small part of this surface is sea ice, which will reflect the light very similarly to snow or ice on land. But the rest of the light will encounter the aquatic surface. Some of this light can also be reflected at the ocean surface; direct reflections are called *sun glint* by remote-sensing scientists, and sun glint is a problem when a satellite instrument is looking directly down at the surface of the oceans. (There's a way to avoid this – just tilt the sensor a bit.) Alternatively, when the winds whip up, they create breaking waves and whitecaps – and the sea foam of a whitecap is also an effective reflector of the incident sunlight.

The final process that can occur at the aquatic surface is transmission – the light photons pass through the boundary between the atmosphere and the ocean and enter the aquatic realm. For optical oceanographers, that's when things really start to get interesting.

The Aquatic Paths: Light in the Ocean

Now the photons are in the water. It's very likely the first substance that they will encounter is water; and the molecules of water are also capable of absorbing or scattering light photons. Rayleigh scattering of light in water contributes to the deeper blue color of water as compared to sky; the "blueness" of water can even be noted in a swimming pool of only moderate depth that is painted white on its bottom and sides. Another factor contributing to the blue of the pool is the strong absorption of the red and yellow wavelengths by water; scuba divers have been taught the name "Roy G. Biv" to describe how colors disappear with increased depth. A small portion of the light will be absorbed by water molecules and converted to heat. As an aside, this is why there are concerns about diminishing Arctic sea ice due to climate change; ice reflects light, but water will absorb it and get warmer, which makes it harder for sea ice to form in subsequent Arctic winters.[22]

If the photons don't interact with water molecules – or even if they do, because one mode of interaction is to pass right through them unaltered, i.e., transmission – the photons will likely interact with a substance in the water. Just as in the atmosphere, there are basically two classes of substances. In the case of water, there are dissolved substances and particles that are suspended in the water – particles may be buoyant so that they stay near the surface most of the time, and some may be non-buoyant with a tendency to sink. Aquatic organisms are an example of the first kind of particle; inorganic sedimentary particles are an example of the second kind. The dissolved substances can absorb light, but usually occur in concentrations that are too small to play a significant role in scattering. The boundary between dissolved and particulate material is not always easy to define; in many cases it is determined by the size of something that can fit through the holes of particular filter sizes.[23]

Perhaps repetitiously, the particles and substances in the water will either scatter, reflect, or absorb light. (Note that reflection can be thought of as a specific direction of scattering.) Inorganic particles, frequently classed as sediments, will mainly reflect light, but darker sediments will also absorb some light. Extremely fine-grained sediments, as well as very small organisms that are made of reflective calcium carbonate ($CaCO_3$) will also scatter light. A familiar example of this is the stunning blue-green aqua color of some mountain lakes, which is due to a suspension of glacial "flour" – very fine-grained sediments – in their

---

[22] NASA, "Recent warming of Arctic may affect worldwide climate", http://www.nasa.gov/centers/goddard/news/topstory/2003/1023esuice.html, 23 October 2004, accessed 10 June 2009.

[23] Monica K. Bruckner, "Measuring Dissolved and Particulate Organic Carbon (DOC and POC)", http://serc.carleton.edu/microbelife/research_methods/biogeochemical/organic_carbon.html, accessed 10 June 2009.

waters. When this same color is encountered at sea, it indicates the presence of *coccolithophorids*; these are described in more detail in the main text. Note that any particles in the water, organic or inorganic, will cause some light scattering.

The colored dissolved substances are predominantly organic, and for that reason they are termed *chromophoric dissolved organic matter* (CDOM). CDOM absorbs light quite well, to the point that some rivers rich in CDOM (from the decay and dissolution of vegetative matter) can appear dark as tea, and black from the viewpoint of space. CDOM is ubiquitous in the oceanic environment, and distinguishing CDOM absorption from chlorophyll absorption is not always easy. Another term for CDOM is *Gelbstoff* (plural *Gelbstoffe*, German for *yellow substance*) – an amorphous, somewhat polymerized, dissolved organic substance that is known to absorb strongly, particularly in the blue range of the spectrum.[24]

Chlorophyll absorption is the other primary fate of photons that enter the ocean's waters and encounter phytoplankton. Chlorophyll is contained in the cells of phytoplankton, frequently in structures called chloroplasts; the size and shape of the chloroplasts inside the cell, as well as the size and shape of the cell itself, will determine the absorptive, and scattering optical properties of a particular phytoplankton species. Overall, though, chlorophyll absorbs light, most predominantly in the blue and red-yellow-orange wavelengths, and reflects the green wavelengths in a manner that is analogous to terrestrial plants. So the more phytoplankton present in the water, the more the water will appear green, rather than the deep blue of the clearest, low-phytoplankton waters of the ocean.

Earlier it was noted that there is one other effect that can occur when light interacts with chlorophyll. This effect is *fluorescence.* When the chlorophyll in the chloroplasts absorbs light energy, the chlorophyll molecules are excited to a higher energy state. Some of the added energy gets converted by the wonders of the chlorophyll molecule and the physiology of the cell into usable biochemical energy, a larger amount is simply degraded into heat, i.e., the cell's temperature increases, but a small portion of it gets re-emitted as light.[25] This particular type of light is very characteristic of chlorophyll, and will be at a longer wavelength than the absorbed light. The peak of the fluorescent light emission from chlorophyll occurs at 683 nm, while the absorption peak is near 670 nm.[26]

Before proceeding further to consider the paths of light in the ocean, it's important to introduce a few vital concepts and their associated terminology. The term *solar irradiance*

---

[24] R. Sudarshana, "Gelbstoff – the interfering yellow substance in cholorophyll remote sensing", *Journal of the Indian Society of Remote Sensing*, 13(2), 53-60, December 1985; Chuqun Chen, Ping Shi, Haigan Zhan, E.P. Achterberg, Samantha Lavender, M. Larson, and L. Jeonsson, "Retrieval of gelbstoff absorption coefficient in Pearl River estuary using remotely-sensed ocean color data", in *Proceedings Geoscience and Remote Sensing Symposium, IGARSS '05*, July 25-29, 2005, IEEE International Volume 4, 2511 – 2514.

[25] "Fundamentals of chlorophyll fluorescence", http://www.optisci.com/cf.htm, (accessed 12 February 2009)

[26] Mark R. Abbott and Ricardo M. Letelier, "Algorithm Theoretical Basis Document Chlorophyll Fluorescence (MODIS Product Number 20)", http://modis.gsfc.nasa.gov/data/atbd/atbd_mod22.pdf (accessed 12 February 2009)

was mentioned earlier; this term includes the word *irradiance.* Irradiance, with the symbol E, essentially describes the energy (power) of light coming from all directions onto a given area on a surface; that's why the units of irradiance are Watts per square meter. The easiest way of visualizing irradiance is to consider a horizontal sheet of paper with a small hole of area ΔA cut in the paper (we will refer to the sheet as the "surface"). During a time interval Δt, a number of photons ΔN, from all directions above the hole pass through. The power, ΔP, associated with the photons that pass through the hole, i.e., the energy carried by all of the photons divided by Δt, divided by ΔA is the irradiance: $E = \Delta P/\Delta A$. The solar irradiance mentioned earlier refers to the irradiance measured on a surface (the sheet) located at the top of the atmosphere, oriented perpendicular to the sun's rays (i.e., for a horizontal surface, the sun would be directly overhead). Note that if the surface (sheet) is not perpendicular to the beam, but oriented at an angle, then fewer photons will pass through the hole and the irradiance will be less. (Think of a sheet of paper in the rain, with no wind: when the surface of the sheet is horizontal rain drops strike it at a certain rate, but when it is oriented in a vertical position, very few rain drops strike it.)

If one places a spectral filter over the hole (e.g. a sheet of colored glass) that transmits light only in a narrow band of wavelengths ($\Delta\lambda$), as shown in Figure A1-6, then the power that passes through the hole is characteristic of the power contained in the wavelength interval $\Delta\lambda$. We call this power $\Delta P(\lambda)$, and define the *spectral irradiance* incident on the surface to be

$$E(\lambda) = \frac{\Delta P(\lambda)}{\Delta A \times \Delta\lambda}.$$

The units of spectral irradiance are Watts/m^2 nm, where nm (nanometers) is the unit for wavelength.

In the ocean, it is necessary to distinguish between light moving upward toward the water surface and light propagating downward toward the abyss. The irradiance of the light propagating downward is called the downwelling spectral irradiance ($E_d(\lambda)$) and the irradiance of the light propagating upward called the upwelling spectral irradiance ($E_u(\lambda)$). A closely-related term to irradiance is *radiance*, and this is the truly vital quantity, the quantity that optical oceanographers and their light-sensing instruments concentrate on measuring. If irradiance can be envisioned as light from all directions, radiance is light traveling in a certain limited range of directions.

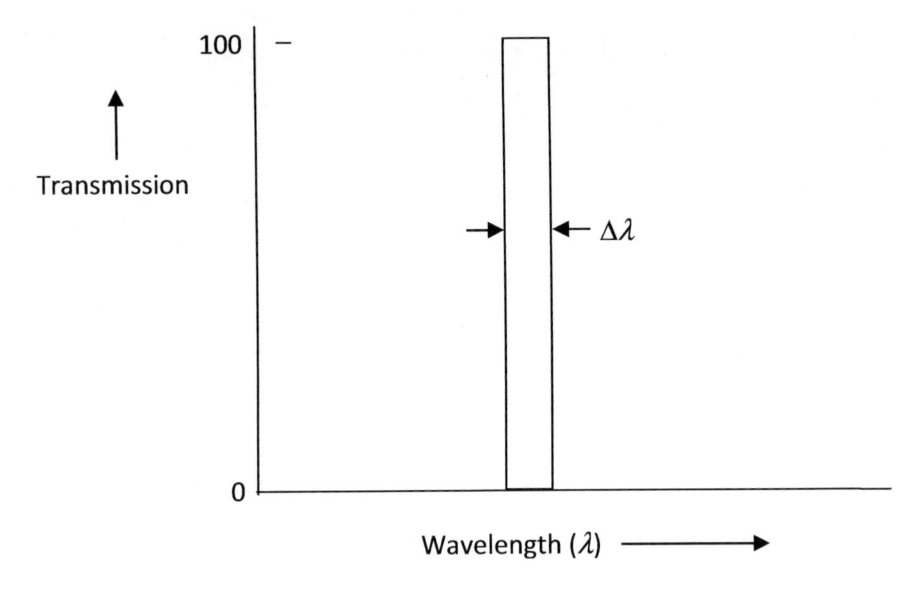

Figure A1-6. The transmission of a spectral filter. The light with wavelengths in the range $\Delta\lambda$ pass through the filter. All others are rejected.

Figure A7 can help explain the concept of radiance. In this figure, the light detector ("Detector") is placed at the end of a hollow cylindrical tube. (In the figure the detector is shown behind the end of the tube for clarity.) The purpose of the tube is to restrict the field of view (FOV) of the detector, i.e., it can only receive light from a range of directions defined by the opening at the end of the tube. One way to describe the restriction of the FOV is the concept of solid angle that we used in conjunction with the volume scattering function. The solid angle (Ω) viewed by the detector is A/r^2 in the figure. Note that this is the same as the area of the end of the tube divided by the square of the length of the tube. Although dimensionless, the solid angle is often given a unit called a *steradian* (*Sr*). This is much like the angle formed by the arc of a circle, which is given the name *radian* (or *degrees*). If the detector could receive light coming toward it from all directions, the the area "*A*" would be the area of a sphere ($4\pi r^2$) and the solid angle would be 4π steradians (*Sr*).

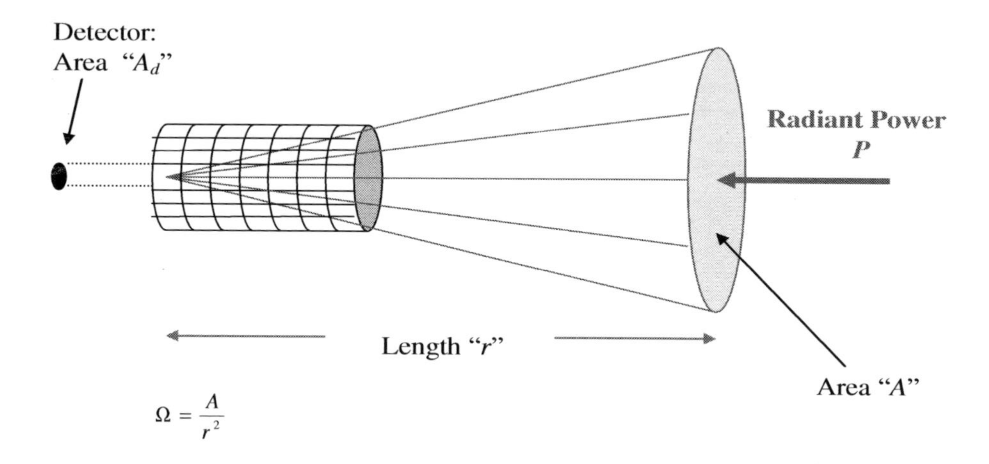

Figure A7. Schematic of a device for measuring radiance.

If the detector is equipped with a spectral filter so it only detects photons within a wavelength range $\Delta\lambda$, then the spectral radiance $L(\lambda, D)$ in a direction that is indicated in some manner by "*D*" is defined by

$$L(\lambda, D) = \frac{\Delta P(\lambda)}{A_d \times \Omega \times \Delta\lambda}$$

The units of radiance are then Watts per square meter per Steradian per nanometer. The shorthand expression for the units of radiance is W m^{-2} Sr^{-1} nm – this shorthand expression is seen frequently in ocean optics papers. Your eye actually measures radiance: its FOV is restricted by the iris, and the individual detector elements are the individual rods and cones. Devices that measure spectral radiance are called *radiometers* or *spectroradiometers*.

There are many ways to indicate the direction *D*, the simplest is to provide the angles θ and φ shown in Figure A1-8. Of particular interest to us is the sub-surface radiance propagating

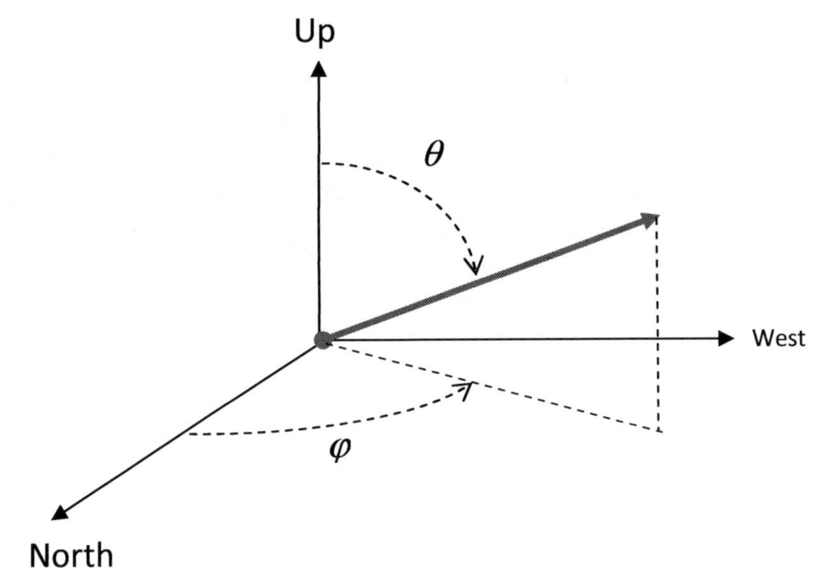

Figure A1-8. One possible method for describing the direction of radiance: providing the angles θ and φ.

straight toward the surface, i.e., in the direction specified by "Up" ($\theta = 0$) in Figure A1-8. This *upwelling radiance* is given a special symbol: L_u.

Equipped with these terms, it is possible to examine optical properties of ocean water and how it is influenced by substances (dissolved or particulate, living or non-living) within them. There are two different types of optical properties: *inherent* optical properties (IOPs) and *apparent* optical properties (AOPs). IOPs are the properties of individual small volumes of water, measured with a specifically defined incident light field. IOPs do not depend on the ambient light. That the properties depend on how the incident light is specified should be fairly obvious; compare the schematic diagram (Figure A1-9) with

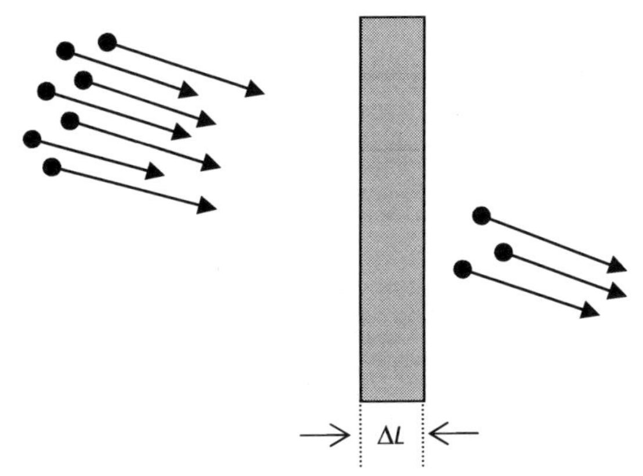

Figure A1-9. More light incident at an angle on a slab is absorbed than would be if it were incident perpendicular to the surface.

the similar one (Figure A1-3) used in the definition of absorption earlier. In this case, more photons are absorbed because the paths of the photons in the medium are longer than before, i.e., more photons are absorbed if the illumination is at an oblique angle. The three IOPS that characterize a water sample are the absorption coefficient a, the scattering coefficient b, and the volume scattering function, $\beta(\theta)$.[27] An important aspect of IOPs is that they can be defined for pure water and for all of the constituents contained in the water. For example, if the water ("*w*") contains phytoplankton ("*ph*") and non-scattering CDOM ("*cdom*") then

$$a = a_w + a_{ph} + a_{cdom},$$
$$b = b_w + b_{ph},$$
$$\beta(\theta) = \beta_w(\theta) + \beta_{ph}(\theta), \quad \text{etc.}$$

In remote sensing, an additional IOP, the backscattering coefficient b_b, which can be derived from $\beta(\theta)$, is of paramount importance, and like $\beta(\theta)$,

$$b_b = (b_b)_w + (b_b)_{ph}.$$

If the oceans were completely homogeneous, then the IOPs would be the same everywhere, i.e., they would be constant, and ocean remote sensing would be much easier (and much less interesting). However, in natural waters, the concentrations of particulate and dissolved matter vary with depth. Thus, all of the IOPs at a given location are dependent on their vertical position within the water column.

The AOPs are more difficult to understand. In fact they are not really optical properties at all. AOPs are properties that depend on the ambient light (technically on the angular distribution of the radiance), but that also display enough regularity and stability to be useful descriptors of the water body. AOPs cannot be measured on small volumes of water; they must be measured *in situ*. As an example consider the downward irradiance on a horizontal surface in the water. Let z be the depth at which the irradiance is measured. It is found through experimentation that the quantity K_d, called the *diffuse attenuation coefficient* or more accurately the *irradiance attenuation coefficient*, can be represented by

$$K_d(z,\lambda) = \frac{-1}{E_d(z,\lambda)}\left(\frac{E_d(z+\Delta z,\lambda) - E_d(z-\Delta z,\lambda)}{2\Delta z}\right),$$

[27] John T.O. Kirk, "The relationship between the inherent and the apparent optical properties of surface waters and its dependence on the shape of the volume scattering function", in Richard Spinrad, Kendall L. Carder, and Mary Jane Perry, *Ocean Optics*, pp. 40-58, New York, Oxford University Press (1994).

where $E_d(z+\Delta z, \lambda)$ means the downwelling irradiance at a depth $z+\Delta z$ and wavelength λ, etc., is almost independent of depth in homogeneous water. Thus it is tempting to say that $K_d(z, \lambda)$ is a property of the water body. However, it is found that the almost-constant $K_d(z, \lambda)$ is different at different times of the day (because the position of the sun varies in the sky during the day). It is smallest near noon and largest near sunset and sunrise. In addition it is different still if the sky is overcast. In other words, $K_d(z, \lambda)$ depends on how the water is illuminated by the sun and sky (or clouds), so even if it were perfectly independent of z, it is a property of not only the water but also the way in which the water is illuminated. Because if this it is called an "apparent" optical property (AOP): it is not an optical property of the water body, it just *appears* to be one. In the same manner the irradiance reflectance $R(z, \lambda)$ and the radiance reflectance $R_L(z, \lambda)$ given by

$$R(z, \lambda) = \frac{E_u(z, \lambda)}{E_d(z, \lambda)} \quad \text{and} \quad R_L(z, \lambda) = \frac{L_u(z, \lambda)}{E_d(z, \lambda)},$$

are nearly independent of z, but depend on the mode of illumination. They are also AOPs.

The basic rationale for AOPs is provided in Figures A1-10 and A1-11. Figure A1-10 shows the downwelling irradiance E_d as a function of depth (z) as well as the radiance propagating in a direction toward the zenith L_u. Notice that both E_d and L_u decrease rapidly with depth. In fact, the decrease is close to *exponential.* In contrast, Figure A1-11 shows the depth behavior of the radiance reflectance R_L and the irradiance attenuation coefficient K_d. Note that they depend very weakly on depth, changing by only a few percent from the surface ($z = 0$) to a depth of 40 meters ($z = -40$). One might then be led to believe that R_L and K_d are properties of the water. But the graphs also show that when the position of the sun is moved from directly overhead (the open symbols) to just 30° above the horizon (the solid symbols) R_L decreases and K_d increases. **Thus, R_L and K_d also depend on variables that have nothing to do with the water body!**

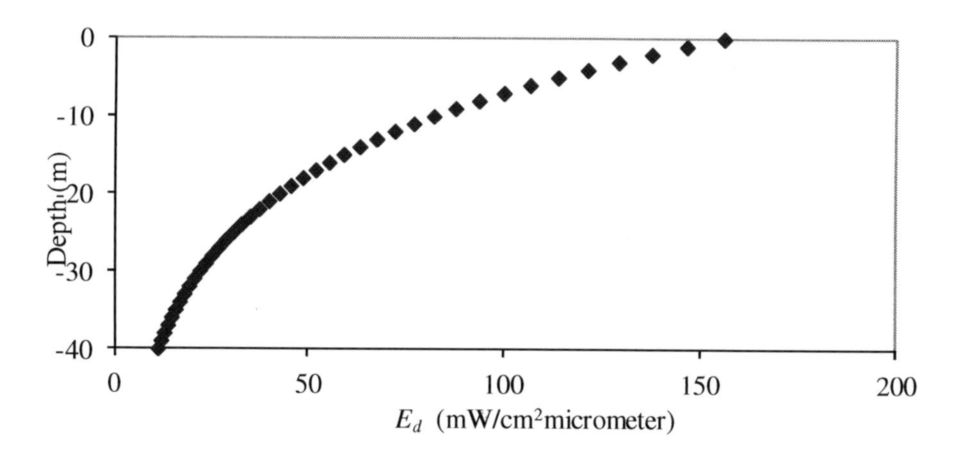

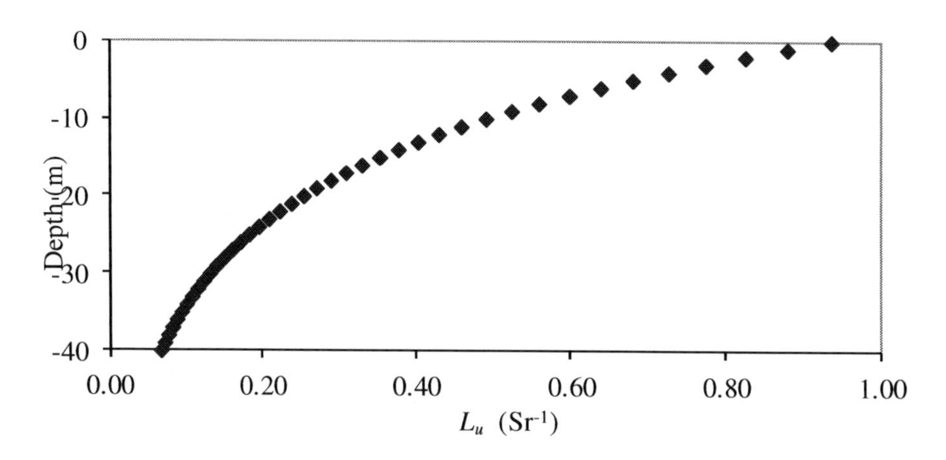

Figure A1-10. Examples of downwelling irradiance (upper panel) and upwelling radiance (lower panel) for a water body with particle absorption and scattering both equal to 0.05 m^{-1} at 440 nm. The sun is at zenith.

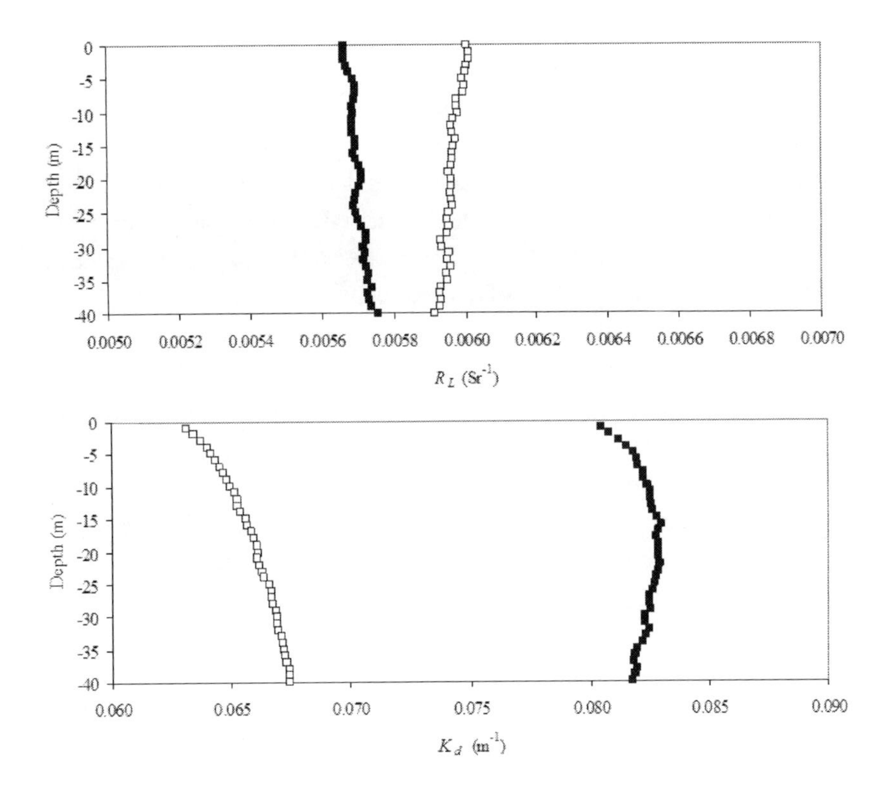

Figure A1-11. Examples of $R_L(z)$ (upper panel) and $K_d(z)$ (lower panel) . The open symbols are for the curves displayed in Figure A1-10. The closed symbols are for a situation where the solar zenith angle is 60°, i.e., the sun is 30° *above* the horizon.

Again, they are called "apparent" optical properties because at first glance they seem to be properties of the water, but in reality they aren't. However, for a specified illumination from the sun, sky, and clouds, they are *almost* properties of the water. They are vitally important in remote sensing, because L_u upon passing through the air-water interface becomes L_w, the *water-leaving radiance* — the radiance that that provides the ocean color "signal." (Figure A1-2). Also, K_d determines the depth range over which L_u is generated. In fact, the depth over which 90% of the water-leaving radiance originates is $1/K_d$.[28] (Note that L_u is not an AOP.)

It is possible, through careful measurements, to estimate AOPs from IOPs. This means that if a volume of water is extracted from its parent body of water at various depths, and *a, b*, and $\beta(\theta)$ are measured at each depth, it is then possible to provide values of the resulting variation of light in the water to values of K_d and R and R_L[29]. For example, in a

[28] Howard R.Gordon and W.Ross McCluney, "Estimation of the Depth of Sunlight Penetration in the Sea for Remote Sensing", *Applied Optics*, 14, 413-416, 1975.
[29] John T.O. Kirk, "The relationship between the inherent and the apparent optical properties of surface waters and its dependence on the shape of the volume scattering function".

water body with homogeneous IOPs (at least in the upper 30 meters or so), R_L just beneath the surface at any wavelength is proportional to the ratio b_b/a. This is extremely important in remote sensing because of the relationship of the upwelling radiance L_u to the water-leaving radiance.

The attenuation of light (irradiance, etc.) with depth is a fundamental characteristic of all oceanic waters. This depth to which light penetrates in the ocean defines the lighted portion of the ocean – the region in which photosynthesis and thus primary productivity can take place, and the region in which most of oceanic life resides. This region is termed the *euphotic zone*, and the depth to which the light penetrates is called the *euphotic depth*.[30] Specifically, if one measures the total number of photons from the sun and sky that penetrate (say, 1 square meter) of the water's surface for the whole visible spectrum, and does the same as a function of depth in the water, then the euphotic depth is can be approximated as the depth over which the number of photons falls to 1% of the number just beneath the surface. The euphotic zone is critically important to the biological remote sensing of the oceans because this zone is the origin of the light that is detected by instruments and analyzed by optical oceanographers to determine (as much as possible) what is present within those waters.

The attenuation of light with depth is also important in determining how much (or more accurately, how little) light returns from depth, because the same optical effects that influence how far light can penetrate into the ocean affect how much light returns from those depths. Therefore, for very clear waters, the upwelling radiance (path a and L_w in Figure A1-2) is an integrated signal of light returning upward from considerable depth (as much as 30 meters), but with most of the signal emanating from much shallower depths. In turbid waters, light may not penetrate more than a few meters, and most of the returning light is from only a few centimeters depth. The clarity of the waters also influences the amount of light that returns toward space; because of scattering and absorption primarily by water molecules, the deep blue ocean waters are quite dark, with an albedo of 0.03-0.10 (i.e., only 3% to 10% of the irradiant intensity on the water surface is returned, compared to land surfaces ranging from 0.04 (dark soil) to 0.45 (light soil and sand). Crops and forests have albedoes in the 0.15 to 0.25 range, though coniferous forests are darker than that. Ice and snow albedo ranges from 0.40 to as high as 0.95. [31]

The Return Path: From the Ocean Surface to Space

To this point, we have considered the light from the Sun as it traveled virtually

---

[30] ZhongPing Lee, Alan Weidemann, John Kindle, Robert Arnone, Kendall L. Carder, and Curtiss O. Davis, "Euphotic zone depth: Its derivation and implication to ocean-color remote sensing", *Journal of Geophysical Research*, 112, C03009, 2007, doi:10.1029/2006JC003802.

[31] Dagmar Budikova, "Albedo" in Cutler J. Cleveland, editor, *Encyclopedia of Earth*, Washington, D.C.: Environmental Information Coalition, National Council for Science and the Environment. [First published in the Encyclopedia of Earth 21 November 2006; last revised 19 March 2008.] http://www.eoearth.org/article/Albedo, accessed 18 June 2009.

unchanged through space; as it entered Earth's atmosphere, where it was subject to some reflection, absorption and scattering; as it encountered the interface between the atmosphere and the surface of the ocean; and then as it interacted with the components of the marine realm, subject to more reflection, scattering, and absorption.

And now, to put it simply, things get even more interesting.

A large number of the cohort of photons that have traveled these paths have already been diverted onto other paths. Now, what must be considered are the photons that remain from the original solar source; the photons that assume a considerably important role because they have been directed into a returning, upward path. They have not been absorbed, or scattered downward and downward again into the darkness of the abyss; they are **survivors**. This group of valuable and distinguished photons represents the properties of the water that has filtered and organized them into a light signal with particular characteristics. This is the light that optical oceanographers are most interested in – the *water-leaving radiance*; the remnant glow of oceanic waters that are illuminated by the Sun. This remnant constitutes a radiant power that is much, much less, only a few percent, of the power of the incoming solar irradiance. This is the light that sophisticated instruments have been designed specifically to detect.

To analyze the water-leaving radiance, however, it must be further refined. In order to use this light signal for quantitative diagnoses of the constituents of the water that it is leaving, it must be *normalized*. Normalization accomplishes a couple of things; it accounts for the attenuation of light on its flight through the atmosphere to the ocean surface, and it also accounts for the changing illumination angle of the Sun on the water surface (the *solar zenith angle*). Thus, the quantity that is even more interesting to optical oceanographers is the normalized water-leaving radiance, which is symbolized by "*nLw*". [32] The purpose of defining *nLw* is to create a quantity that represents the upwelling light energy exactly at the surface of the ocean, with the Sun shining down from directly overhead with the atmosphere removed.

There is now still one more hazardous path for the water-leaving radiance photons to travel. This path is the flight through the atmosphere toward space. As before, the molecules of air and substances floating in the air, such as dust or pollutants, as well as clouds, will reflect, scatter, or absorb the vital photons that are needed by the satellite sensor to quantify the ocean. If these photons encounter clouds on their upward path, then they will be reflected back downward and lost. They may be scattered sideways and back downward, and also never leave the atmosphere. And they may still be absorbed by water vapor and ozone molecules lying in their path. These losses mean that the water-leaving radiance signal is further attenuated before the light of the water-leaving radiance exits the

---

[32] Howard Gordon and Dennis Clark, "Clear water radiances for atmospheric correction of coastal zone color scanner imagery," *Applied Optics*, 20, 4175-4180 (1981); Howard Gordon and Kenneth Voss, "MODIS Normalized Water leaving Radiance Algorithm Theoretical Basis Document (MOD 18), Version 4", http://modis.gsfc.nasa.gov/data/atbd/atbd_mod17.pdf (accessed 12 February 2009); Howard Gordon, "Normalized water-leaving radiance: revisiting the influence of surface roughness," *Applied Optics* 44 (2), 241-248 (10 January 2005).

atmosphere and heads into space, to its final destination, the light collecting and detecting optics of a satellite instrument. The satellite instrument is, of course, also detecting photons that never even reached the water (L_p in Figure A1-2). In fact, typically, over the open ocean L_p is responsible for about 90% of the radiance measured by the sensor. For this reason, the light that reaches the top of the atmosphere is not water-leaving radiance; it is water-leaving radiance that has been modified by its interaction with the atmosphere and augmented by the atmospheric backscattering of photons that never reached the water. This atmospheric interference is the reason that one ocean color scientist remarked that ocean color satellites don't actually observe the color of the ocean; they actually observe the color of the atmosphere modified by the presence of the ocean waters below the atmosphere.

To determine the actual (at least approximate) value of the water-leaving radiance, the large influence of the atmosphere must be removed – the process of *atmospheric correction.* It is fair to say that the optical analysis that led to the atmospheric correction method enabled the science of ocean color remote sensing. Only the basics of this process are presented here; the actual underlying algorithms are complex and are still being refined to produce better estimation of the water-leaving radiance, which allows better estimation of the constituents of the ocean that are of interest to biological oceanographers.

In a sense, atmospheric correction is like watching a magic show. The skillful magician makes what he's doing look easy (and hopefully astonishing) – the audience doesn't see the mechanics and details "behind the scenes" that make the illusion work. Atmospheric correction is not magic and is based on rigorous physics, but to a non-physicist, the transformation of the raw radiances detected by the sensor into the "useful" water-leaving radiances employed by the algorithms is nonetheless a perceptibly remarkable achievement.

The essence of atmospheric correction is distilled into this equation:

$$L_{toa} = L_r + (L_a + L_{ra}) + tL_{wc} + TL_g + t\,L_w$$

in which the quantities stand for the following:

- L_{toa} is the radiance at the top of the atmosphere (the radiance detected by the sensor)
- L_r is the Rayleigh scattering term
- $L_a + L_{ra}$: L_a represents scattering by aerosols, and L_{ra} is the interaction (multiple scattering) between molecules and aerosols. (The scattering by molecules is modeled by Rayleigh theory and that by aerosols by Mie theory).
- L_{wc} represents the radiance due to reflection off of whitecaps
- L_g represents the radiance due to sun glint (direct reflection off the sea surface), and

- L_w represents the quantity desired, the water-leaving radiance![33]

So all that needs to be done is to make accurate estimates of the quantities other than the water-leaving radiance, subtract these from the top-of-the-atmosphere radiance, and that leaves the water-leaving radiance. That was easy![34]

The previous equation makes it look easy (the magic part), but now we'll go into some of the details. First of all, the Rayleigh scattering term can be estimated fairly simply from the atmospheric air pressure – knowing the approximate number of air molecules between the ocean and the top of the atmosphere indicates how much light will be scattered by them. The whitecap term can be estimated based on wind speed. Sun glint can be avoided – the history of the development of ocean color instruments will discuss how. If the sun glint is not avoided, there will be some areas where the sun glint overwhelms all of the other light, and data from those areas will not be usable.

That leaves the aerosols, and that's where there's significantly more work. Essentially the first thing that needs to be done is to build a model atmosphere. The model atmosphere will include the influence of Rayleigh scattering, absorption by ozone (for which data from other satellites is needed) and a model of typical aerosols and how these typical aerosols interact with light. The first models used a single-scattering method (photons only scatter once) to estimate the scattering terms (Rayleigh and aerosol) – in many cases this is entirely sufficient to get reasonably accurate atmospheric correction; however, if the aerosol concentration is too high it is necessary to include multiple scattering (photons scatter more than once).[35] In addition, there are cases, particularly mixed cases where air masses from the continents mix with air masses over the ocean, and where there are different kinds of aerosols present, where multiple scattering effects have to be employed to get an accurate atmospheric representation.[36]

The next step is to figure out how much the light from the ocean has been influenced by the atmosphere. The way that this is done is to use bands on the boundary between the visible spectrum and the infrared – the near-infrared (NIR) bands, which don't provide any contribution from the ocean because the ocean is dark at those wavelengths. The NIR bands therefore provide information only on the optical characteristics of the atmosphere. The values of the NIR radiances thus provide the $L_a + L_{ra}$ terms in the simple atmospheric

---

[33] NASA, SeaWiFS Atmospheric Correction, http://oceancolor.gsfc.nasa.gov/SeaWiFS/TEACHERS/CORRECTIONS/, (accessed 13 February 2009)

[34] Howard R. Gordon, "Atmospheric correction of ocean color imagery in the Earth Observing System Era", *Journal of Geophysical Research*, 102 (D14), 17,081-17,106, (27 July 1997); John Porter and Torben Nielsen, "Atmospheric Correction Near Hawaii: Clear Sky and Volcano Plumes", *Pacific Ocean Remote Sensing Conference*, Bali, 2002, http://www.soest.hawaii.edu/porter/Bali_Atmos_Corr.pdf (accessed 13 February 2009).

[35] Howard Gordon, "Atmospheric correction of ocean color imagery in the Earth Observing System era".

[36] Howard Gordon, "Atmospheric correction of ocean color imagery in the Earth Observing System era".

correction equation.

There is still much work to be done. More advanced aerosol models address the aerosols which absorb light, as well as scatter it. Furthermore, the complexity of the atmosphere must also be addressed; the atmosphere can be modeled as having a uniform composition from top to bottom (very simple), as having a tropospheric layer and a stratospheric layer (more difficult) and as having multiple layers with different kinds of aerosols (quite complex). Finally, the details of how molecules and particles in the atmosphere scatter light are continually being evaluated and better understood, and this increased understanding can be utilized to tweak the coefficients of the equations that represent the constituents and light interactions in the model atmosphere.

Ultimately, there is a remaining small percentage of the original phalanx of solar photons that have not been diverted to one of the myriad of paths that do not lead back to the near-emptiness of space. These photons have interacted with the atmosphere and ocean and have re-emerged from the Earth system, and thus they carry information about the Earth system; their modified aggregate spectrum is the result of the interactions. These are the truly vital photons that enable remote sensing of the oceans. These photons make the final stage of the journey through space and into a window that leads to electronic technology that directs, sorts, separates, and detects them, and this information about the quantity of each, the radiant power of light which is detected in space, is then transmitted back to Earth, where it can be analyzed.

Detecting and Measuring the Light from the Ocean

Now that we have seen how light travels from the Sun through space to the Earth, through various paths within the Earth system and then back to space, we can now describe how this light can be quantitatively observed by various forms of optical instrumentation. This discussion will not go into the details of how optical instrumentation actually translates light intensity falling on a detector element into electronic information, whether in an instrument used at sea or an instrument orbiting in space; this discussion will concentrate on what is being measured and how it is measured.

There are two basic classes of instruments that measure light: *radiometers* (also called *spectroradiometers*) and *spectrometers*. Radiometry is the process of measuring radiation, which covers the entire range of the electromagnetic spectrum; so there are visible-light radiometers, microwave radiometers, infrared and ultraviolet radiometers, gamma-ray radiometers, etc.

The difference between a radiometer and a spectrometer is that the radiometer measures the light intensity in specific discrete sections of the electromagnetic spectrum, called *bands*, while a spectrometer measures the entire continuous spectral curve of light intensity. The spectrometer, by generating the continuous spectrum, is thus able to separate what is being measured into its spectral parts. Spectrometers are valuable tools because the light signal from the ocean is a mixture of several spectral curves (Figure A1-12).

A *photometer* is a special class of radiometer that measures light in the visible spectrum in a manner that simulates the human eye. Radiometry and photometry measure different quantities; radiometry measures the power of electromagnetic radiation in a narrow, well defined spectral band, while photometry measures the totality of radiant power in a wide spectral band with a spectral sensitivity that is similar to the human eye — the first light detector used by scientists to study light.

Figure A1-12 shows where the bands of several different remote-sensing instruments are located (and how wide they are) in the visible light spectral range.[37] Below that is the spectrum of absorption of pure water, chlorophyll, and CDOM.[38]

If a spectrometer were being used to measure the light intensity, it would produce a curve expressing the light intensity continuously. In contrast, the radiometer can only measure the radiant intensity in the specific range of each band that it possesses.

To understand and quantify the light from the ocean that can be detected from space, optical oceanographers deploy a variety of instrumentation at sea. In-water (submersible) instruments are lowered into the oceanic water column to measure both the downwelling spectral irradiance (E_d) and the upwelling spectral radiance (L_u). By integrating the measurements taken at various depths, the expected water-leaving spectral radiance at the surface of the ocean can be calculated from these measurements. There are also instruments that can be aimed at the surface of the ocean, which directly measure the water-leaving radiances. Instrumentation on the ship can also measure the solar irradiance, which can be related to the radiances measured in the water.

Other types of instrumentation deployed in the water measure the scattering effects of particles (backscatter), the fluorescence of chlorophyll (a very accurate way to determine the concentration of chlorophyll) and the transmission of light through a given distance (transmissometers for measuring *c*). Oceanographers will also measure many other variables related to the optical characteristics of the environment; for example, atmospheric instruments can measure the absorption of light by aerosols and ozone, and lasers can be used to induce fluorescence in chlorophyll and related photosynthetic pigments.

---

[37] NASA Goddard Earth Sciences DISC, http://daac.gsfc.nasa.gov/oceancolor/images/ocean_color_wavelengths.gif; Top figure, personal files of James G. Acker, from Goddard DAAC dataset guide documents. Bottom figure, Wetlabs, Inc., "More about absorption", http://www.wetlabs.com/iopdescript/abslvl2.htm, accessed 12 February 2009.

[38] Wetlabs, Inc., http://www.wetlabs.com/iopdescript/absimages/11a.gif

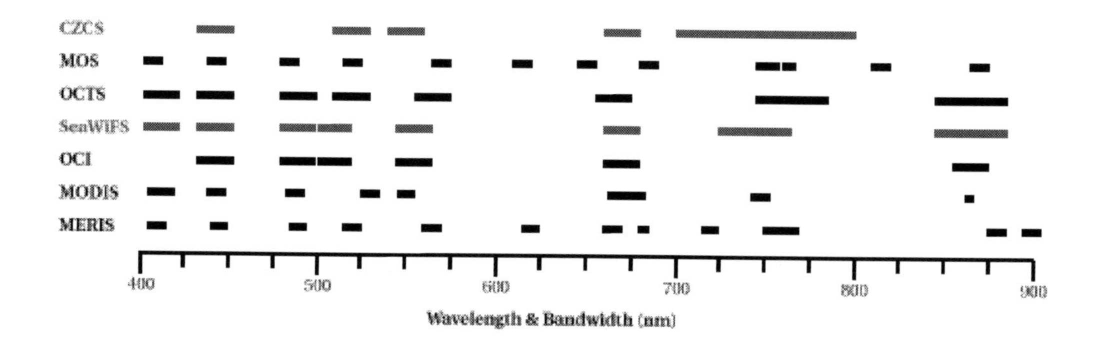

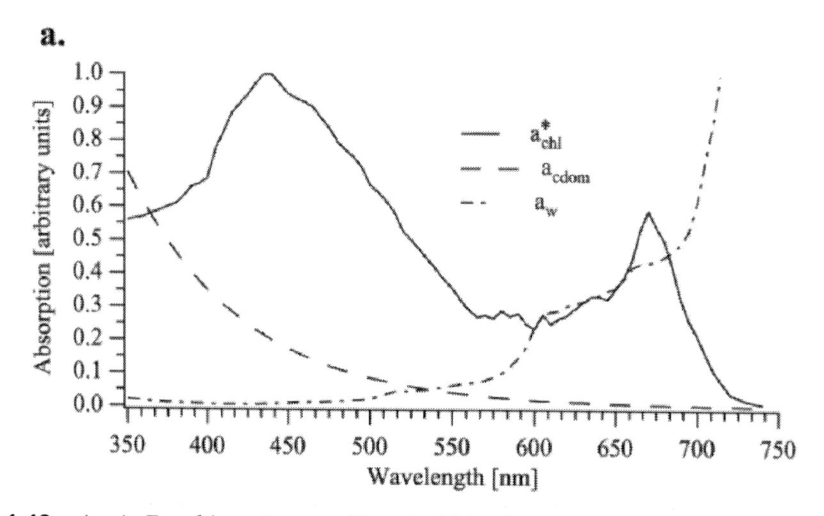

Figure A1-12. (top) Band locations and bandwidths for several satellite ocean color remote-sensing instruments. (bottom) Absorption spectra for chlorophyll (solid line), CDOM (dashed line) and water (dash-dot line).

All of the measurements taken at sea provide a full characterization of the marine optical environment. Such information is vital to understand all the details of how light behaves in the ocean, even done to the characteristics of how the various morphologies of single cells of phytoplankton influence the marine light field.[39]

Another reason to make ocean optical measurements at sea is to support and calibrate the measurement of ocean optical quantities from space, which is the subject of this history. There are many important design considerations that influenced the development of spaceborne oceanographic remote sensing instruments. While there are certainly vital technical details regarding how the instruments actually work, here we will be concerned with what they are attempting to measure.

A basic concept of a remote-sensing instrument that is intended to examine the oceanic

[39] André Morel, "Optics from the Single Cell to the Mesoscale," in Richard Spinrad, Kendall L. Carder, and Mary Jane Perry, *Ocean Optics*, 93-117, New York, Oxford University Press (1994).

light spectrum has several fundamental elements. The first element is the device which receives and directs the light into the instrument. While this component doesn't have to be a mirror, it frequently is, because there are several different ways that mirrors can be mechanized to enable repeated scanning of the Earth surface. Light from the Earth will enter an aperture and then be gathered and directed by the collector into the instrument optics.

There is one particular factor that significantly influences the design of sensors which view the light emanating from the surface of the ocean. This factor is *polarization.* As described earlier, polarization of an electromagnetic wave refers to the direction of oscillation of the electric vector. Light from the sun is unpolarized, meaning that sunlight is propagating with the planes of the electric field randomly oriented from photon to photon. If the polarization is resolved vertically and horizontally, half of the power is associated with electromagnetic waves with the electric oscillation in a vertical plane, and the other half with the electric oscillation in a horizontal plane (these two planes intersect along the line of propagation of the light, Figure A1-1). When light scatters from a particle, the power scattered by these two directions of oscillations will not be the same, and the scattered light therefore has more of one polarization than the other — the scattered light is *partially* polarized. Since the light that leaves the water (L_w in Figure A1-2) and the light that is backscattered by the atmosphere (L_p in Figure A1-2) are both scattered, that light is partially polarized.

The remote sensor in space is equipped with instrumentation that measures power associated with the radiance, but cannot discern its polarization. It is an unfortunate fact (which must be dealt with) that most optical detection systems are somewhat sensitive to the polarization of the light they are measuring. Thus, to get an accurate measurement of the radiance, remote sensors are usually equipped with depolarizers or polarization scramblers that depolarize the incoming light and allow an accurate measurement.

After accomplishing the absolutely necessary step of depolarization, the beam of valuable light is then split into its spectral components by means of more mirrors, and devices which create a spectrum, such as prisms or diffraction gratings. When the light has been separated into these components, it is then directed onto the detector elements; these are electronic devices that are light sensitive and which transform the light intensity that falls on them into an electric current. The strength of the current is directly related to the light intensity, so the electronics of the instrument take this information and either transmit or record it (more on that process in a moment).

Because of the importance of the light that is being detected, the detectors must be remarkably sensitive to the specific wavelengths of light they are designed for. The more sensitive a detector and its associated electronics, the more likely that there will be sources of interference (noise), both optical and electronic. So for the light measured to be meaningful, the detector must have sufficient capability to quantify the light "signal" and to introduce a minimum amount of false variations ("noise") that would be interpreted as variations in the water-leaving radiance, i.e, it must have a sufficiently high signal-to-noise ratio (SNR). The paramount problem of visible light ocean remote sensing is that the actual signal is low compared to the many sources of noise. There are several methods by which the signal can be amplified compared to the noise. The amount of signal received is directly related to the amount of time available to collect the light stemming from the Earth, so the longer that an instrument can look at a specific area, the more light that can be collected. Unfortunately, instruments scanning the Earth only have a finite, and very short, amount of time to look at any one particular spot and to collect the light from it. So there are both mechanical ways to amplify the light, such as the use of focusing lenses (i.e., a telescope) to collect more photons, and electronic ways such as time delay and integration (TDI). TDI utilizes multiple detector elements, all of which gather light from the same spot for a designated period of time, and then the instrument electronics sum up (integrate) the signal from each detector element.

Those are the very basic considerations of the instrument operations. Another significant consideration is where to locate the various spectral bands that the instrument will have. The earlier figure showed where the bands for various ocean-remote sensing instruments are located, in relation to the absorption features of water, chlorophyll, and CDOM. The band location for ocean remote sensing instruments is intended to "sample" the spectrum of light and to use this information to derive various quantities of interest in the ocean (see below). But there are some places that have to be avoided. Figure A1-13 shows where the absorption bands of ozone, oxygen, and water vapor are found in the visible light spectrum.[40]

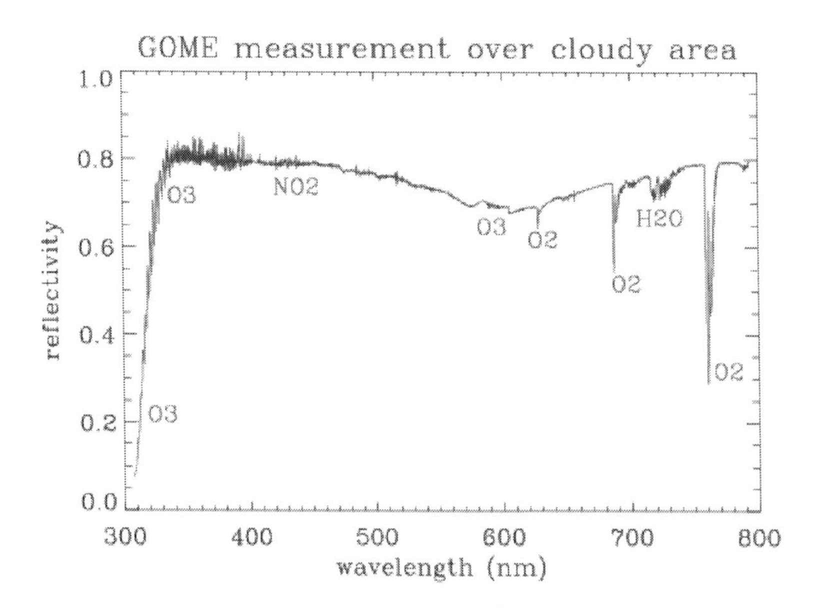

Figure A1-13. Atmospheric spectrum from Global Ozone Monitoring Instrument (GOME).

[40] Spectrum courtesy of Dr. R.B.A. (Robert) Koelemeijer, Royal Netherlands Meteorological Institute.

So when choosing the desired bands of an orbiting radiometer, the optical features of interest in the ocean must be considered, as well as the optical features of the atmosphere that need to be avoided.

Having considered the basics of instrument design and band location, the other necessary consideration for a satellite instrument regards its imaging capabilities. (Note: these considerations are considerably expanded on in the main text of the history, in the same manner that engineers and oceanographers comprehended and considered them as they designed and built instruments and analyzed the data from those instruments.) While we are growing accustomed to seeing high resolution imagery of the Earth at somewhat astonishing spatial resolutions of a meter or less, scientists working on the design of remote-sensing instruments have to balance several competing factors. The amount of light (or electromagnetic energy) that it is necessary to acquire from a certain area in order to have useful information is one of the most important factors. The optical, mechanical, electronic characteristics of the instrument are also significant factors. The altitude of the satellite orbit, and the speed that it travels over the Earth, must also be considered. These factors result in a specific set of data characteristics: the spatial range of the area viewed by the instrument (the *scan swath* or *footprint*); the spatial resolution, i.e., the size on the Earth surface of each picture element (*pixel*); and the translation of the electronic signal into "numbers" that can be transmitted from the satellite to Earth, the process of digitization. The actual electronic signal is an *analog* signal, but the form of the data sent to Earth is digital, and this transformation also changes the nature of the data that are received from the satellite.

Envisioning the Light in the Ocean

Thus, the photons from the Sun have traveled through space, within the Earth system, back to space, and they have encountered a satellite with instruments designed to collect and detect them. The instrument has performed its duty, quantifying the light intensity from each small segment of its scanning region in each of its detecting bands, and then storing (or broadcasting) this information to a receiving station and data system on the ground. This information – the satellite telemetry – has its own particular characteristics that will allow it to be translated into information that will be useful to oceanographic scientists. So now we are no longer considering a physical path of light; rather, we are considering an analytical and computational path that leads to knowledge of the oceans. There are many pitfalls along this particular path that must be recognized.[41]

[41] The presentation of remote sensing data processing utilizes the following references: GES DISC, "Chapter 11: On the Level from Radiation to Scientific Imagery"; Nicholas Short, "Remote Sensing Tutorial", sections entitled "Sensor Technology: Types of Resolution", http://rst.gsfc.nasa.gov/Intro/Part2_5a.html, accessed 11 June 2009; and "Processing and Classification of Remotely Sensed Data; Pattern Recognition; Approaches to Data/Image Interpretation", http://rst.gsfc.nasa.gov/Intro/Part2_6.html, accessed 11 June 2009; Donna Thomas

Inside the Instrument: Processing the Signal

The precious photons that carry information about the oceanic environment have now returned to space and have been collected, reflected, directed, and then detected by the sensitive optical elements within the satellite sensor. The detectors collect the light for their designated microseconds of time, then discharge the accumulated electrical signal to the sensor's integrating device. The integrator takes the signal from each element of the detector, assigns it to its designated place, and then converts the electrical signal to a digital number, the process of digitization.

Because computing systems are at their heart binary systems (the information consists of "1's" and "0's", digitization is the process of converting the power of the electrical signal into a number, or byte. In the early days of remote sensing, 8-bit digitization was common, which meant that the byte value could range from 0 to 2^8, which means 0-256. So the minimum signal detected would be 0, and the maximum signal would be 256. Simply increasing the digitization capability 8-bit to 10-bit, from 2^8 to 2^{10}, means going from 256 discrete values to 1,024 discrete values, a much finer resolution of the light intensities stemming from the Earth surface. And going from 10-bit to 16-bit digitization is the difference between 1,024 and 65,536 discrete values. The problem is: storing so much more data, i.e., so many more 1's and 0's, requires a lot more instrumental memory.

The analog-to-digital converter (ADC) accomplishes the conversion, so that now each detector element yields a digital number. These digital numbers are linked together in the order that they were detected as the light from the scanning system impinged the detectors – thus creating a single *scan line* of digital values. As the satellite moves along its orbit and as the scanner keeps collecting each consecutive scan line of data, a data object takes shape: simply put, this is a data file. The data file can be stored in space in the computer memory of the satellite, or it can be created and stored on the ground, the file constructed by a computer at a ground station that receives the signal broadcast continuously from the satellite (the data broadcast from the satellite is called telemetry; that term covers more than just the data in the scan file – it also contains information on the satellite position, the operating status of the satellite, associated information such as time and temperature, etc., all of which are very important information to the operators of a satellite remote-sensing mission. The numbers stored on the satellite (for later transmission to a ground station) or directly broadcast are somewhat inelegantly called *raw data* – an array of numbers. It takes more information to create something that will be meaningful to scientists.

---

and Bryan Franz, "Overview of SeaWiFS data processing and distribution", http://oceancolor.gsfc.nasa.gov/DOCS/SW_proc.html, accessed 11 June 2009.

Converting Telemetry to Radiances

Once the raw data have been received, data processing begins. The first step is to assign the appropriate location to the data, using the associated navigational data – which is why it's critical for the satellite to know where it was when the data was collected, and what time it was collected. Ultra-precise time-keeping and location determinations are vital and unsung elements of remote sensing. Once the navigational data have been utilized (and a few other pieces of information, such as where one file ends and another begins), a data file can be created. This is a *Level 0* data file; raw data along with the location on the Earth corresponding to each pixel of raw data. Note that in case of data collected by a spectroradiometer, the data are also organized according by each spectral band of the detectors in the sensor.

The next step is to convert the raw numbers into radiance values. This process is accomplished by using some of the associated information from the satellite, which may be available in the scanning data or already stored in the data processing system. According to the calibration data for the instrument, each digital byte number corresponds to a radiance value. Converting the Level 0 data file to a *Level 1* data file thus primarily consists of converting the byte values to radiance values. There is a very important distinction to be noted here; this conversion ONLY converts the byte values to *raw* radiances, which is shorter way of saying the radiances measured at the satellite, above the atmosphere of the Earth. So the Level 1 data file consists of uncorrected radiance values, corresponding to each band of the instrument (if the instrument has bands – some instruments just measure light intensity across the spectrum, but ocean color instruments are spectroradiometers and therefore have band-by-band data).

Converting Radiances to Geophysical Products

Everything that has happened to the data before this next step has been pretty technical; most scientists interested in using the data from the instruments aren't all that interested in the first couple of steps, other than expecting that the mission scientists and engineers know what they're doing to deliver accurate raw radiance data from the satellite. As will be seen, this is a challenging task; instruments and satellites constantly change in the environment of space, so insuring data quality starts at the very beginning of the mission and the very beginning of the data acquisition process – and never ends until long after the mission is over.

The next milestone in the computational process is where it gets considerably more interesting from the scientific perspective. It is here at this stage that computational algorithms (achieved by research) are applied to the data to convert the raw radiance values into ocean-relevant radiance values – for ocean color remote sensing, this means applying the necessary atmospheric corrections for the processes occurring in the atmosphere (including absorption by water vapor and ozone, and scattering by aerosols and air molecules); testing the data for a variety of conditions which make it unusable or at least suspect; correcting for the angle of the sun and the viewing angle of the satellite; and applying calibration data. Once this done, the result is *Level 2* data in a Level 2 data file. For ocean color science, Level 2 data means the water-leaving radiances for each band. The water-leaving radiances can then be used to calculate other quantities, such as the highly-desired concentration of chlorophyll in ocean waters. These are collectively referred to as *geophysical values.*

It should also be noted that both Level 1 and Level 2 data files are commonly organized into files corresponding to a discrete period of scanning time or a segment of the satellite orbit. The Level 1 and Level 2 data files may therefore be referred to as *swath* files or *scan swath* files. The files consist of a set of scan lines with the radiance values in each pixel of the each scan line. This data is thus segmented as it was collected by the satellite.

Level 1 and Level 2 data are also (usually) data at the maximum spatial resolution collected by the satellite. But there is another distinction to be made here; for some missions or data processing systems, in order to have sufficient memory to store the data, the data may be subsampled to reduce the total amount of data stored. In these cases, the maximum resolution data (without subsampling) is called LAC, for Local Area Coverage; the subsampled data is termed GAC, for Global Area Coverage. The importance of this distinction is that there can be both Level 1 and Level 2 LAC data files and Level 1 and Level 2 GAC data files. So a GAC swath file will have less data in it than a LAC swath file, even though they cover the same area on the Earth's surface.

Displaying and Mapping the Geophysical Products

With the geophysical values now contained in Level 2 swath files, the final step is to indulge in the historically-honored process of cartography – which is to say, make maps. Map-making in this context is not quite the same as carefully delineating the location of geographical, geological, or geo-political features; in this case, it means taking the data from their existence in a satellite observational swath and locating the data on a more familiar map of the Earth. One of the interesting aspects of this process it that it is not simply a spatial process; the data also have a place in time, so that mapping of the data involves location in both time and space. The oceans are always changing; thus, this mapping will capture the process of change as the satellite sensors observe it.

So the next procedure is to take the data from the swaths and put it into a location that has both spatial and temporal significance. For the purpose of remote sensing, this location is called a "bin" – conjuring up a vision of a robot tossing data pixels with incredible rapidity into their appropriate bins. Such a vision is not too far removed from what the data systems actually do. As the data from each swath is received, it is assigned to its appropriate spatial bin, and every pixel value for that location is added to the bin. For the purpose of global mapping, the spatial resolution is reduced compared to the LAC (and sometimes the GAC) resolution – so there will be many data values added to a single bin.

Even for a single day of scanning, some satellite observations will overlap, so there can be more than one data value for a given bin in a single day. The spatial location of the bins won't change, but the contents will vary depending on the temporal resolution of the bin: a bin can be a daily, weekly, monthly, or annual bin, containing all of the valid data pixels acquired by the satellite sensor over that time interval.

The data in the bins accumulates, and are then averaged according the bin resolution. For example, if the data are collected at 1 km resolution and the bin is 25 x 25 km (which is roughly a quarter of a geographical degree, given the curvature of the Earth), then there will be 625 pixels in each bin. To calculate the bin value for one day, provided there was no overlapping swath coverage for this bin, all of the valid pixel values from the 625 pixel locations in the bin will be averaged (remember that pixel values might not be counted if they were suspected to be inaccurate, if they were covered by clouds, or if they were on land).

If the data were collected over one week, all of the valid pixel values from the 625 pixel locations in the bin, collected each day for seven days, will be averaged. Then the values are projected onto a map projection, and a global map of the binned and averaged data values is constructed. (Not surprisingly, this process is called *binning and averaging*!)

There are potential pitfalls in this process, and the methods used to calculate the values for such maps must be carefully considered. Averages can be calculated according to different methods (such as the *arithmetic mean* or the *geometric mean*), and there are other values that could be calculated (such as the *median* or the *Maximum Likelihood Estimator*, MLE). The actual period of data collection has to be considered; the length of time it takes for satellites to orbit the Earth is rarely amenable to neat divisions according to calendrical days, weeks, or months.

Furthermore, data from the oceans are not distributed according to the classic "bell curve" statistical distribution; rather, the data have a bimodal distribution of a small number of high values and a very large number of low values. This distribution also affects the calculation of mean values over time.

The outcome of these processing steps is a global data array. It's not quite a map yet, because a map is a visual object. One final procedure remains to make what is familiarly called a map – converting the values into a visual display. This conversion is accomplished by assigning the data values to colors in a color palette, and then placing the colors corresponding to the values in their geographical location. In order to depict small gradations in values, frequently a color palette considerably removed from reality is used, which is called a *false color* palette to produce a false color map. The familiar "Rainbow" palette depicts low values with violets and blues, moderate values with shades of green and yellow, and high values with hues of orange and red, and is used for many different kinds of remotely-sensed data, not just the data produced by ocean color sensors.

It is also possible to mathematically combine data from the bands in the red, green, and blue (RGB) parts of the spectrum to create a map that looks fairly similar to the actual colors of the Earth as seen from space; while these are sometimes called *true color* maps, because the correspondence of the mathematically combined bands to actual visually-perceived colors is not exact, they are more accurately called *pseudo-true color* maps.

Conclusion – at the beginning

Now the process is complete; the data are available in files, or data arrays, or maps – and it can be utilized by scientists for oceanographic research. It is remarkable that these maps accurately depict geophysical values calculated by algorithms from radiances deduced from electrical signals resulting from the detection of a few photons of light by a satellite sensor in orbit above the Earth.

If that sounds like a technically miraculous outcome, that impression is reasonably valid – and that is why the history of NASA ocean color remote sensing is such a compelling and somewhat unknown story. The reason that it is somewhat unknown is partly because it has been successful, and that success has tended to obscure the myriad of technological advances, research campaigns, late nights and long days, theoretical insights, close calls and near-misses, frustrating delays, second- and third-guessing, funding problems, lucky happenstance and awful fate, human foibles, and above all, the truly stunning dedication to scientific accuracy motivated by the belief that this branch of oceanographic science could be done – and that it would be extraordinarily meaningful to the human understanding of the how the oceans work. It is not just a story about how maps of chlorophyll and stunning images of the oceans from space are created; it is a history of scientific achievement, accomplished primarily by the combined human powers of intellect and perseverance. The pages of this history have endeavoured to tell that story, and to show how such a scientific success was dependent on painstaking and sometimes painfully incremental steps.

Science progresses most vigorously when each new discovery generates new questions. Lacking such continual revitalization, science tends to succumb to sterile and doctrinaire dormancy. **– Gifford Ewing**

Appendix 2: The Nimbus Program: Nimbus 1 through Nimbus 5

The researchers and the activities of the researchers described in Chapter 1 and 2 contributed to the theoretical basis and the engineering knowledge required to envision a satellite-borne ocean color instrument. However, a lot more work was required on the engineering and feasibility aspects of such a mission. The deployment of the airborne spectroradiometer during HAOCE was an important step. As these activities were taking place in the 1960s and early 1970s, NASA had already initiated the Nimbus program. The Nimbus program was NASA's pioneering program to observe the Earth from space.

Satellite	Launch Date
Nimbus 1	1964-08-28
Nimbus 2	1966-05-15
Nimbus 3	1969-04-14
Nimbus 4	1970-04-08
Nimbus 5	1972-12-11

Largely unnoticed by the public due to the simultaneously occurring human spaceflight activities leading to the landing on the Moon, Nimbus nonetheless served as a testing ground and instrumental demonstration program leading to standard operational observations of important variables in the Earth's climate and environment.[42] All of the Nimbus satellites were launched from Vandenberg Air Force Base in California. The Nimbus program and much of the instrument fabrication and satellite construction and preparation took place at Goddard Space Flight Center in Greenbelt, Maryland. The Mission Director for Nimbus 1-5 was William Brotherton Huston.[43]

The Nimbus program utilized a common satellite design, which frequently appears in depictions of satellites in orbit due to its unique appearance. The observational instruments were mounted on the Earth-facing side of a circular platform. The opposite side featured a guidance and power structure with a vaguely alien or robotic appearance, with the most notable feature being the pair of wing-like solar panels[44] (see Figure A2-1).

---

[42] "Nimbus 1963-2004", http://atmospheres.gsfc.nasa.gov/nimbus/ (accessed 31 January 2009); "Nimbus Program", http://nssdc.gsfc.nasa.gov/earth/nimbus.html (accessed 31 January 2009); Rebecca Lindsey, "Nimbus 40th Anniversary", http://earthobservatory.nasa.gov/Features/Nimbus/nimbus.php (accessed 31 January 2009)
[43] "Wilbur B. Huston — 'The smartest boy in America' (in 1929) is dead", http://www.bookofjoe.com/2006/06/wilbur_b_huston.html (accessed 31 January 2009); "Wilber B. Huston", http://www.nationmaster.com/encyclopedia/Wilber-B.-Huston (accessed 31 January 2009).
[44] Rebecca Lindsey, "Nimbus 40th Anniversary".

Figure A2-1. Artist's drawing of the general design of the Nimbus series of satellites. Image taken from Madrid, C.R., ed. (1978) *The Nimbus 7 Users' Guide.* Goddard Space Flight Center: National Aeronautics and Space Administration.

Nimbus 1 carried three instruments: the Advanced Vidicon Camera System, the Automatic Picture Transmission System (instrument?), and the High-Resolution Infrared Radiometer. Dr. William Nordberg was the Project Scientist, Dr. Morris Tepper was the Program Scientist, and Mr. Harry Press was the Project Manager for Nimbus 1.[45]

Nimbus 2 added one new instrument, the Medium-Resolution Infrared Radiometer. Nordberg, Tepper, and Press continued in their same positions for the Nimbus 2 mission.[46]

Nimbus 3 carried an expanded payload of instruments and new systems for power and communications. The notable new addition to the instrumental payload was the Satellite Infrared Spectrometer (SIRS), designed to obtain vertical atmospheric temperature profiles. Another new spectrometer on this mission was the Infrared Interferometer Spectrometer (IRIS). The trio of Nordberg, Tepper, and Press continued to perform the same duties that they had performed for the previous Nimbus missions.[47]

Nimbus 4 featured several new instruments and project personnel.[48] A second SIRS and a second IRIS orbited on Nimbus 4, along with the Backscatter Ultraviolet Spectrometer (BUV); the Image Dissector Camera System; the Selective Chopper Radiometer (SCR), which observed radiative emissions from atmospheric carbon dioxide to determine the temperature of various atmospheric layers; the Solar UV Monitor; the Temperature-

---

[45] "Nimbus 1", http://nssdc.gsfc.nasa.gov/nmc/masterCatalog.do?sc=1964-052A (accessed 31 January 2009), "Nimbus 1 experiment search result", http://nssdc.gsfc.nasa.gov/nmc/masterCatalog.do?sc=1964-052A&ex=* (accessed 31 January 2009)
[46] "Nimbus 2", http://nssdc.gsfc.nasa.gov/nmc/masterCatalog.do?sc=1966-040A (accessed 31 January 2009); "Nimbus 2 experiment search result", http://nssdc.gsfc.nasa.gov/nmc/masterCatalog.do?sc=1966-040A&ex=* (accessed 31 January 2009)
[47] "Nimbus 3", http://nssdc.gsfc.nasa.gov/nmc/masterCatalog.do?sc=1969-037A (accessed 31 January 2009); "Nimbus 3 experiment search result", http://nssdc.gsfc.nasa.gov/nmc/masterCatalog.do?sc=1969-037A&ex=* (accessed 31 January 2009)
[48] "Nimbus 4", http://nssdc.gsfc.nasa.gov/nmc/masterCatalog.do?sc=1970-025A (accessed 31 January 2009); "Nimbus 4 experiment search result", http://nssdc.gsfc.nasa.gov/nmc/masterCatalog.do?sc=1970-025A&ex=* (accessed 31 January 2009)

Humidity Infrared Radiometer (THIR); and the Filter Wedge Spectrometer. The FWS is notable because the Principal Scientist was Warren Hovis, who would become the leader of the Nimbus program's ocean color mission.[49] Unfortunately, the FWS instrument did not return useful data. The BUV instrument was the first instrument orbited in space that measured the distribution of atmospheric ozone.[50] For Nimbus 4, the program managers and scientists changed, with Ray Arnold taking over as Program Manager, Charles Mackenzie becoming the Project Manager, and Al Fleig the new Project Scientist.

Nimbus 4 also carried an interesting experiment, the Interrogation, Recording, and Location System (IRLS).[51] The objective of this experiment was to communicate with various instrumented recording devices like weather balloons, ocean buoys, and research vessels. When the device communicated with the satellite, the satellite determined the location of the device on the Earth surface to within 2 km by determining the distance from the satellite to the device using the round-trip travel time for the radio signal. The IRLS also received the data recorded by the device for later retransmission. Given that we are now accustomed to 1 meter or better locations from the Global Positioning System, the 2 km objective of the IRLS seems rather quaint. The IRLS was a notable precursor to global ocean drifter programs, where an instrumented float moves at various depths with ocean currents until a specified time, then surfaces to transmit location and collected data to an orbiting satellite.[52] More than 3,000 ocean drifters have been deployed since the start of the program.

Nimbus 5 carried a reduced payload of six instruments.[53] A second THIR and another SCR were deployed on Nimbus 6, along with the Electrically Scanning Microwave Radiometer (ESMR), the Infrared Temperature Profile Radiometer, and the Microwave Spectrometer, called NEMS because the Nimbus satellites were designated with letters before launch and numbers after they were orbited successfully, so NEMS stands for Nimbus E Microwave Spectrometer. NEMS measured both atmospheric and Earth surface temperatures as well as atmospheric water vapor and cloud liquid water. ESMR was noteworthy because it attempted the first observations of sea ice cover and extent, as well as

---

[49] "Filter Wedge Spectrometer (FWS)", http://nssdc.gsfc.nasa.gov/nmc/experimentDisplay.do?id=1970-025A-09 (accessed 31 January 2009).

[50] "Backscatter Ultraviolet (BUV) Spectrometer", http://nssdc.gsfc.nasa.gov/nmc/experimentDisplay.do?id=1970-025A-05 (accessed 31 January 2009).

[51] "Interrogation, Recording, and Location System (IRLS)", http://nssdc.gsfc.nasa.gov/nmc/experimentDisplay.do?id=1970-025A-07 (accessed 31 January 2009).

[52] "The Global Drifter Program", http://www.aoml.noaa.gov/phod/dac/gdp_information.html (accessed 31 January 2009).

[53] "Nimbus 5", http://nssdc.gsfc.nasa.gov/nmc/masterCatalog.do?sc=1972-097A (accessed 31 January 2009); "Nimbus 5 experiment search result", http://nssdc.gsfc.nasa.gov/nmc/masterCatalog.do?sc=1972-097A&ex=* (accessed 31 January 2009).

examining surface composition and soil moisture.[54]

The sixth instrument on Nimbus 5 was the Surface Composition Mapping Radiometer (SCMR), with Warren Hovis as Principal Scientist. The "skimmer" was designed to analyze Earth surface composition based on temperature, and also to measure sea surface temperatures. The ground resolution of the SCMR was 660 meters. Though SCMR did produce some usable data, unlike Hovis' previous instrument on Nimbus 4, it started being recalcitrant soon after the December 11, 1972 launch and stopped transmitting data entirely on January 4, 1973 – only about four weeks.[55]

Mackenzie and Fleig continued in their roles for Nimbus 5, with a new Nimbus Program Manager, George Esenwein, Jr.

Thus, during the late 1960s and early 1970s, NASA Goddard Space Flight Center was a fertile proving ground for design, development, and fabrication of the first wave of Earth observing instruments launched into space. With Nimbus 6 and Nimbus 7 in the planning stages, the skilled engineers were hungry for new challenges. Warren Hovis in particular may have been seeking a new project that would fare better than his previous two Nimbus instruments – perhaps a project based on the spectroradiometer that he had flown in a Learjet over the California coast in 1971.

---

[54] "Electrically Scanning Microwave Radiometer (ESMR)", http://nssdc.gsfc.nasa.gov/nmc/experimentDisplay.do?id=1972-097A-04 (accessed 31 January 2009).

[55] "Surface Composition Mapping Radiometer", http://nssdc.gsfc.nasa.gov/nmc/experimentDisplay.do?id=1972-097A-05 (accessed 31 January 2009).

Appendix 3: Scientific results of the CZCS and SeaWiFS missions

CZCS

Major Current Systems

The CZCS provided images of the biological characteristics of Earth's major ocean currents, augmenting the coverage that had been provided by SST remote sensing, and demonstrating how the frontal zones defined by currents frequently separated regions of generally higher productivity from regions of lower productivity. The global view from CZCS indicated how the major ocean currents also define the edges of the major ocean basins, the centers of which usually host low-productivity oligotrophic water masses. And the CZCS famously showed the development of eddy and ring systems, which can last for months and years and which can transport higher productivity waters through regions of low productivity over remarkably long distances.

Some of the most striking and memorable ocean images from the CZCS, images which have been repeatedly published, feature the Gulf Stream off the U.S. East Coast. While genuinely impressive, these images don't demonstrate what was learned about the Gulf Stream from CZCS data. Both during and soon after the mission, observations of the ring systems were published.[56] The CZCS augmented the existing knowledge of the rings by showing how the rings aged over time, and provided an additional way to inventory the population of the rings to determine how they were distributed in the northwestern Atlantic.

Joji Ishizaka of Texas A&M University evaluated frontal dynamics in the Gulf Stream, and showed that the main chlorophyll "events", which were responsible for the highest degree of variability, were due to the movement of Gulf Stream frontal eddies. Frontal eddies are a distinct feature of the Gulf Stream; the current characteristically forms meanders with the appearance of waves, which tend to move northward. In the trough of the wave, colder waters from the continental shelf circulate cyclonically (counter-clockwise in the Northern Hemisphere). If the wave extends and pinches off, it forms a cold-core ring (Figure 4.1). Ishizaka published a set of three papers which described model comparisons to CZCS observations.[57]

---

[56] Howard R. Gordon, Dennis K. Clark, James W. Brown, Otis B. Brown, and Robert H. Evans, R.H. "Satellite measurement of the phytoplankton pigment concentration in the surface waters of a warm core Gulf Stream ring", *Journal of Marine Research*, 40, 491-502 (1982); Raymond C. Smith, Otis B. Brown, Frank E. Hoge, K.S. Baker, Robert H. Evans, Robert N. Swift, and Wayne E. Esaias, "Multiplatform sampling (ship, aircraft, and satellite) of a Gulf Stream warm core ring".

[57] Joji Ishizaka, "Coupling of coastal zone color scanner data to a physical-biological model of the southeastern U.S. continental shelf ecosystem 1. CZCS data description and Lagrangian particle tracing experiments", *Journal of Geophysical Research*, 95 (C11), 20,167–20,182 (1990); Joji Ishizaka, "Coupling of coastal zone color scanner data to a physical-biological model of the southeastern U.S.

In 1984, McClain and Atkinson analyzed a single CZCS image of the Charleston Bump region located, as might be guessed, off the South Carolina coast near Charleston.[58] The Charleston Bump is a small bottom feature, a benthic hillock, near the edge of the continental shelf. The Bump influences the flow of the Gulf Stream sufficiently to cause the formation of a semi-permanent eddy; deeper water forced to the surface in this circulation provides nutrients that enhance productivity and provide a well-known deep-sea fishing hot spot.

The influence of the Gulf Stream can extend well beyond its actual flowing boundaries. A study published in 1999 demonstrated that the Gulf Stream is a major influence on the productivity of the waters of the "Slope Sea", the region lying between the continental shelf off of the Mid-Atlantic States and the northern wall of the Gulf Stream.[59] This team examined the critical transitional period between winter and summer conditions, when surface temperatures warm markedly, causing mixing and rapid hydrographic changes. The enhancement of productivity in the Slope Sea can last over 2 ½ months of spring. One of the most intriguing aspects of this period is that the warm core rings spinning off the Gulf Stream can encroach on the waters of the shelf, and their strong circulation draws the high productivity shelf waters seaward, causing blobs and filaments of productivity to extend much further offshore.

Showing that CZCS data can continue to provide oceanographic insight even in the era of wide availability of SeaWiFS, MODIS, and other ocean color satellite sensors, Martins and Pelegri investigated the South Atlantic Bight and Gulf Stream during the low-stratification period, which for the oceans is frequently called winter.[60] They examined the reduction in pigment concentration from the coast to the open waters, and also noted an increase in the scale of pigment variability.

The Pacific counterpart of the Gulf Stream is the Kuroshio Current, which flows from its origin in the South China Sea past the Phillippines and Taiwan, up to the islands of Japan, where it turns to the east from Honshu and extends its influence hundreds of kilometers into the northern Pacific Ocean. Ning and colleagues characterized the southern flow of the Kuroshio and the waters of the East China Sea, documenting significant features caused by river outflow.[61] The Kuroshio current here is very similar to the Gulf Stream, carrying

---

continental shelf ecosystem 2. An Eulerian model", *Journal of Geophysical Research*, 95 (C11), 20,183–20,19 (1990); Joji Ishizaka, "Coupling of coastal zone color scanner data to a physical-biological model of the southeastern U.S. continental shelf ecosystem 3. Nutrient and phytoplankton fluxes and CZCS data assimilation", *Journal of Geophysical Research*, 95 (C11)20,201–20,212, (1990).

[58] Charles McClain, and Larry P. Atkinson, "A note on the Charleston Gyre", *Journal of Geophysical Research*, 90 (C14), 11857-1186, (1984).

[59] John P. Ryan, James A. Yoder, and Peter C. Cornillon, "Enhanced chlorophyll at the shelfbreak of the Mid-Atlantic Bight and Georges Bank during the spring transition", *Limnology and Oceanography*, 44(1), 1–11, (1999).

[60] Ana M. Martins and J. Pelegri, "CZCS chlorophyll patterns in the South Atlantic Bight during low vertical stratification conditions", *Continental Shelf Research*, 26(4), 429-457, (2006).

[61] Xiuren Ning, Ming Fang, Zilin Liu, Jiezhong Chen, and Yuming Cai, "Physico-biological oceanographic remote sensing of the East China Sea: satellite and *in situ* observations", *Journal of Geophysical Research-Oceans*, 103(C10), 21,623-21,635, (1998).

warm, low productivity waters northward. Due to the distances involved in conducting research in the vast Pacific, very few investigations attempted to characterize the seaward extension of the Kuroshio. Coverage was also limited due to the observational time budget. The few available CZCS pigment images demonstrated that the Kuroshio off Japan looks very similar to the Gulf Stream, with low productivity warm water to the south and successively higher concentrations to the north.

Due to the participation of South African scientists on the CZCS NET, the oceanic waters around South Africa received considerable CZCS observational time. These waters have two major current systems: the Benguela, which creates a large upwelling zone (which will be discussed subsequently) and the Agulhas, which creates one of the most interesting and picturesque (in the remote-sensing sense) zones in the entire world ocean. The Agulhas is the only equivalent in the Southern Hemisphere to the dominant warm-water western boundary currents in the Northern Hemisphere, the Gulf Stream and the Kuroshio. The Agulhas flows southward along the eastern African Coast, bringing warm water from the Indian Ocean past Madagascar through the Mozambique Channel, and rounding the southern coast toward Cape Town. But just a bit past Port Elizabeth, something unusual happens. The Agulhas Current abruptly turns south and heads out to sea, flowing southward until it collides with the eastward flow of the Southern Circumpolar Current (SSC). The SSC seizes the Agulhas current and twists it nearly back on itself, in a zone of ocean interaction called the Agulhas Retroflection.[62] Soon after it turns, the current encounters the Agulhas Plateau, a shallow region of the ocean that adds additional turbulent energy to this zone. The result of this turbulent collision of strong oceanic currents is a remarkable zone of huge meanders, colored brightly in CZCS pigment data as the interaction of mixing waters forces nutrients to the surface. This is a region that forms huge rogue waves that can threaten oil tankers. Another consequence of this interaction is the formation of a multitude of spinning Agulhas ring systems, which actually gravitate westward, toward South America.[63]

Scarla Weeks and Frank Shillington described several aspects of this region utilizing both CZCS pigment data and SST data.[64] They observed considerable year-to-year differences in the position of the Agulhas Front, and determined that the front was one of the main factors determining the distribution of phytoplankton pigment in this region. The Agulhas system has also been discussed in conjunction with investigations of the Benguela upwelling system.

---

[62] NASA GES DISC, "Classic CZCS Scenes: Chapter 9: The Agulhas Retroflection", http://daac.gsfc.nasa.gov/oceancolor/scifocus/classic_scenes/09_classics_agulhas.shtml, accessed 28 April 2009.

[63] Arnold L. Gordon and W.F. Haxby. "Agulhas eddies invade the South Atlantic – evidence from GEOSAT altimeter and shipboard conductivity–temperature– depth survey", *Journal of Geophysical Research*, 95 (C3), 3117–3125, (1990).

[64] Scarla Weeks and Frank A. Shillington, "Phytoplankton pigment distribution and frontal structure in the subtropical convergence region south of Africa", *Deep-Sea Research. Part 1*, 43(5), 739-768, (1996).

Major Upwelling Zones

On the ocean color stage, there are just a few oceanic regions that deserve superstar status. These are the lead singers of the chlorophyll concert, the areas where the solar spotlight shines brightly and the chlorophyll concentrations reach their highest levels in the entire world ocean. These are the areas that draw the eye first on the false-color maps of pigment and chlorophyll concentration: the fiery reds and bright oranges and yellows indicate where phytoplankton proliferate, utilizing abundant nutrients and strong sunlight to create ocean waters that are literally green with life, waters so rich in chlorophyll that they test the limits of the algorithms designed to measure them.

So first we must address the question: *what is upwelling?* Oceanographers reading this history likely know; readers of this history who have encountered the word in earlier chapters may have already looked it up. Because upwelling is such a vital process for ocean color research, it deserves a slightly longer explication here.

Upwelling refers to the oceanic process by which the flow of a wind-driven surface current causes the horizontal advection of a surface water mass, which must be replaced by deeper water rising to the surface. When winds blow along the surface of the ocean, they induce the movement of water at the surface. This situation gives rise to a process called "Ekman transport". Essentially, the transfer of momentum from the surface wind to the water causes the movement of the water to be aligned approximately 90° (perpendicular) to the direction of the wind. When the winds are aligned along the coast, as is the case for several major upwelling zones, the resulting transport of water is seaward, and deeper nutrient-rich water flows up to take the place of the surface water moving offshore (Figure A3-1).[65]

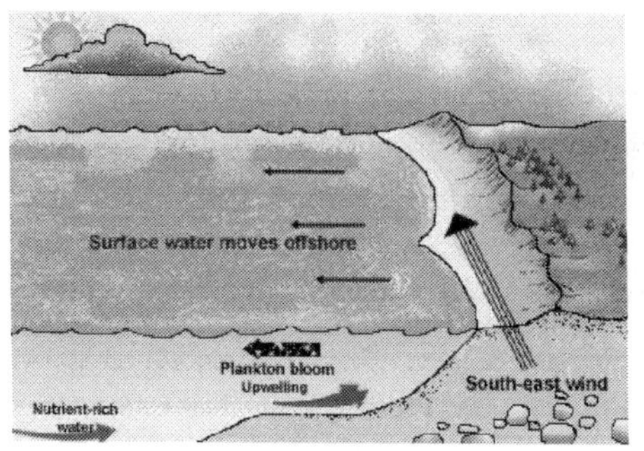

Figure A3-1. Schematic diagram of an upwelling zone similar to the Benguela system.

[65] The animation of the Benguela upwelling zone was provided by Dr. Derek Keats of the Botany Department of the University of the Western Cape in South Africa and is also found on the GES DISC Web page "Chapter 2, The Benguela Upwelling Zone", http://disc.sci.gsfc.nasa.gov/oceancolor/additional/science-focus/classic_scenes/02_classics_benguela.shtml, accessed 8 January 2010.

Upwelling doesn't always take place due to winds which predominantly blow in a certain direction. Major storms such as hurricanes can set up a temporary occurrence of upwelling; storm winds will also cause mixing of deeper waters with surface waters, which also brings nutrients to the surface. There are also zones of convergence, where two currents flowing in opposite direction interact; this also causes the transport of nutrients to the surface. This is the process occurring along the Equator in the Pacific Ocean in normal years, and also happens along the eastern coast of South America.

There are two particularly notable coastal upwelling zones; the Peru Upwelling along the coast of Peru, and which is notably affected by El Niño; and the Benguela Upwelling Zone, on the Atlantic coast of western South Africa and the Atlantic coast of Namibia. Because South African scientists were represented on the CZCS NET, and because they conducted collaborative research, the Benguela upwelling received very good observational coverage by the CZCS. A representative publication for this region was another work by Weeks and Shillington, which investigated interannual variability of oceanic pigments in the Benguela.[66] Their analyses clearly demonstrated how the repeated observations of the CZCS over time distinguished both spatial and temporal patterns of variability in this region. One observation was that the Benguela could not be considered to be one large system; rather, it had a northern and southern component, with dissimilar seasonal patterns. In the northern Benguela, pigment concentrations had a small decrease in spring, a decrease in summer, and a large increase in autumn; while in the southern Benguela only one concentration maximum, in the summer, was observed. The Benguela also a distinct pattern of interannual variability; surprisingly, examination of the CZCS annual composite images showed a significant increase in pigment concentrations in the period October 1981-September 1982, compared to the period October 1978-September 1979. The Agulhas Retroflection and the Subtropical Convergence Zone south of South Africa also showed distinct increases. Weeks and Shillington noted that the largest differences coincided with the occurrence of the 1982-1983 El Niño, a period when unusual oceanographic and meteorological conditions were also noted in the environs of South Africa. The paper also features one of the remarkable images of the CZCS mission, a monthly image from February 1983, where the wave-like meanders downstream of the Agulhas Retroflection are particularly prominent.[67]

The coastal waters adjacent to Northwest Africa also received good coverage by the CZCS. A team of researchers led by Lieve Van Camp investigated the occurrence of upwelling in this region.[68] In this study, the SST and CZCS pigment data were not examined for the same time periods; the SST data was for 1997, while the CZCS data was for 1980-1985. (This is a small indication of why remote-sensing oceanographers had a wish list for a

---

[66] Scarla J. Weeks, and Frank A. Shillington, "Interannual scales of variation of pigment concentrations from coastal zone color scanner data in the Benguela upwelling system and the subtropical convergence zone south of Africa", *Journal of Geophysical Research*, 99, 7385-7399, (1994).
[67] NASA GES DISC, "Classic CZCS Scenes: Chapter 9: The Agulhas Retroflection".
[68] Lieve Van Camp, Leo Nykjær, Ekkehard Mittelstaedt, and Peter Schlittenhardt, "Upwelling and boundary circulation off Northwest Africa as depicted by infrared and visible satellite observations", *Progress in Oceanography*, 26, 357-402, (1991).

sensor that collected SST and ocean color data at the same time!) Van Camp et al. stated that winter and spring pigment concentrations ranged from 1.5 to over 20 mg/m^{-3}, which was verified by surface observations. The researchers noted the extensive occurrence of filaments, which are long narrow extensions of higher pigment waters, caused by sudden strong offshore transport jets. As Dick Barber had already noted, CZCS imagery showed that coastal oceanic pigment distributions were not uniform sheets with similar concentrations everywhere; they were dynamic, with rapidly changing features and high variability. Van Camp et al. did examine one fortuitous occurrence of near-simultaneous SST and CZCS data acquisition, which occurred on July 2, 1985. On this date, it was fairly easy to identify coastal water masses by their low SST and high pigment concentration characteristics.

The CZCS also allowed examination of other upwelling zones distributed along the world's oceanic coast. Prolific Joji Ishizaka described observations of coastal upwelling near Japan's famous Izu peninsula (prominently located near Tokyo) during May 1982.[69] This paper demonstrated the application of a regional algorithm, because the CZCS and ship-measured pigment concentrations originally differed by a factor of 5. When the algorithm was tuned so that the CZCS pigment values and the ship-measured values agreed, the results were in close agreement over the entire region. This observational method indicated the need for a constant source of accurate radiometric data to maintain accurate sensor calibration. Ishizaka's 1992 paper featured two co-authors worthy of note, Motoaki Kishino and Hajime Fukushima, who would be important figures in Japan's ocean color missions in the 1990s. In a few years, Kishino would design and deploy an in-water radiometric sensor buoy off the coast of Japan.

Polar Oceans

One of the significant accomplishments of the CZCS was that it provided a view of oceanic processes in areas that are markedly inhospitable to traditional at-sea research: the polar regions. At-sea oceanographic research in the frigid waters of the boreal and austral seas commonly requires ice-breaking vessels, extraordinarily warm clothing, and a great deal of patience to await reasonable weather conditions that allow the conduct of research. As some polar oceanographers will relate, research in these areas may be conducted in unreasonable weather conditions, too. Just as it is not easy working there, it is not easy getting to these distant locations, either. So CZCS opened a window on the wintry worlds of the polar seas that was a singular advance in our understanding of their biological systems.

The demarcation of biological patterns is particularly notable at the moving seasonal boundaries of the polar sea ice. Joey Comiso extended Cornelius Sullivan's early research in

---

[69] Joji Ishizaka, Hajime Fukushima, Motoaki Kishino, Toshiro Saino, and M. Takahashi, "Phytoplankton pigment distributions in regional upwelling around the Izu peninsula detected by coastal zone color scanner on May 1982", *Journal of Oceanography*, 48, 305-327, (1992).

an examination of the pigment patterns at the ice edges, determining the large areal extent of the phytoplankton blooms and also determining that they were persistent from spring into autumn, not just occurring in the early spring.[70]

On one side of North America, Frank Muller-Karger and his co-authors compared phytoplankton measured from a research cruise to CZCS observations of the Bering Sea.[71] They determined the normal timing of the spring bloom, for which they had correlated shipboard data, and also observed phytoplankton blooms in the autumn, for which there wasn't a lot of ship-based information. This is understandable, as many a king crab boat captain and crew will attest. On the other side of the continent, Greg Mitchell led a team that combined CZCS data with shipboard observations in the Barents Sea and described differing zones of productivity in that region.[72]

The Mediterranean Sea

The Greek poet Homer described the seafaring adventures of Odysseus in *The Odyssey*, likely in parts of the Mediterranean and Aegean Seas. While in modern times we don't expect to encounter bewitching singing maidens, one-eyed giants, ship-devouring whirlpools, or multi-headed sea serpents, the non-mythical Mediterranean is still one of the most familiar bodies of water in the world and one of the most studied. The CZCS added considerable insight to this portion of the marine realm.

It should not be a surprise that the utilization of CZCS data to study the Mediterranean involved André Morel. Morel and Jean-Michel André observed the patterns of pigment in the western Mediterranean (the Alboran Sea) and based on this data, modeled the primary productivity of this region.[73] Following up a few years later, David Antoine, Morel, and André did the same for the eastern Mediterranean.[74] Morel and André focused on the year 1981, utilizing over 100 CZCS scenes, and noted that the spring bloom in this region was strongest in May. The model they employed provided an estimate of the amount of productivity and the amount of carbon fixation here.

In 1995, Navy researcher Robert Arnone summarized what the CZCS had observed in

[70] Josefino C. Comiso, Nancy G. Maynard, Walker O. Smith Jr., and Cornelius W. Sullivan, "Satellite ocean color studies of Antarctic ice edges in summer and autumn", *Journal of Geophysical Research*, 95, 9481-9496, (1990).
[71] Frank E. Müller-Karger, Charles R. McClain, Raymond N. Sambrotto, and G.C Ray, "A comparison of ship and coastal zone color scanner mapped distribution of phytoplankton in the southeastern Bering Sea", *Journal of Geophysical Research*, 95, 11,483-11,499, (1990).
[72] B. Greg Mitchell, E. A. Brody, E.-N. Yeh, Charles R. McClain, Josefino C. Comiso, and Nancy G. Maynard, "Meridional zonation of the Barents Sea ecosystem inferred from satellite remote sensing and in situ bio-optical observations", *Polar Research*, 10, 147-162, (1991).
[73] André Morel, and Jean-Michel André, "Pigment distribution and primary production in the western Mediterranean as derived and modeled from Coastal Zone Color Scanner observations", *Journal of Geophysical Research*, 96, 12,685-12,698, (1991).
[74] David Antoine, D., Jean-Michel André, and André Morel, A, "Oceanic primary production 2. Estimation at global scale from satellite (coastal zone color scanner) chlorophyll", *Global Biogeochemical Cycles*, 10, 57-69, (1996).

the western Mediterranean.[75] Arnone had previously described the dual gyre circulation of the Alboran Sea in a 1990 paper.[76]

Coastal Currents, Rivers, and Turbid Waters

The word "coastal" is obviously a prominent part of the name of the sensor, and coastal remote sensing was clearly one of the main tasks of the CZCS. It is ironic, then, that it would turn out that while it was easy to *observe* the wide variation of oceanic color near the coast, the wide variety of factors causing the variation of color in the coastal zone would make the extraction of accurate values of phytoplankton pigment concentration very difficult in those waters. The necessary improvements in accuracy to get a true chlorophyll value – and to get this value where the chlorophyll concentrations are higher in the presence of multiple interfering factors – would influence the design of the next-generation sensors, and both beguile and bedevil the dreams and waking hours of algorithm developers. The CZCS provided an unprecedented view of the world's oceanic coastlines, and forced the consideration of the factors that affected the optical environment and the biological processes occurring where these factors intermixed.

Research by Dale Kiefer and his colleagues in 1979 anticipated some of the work in the coastal zone, where there are several factors contributing to the ocean color signal. Utilizing cell cultures to examine particle transmittance and reflectance, these scientists showed that the optical properties of phytoplankton could be related to the carbon content of the cells and their growth rate – indicating that the chlorophyll-to-carbon ratio in the cells varied. The researchers also demonstrated that the optical properties of the cells changed as they aged, which indicated that there would be increasing concentrations of non-living detritus, which still absorbed light, in areas where phytoplankton grow rapidly. These relationships could be utilized to better estimate phytoplankton productivity in the coastal zone.[77]

The West Coast Time-Series enabled examination of the California Current and its related jets and squirts of offshore productivity. Mark Abbott noted that he knew that the science was advancing when he was able to observe the evolution of a coastal upwelling event in July 1981. Using both CZCS data and AVHRR, Abbott said that they were actually able to see what they had previously only envisioned – pixel by pixel, as the cold deep water moved onshore, the phytoplankton responded to the high nutrient concentrations. This

---

[75] Robert A. Arnone, "The temporal and spatial variability of chlorophyll in the western Mediterranean: coastal and estuarine studies", in Paul E. LaViolette, editor, *Seasonal and Interannual Variability of the Western Mediterranean Sea,* American Geophysical Union, Washington DC, pp. 195-225, (1995).

[76] Robert A. Arnone, Denis Wiesenburg, and K. Saunders, "The origin and characteristics of the Algerian current", *Journal of Geophysical Research*, 95, 1587-1598, (1990).

[77] Dale Kiefer, R.J. Olson, and W.H. Wilson, "Reflectance spectroscopy of marine phytoplankton. Part 1. Optical properties as related to age and growth rate", Limnology and Oceanography, 24(4), 664-672, 1979.

event was described in a notable 1985 publication.[78]

In 1990, Ted Strub of Oregon State University and several co-authors examined the complex variability of the California Current.[79] These researchers documented different seasonal cycles moving northward from southern California to Puget Sound, and noted that the strongest non-seasonal signal was due to the 1982-1983 El Niño event. A co-author on that paper, Andrew Thomas, followed up in 1994 with a comparison of the Peru Current and the California Current.[80] This research group supported many of the earlier published results of Strub's earlier paper, but featured one intriguing non-observation; the effect of the 1982-1983 El Niño event was (surprisingly) not strongly in evidence in the Peru Current system in the CZCS data. They speculated that the lack of coverage was one of the reasons that the El Niño-induced variability was largely absent. It would take another mission and another El Niño to test this speculation.

Herschel Hochman and his fellow researchers took another look at the influence of the Orinoco River outflow on the Caribbean Sea.[81] This paper clearly examined the obscurative effects of the riverborne constituents that the Orinoco seasonally poured into the Caribbean Sea during the rainy season. While this paper was not the first – and certainly would not be the last – to determine this, the results of this paper showed that approximately 50% of the remote-sensing signal attributed to chlorophyll was actually due to dissolved organic matter, which in the case of the Orinoco was due to various sources, including deceased organisms, plant matter, and the ever-present, amorphous *Gelbstoff*. The following paper in the same issue of the Journal of Geophysical Research did a modeling study of similar oceanic conditions. Roland Doerffer and J. Fischer attempted to estimate the various concentrations of chlorophyll, suspended sediments, and yes, *Gelbstoff*.[82] Doerffer and Fischer were familiar with the Germanic coast of the North Sea, an area which was an excellent Case II water algorithm test site.

One other paper related to the difficulties of remote sensing of coastal waters was published in 1991 by Richard Stumpf and Jonathan Pennock.[83] Building on their previous work, they used data from the AVHRR on the NOAA polar-orbiting satellite series in

---

[78] Mark Abbott, recorded interview, 28 November 2008; Mark R. Abbott and Philip M. Zion, "Satellite observations of phytoplankton variability during an upwelling event", *Continental Shelf Research*, 4(6), 661-680, (1985).
[79] Theodore P. Strub, C. James, Andrew C. Thomas, and Mark R. Abbott, "Seasonal and nonseasonal variability of satellite-derived surface pigment concentration in the California current", *Journal of Geophysical Research*, 95, 11501-11530, (1990).
[80] Andrew C. Thomas, F. Huang, Theodore P. Strub, and C. James, "Comparison of the seasonal and interannual variability of phytoplankton pigment concentrations in the Peru and California Current systems", *Journal of Geophysical Research*, 99(C4), 7355–7370, (1994).
[81] Herschel T. Hochman, Frank E. Müller-Karger, and John J. Walsh, J.J., "Interpretation of the coastal zone color scanner signature of the Orinoco River plume", *Journal of Geophysical Research*, 99, 7443-7455, (1994).
[82] Roland Doerffer and Jürgen Fischer, "Concentrations of chlorophyll, suspended matter, and gelbstoff in case II waters derived from satellite coastal zone color scanner data with inverse modeling methods", *Journal of Geophysical Research*, 99, 7457-7466, (1994).
[83] Richard P. Stumpf, and Jonathan R. Pennock, J.R.,"Remote estimation of the diffuse attenuation coefficient in a moderately turbid estuary", *Remote Sensing of Environment*, 38, 183-191, (1991).

comparison to *in situ* data collections to evaluate how well the diffuse attenuation coefficient, a value related to the turbidity of the water, could be estimated. [84] The diffuse attenuation coefficient at 490 nanometers (K490) data product from the CZCS ultimately became the sensor's best product to evaluate suspended sediments and turbidity, as it had been determined fairly early during the mission that actually retrieving accurate suspended sediment concentrations was beyond both sensor and algorithm capabilities. Stumpf and Pennock's ongoing research work showed how the AVHRR reflectance product could be used to observe reflectance changes related to sediment suspension, and provided bio-optical oceanographers with a remote-sensing product they could utilize in the years following the end of the CZCS mission. As an example, the AVHRR reflectance product showed marked increases in reflectance in Florida Bay in December 1989 after one of the coldest air masses in history blasted through the state and caused a devastating freeze; the reflectance increase was due to the suspension of sediment in the shallow bay due to strong winds.

Some of Stumpf's research was ground-breaking enough to be initially considered of dubious accuracy; in particular, a paper that he considered "one of his best" concerning the effect of floods on sediment transport in the Chesapeake Bay, authored with Mary A. Tyler, "had to overcome the community resistance at the time".[85] Stumpf believes that his work with AVHRR "helped to get people to start thinking of ocean color data in the coast in new ways."

Atmospheric Corrections

The subject of atmospheric correction, and the effects of differing atmospheric conditions on the CZCS data, continued to be of interest to researchers. One important contribution along these lines came from Ken Carder, collaborating with Watson Gregg, David Costello, Kenneth Haddad, and atmospheric dust expert Joseph Prospero of RSMAS (Carder, Gregg, Costello, and Haddad were at the University of South Florida).[86] Residents of Florida and the Caribbean occasionally see the results of large dust storms crossing the Atlantic Ocean from the Sahara desert, primarily perceived as a distinct yellowing and dimming of the solar orb. The CZCS atmospheric correction algorithm was challenged by desert dust aerosols, so Carder and his team sought to determine if the presence of dust aerosols could be corrected for in the CZCS data. A major dust storm event occurring in

[84] Richard P. Stumpf, and Jonathan R. Pennock, "Calibration of a general optical equation for remote sensing of suspended sediments in a moderately turbid estuary", *Journal of Geophysical Research – Oceans*, 94(C10), 14,363-14,371, (1989).

[85] Richard A. Stumpf, email message received 2 May 2008; Richard A. Stumpf and Mary A. Tyler, "Sediment transport in Chesapeake Bay during floods; analysis using satellite and surface observations", *Journal of Coastal Research*, 4, 1-15, (1988).

[86] Kendall L. Carder, Watson W. Gregg, David K. Costello, Kenneth Haddad, and Joseph M. Prospero, "Determination of Saharan dust radiance and chlorophyll from CZCS imagery", *Journal of Geophysical Research*, 96, 5369-5378, (1991).

June 1980 provided an ideal test case. (The dust from this event was so pervasive that the dust deposited on the sea surface in the Sargasso Sea was actually collected by sediment trap collectors deployed 30 meters underwater.) They determined that a modified algorithm with two components – one representing dust, the other representing a "bluish haze", which could be distinguished from atmospheric pollution emanating from the east coast of Florida – improved the analysis somewhat. Rather than overestimating chlorophyll by either 54-140% or 122-180% when only one aerosol type was used, the two-component model only resulted in overestimates of 64-78%. They indicated the overestimate was due to the assumptions made about the deterioration of the CZCS bands over time.

This particular subject, the ongoing calibration of the CZCS, was the subject of the critical paper written by Robert Evans and Howard Gordon in 1994, "Coastal zone color scanner 'system calibration': A retrospective examination".[87] The availability of the full mission global data set allowed a detailed analysis of the data, which evaluated all of the factors which were changing during the mission, both relatively long-term trends in sensitivity degradation, and shorter-term fluctuations that were most notable each time the instrument was turned off and then on again. Each time this happened, the CZCS essentially "reset" itself, not necessarily in quite the same state as when it had been turned off. The long-term trends were basically due to sensor degradation; the most severe degradation was noted for band 1, the blue band, and the degradation of this band led to the data processing team to note that unrealistically high chlorophyll values were being computed, because the degradation rate accelerated about one year into the mission. Evans and Gordon determined that the El Chichón aerosol did not appear to cause short-term variations in the data. One of the main things conclusively determined in this contribution was that a next-generation ocean color sensor mission should include several different techniques to maintain instrument calibration. One recommended technique was the collection of frequent lunar and solar calibration data; another was the establishment of an *in situ* observation station in clear waters that would provide radiance data regularly as the satellite passed overhead. Such recommendations were already being heard and heeded in 1994 by the mission planners.

Evans and Gordon provide an interesting side note that underscores the increasing computational power available to researchers: for their paper they processed the CZCS data set 12 separate times, which required 6 computer years (on a VAX 3200).

Open Ocean and Primary Production

With the CZCS global data set demonstrating the feasibility (though not the realization) of global ocean color remote sensing, researchers using the data broadened the scope of their investigations. The availability of this data – even given the concerns about accuracy,

---

[87] Robert H. Evans and Howard R. Gordon, "Coastal zone color scanner 'system calibration': A retrospective examination", *Journal of Geophysical Research*, 99, 7293-7307, (1994).

the lack of complete coverage, and a host of other uncertainties – allowed the first characterizations of basin-scale processes, regular seasonal changes, and both regional and global ocean productivity.

Paul Fiedler of the NOAA Southwest Fisheries Science Center in La Jolla, California (adjacent to Scripps Institute of Oceanography) investigated the seasonal and interannual changes in the CZCS pigment data, and their relationship to surface ocean winds, in the eastern tropical Pacific.[88] Carrie Leonard and Charles McClain also analyzed the tropical Pacific over the entire CZCS mission.[89]

Researchers in Canada generated a large number of contributions addressing primary production on the basin-scale in the Atlantic and in the global ocean.[90] Campbell and Aarup also estimated production in the north Atlantic, while global estimates were also provided by Antoine, André, and Morel in 1996.[91] Jim Yoder and the Goddard ocean color group summarized the annual cycles that could be perceived in the global CZCS data set in a 1993 paper.[92] Both the Yoder paper and the Antoine et al. paper were published in *Global Biochemical Cycles*, demonstrating that despite the incomplete coverage of the global oceans that the CZCS had provided, the data were demonstrating the global importance of ocean color data – a factor that was significant to the advocacy within the oceanographic community for the follow-on mission to the CZCS.

Primary production in the oceans means, at its core, the living activities of phytoplankton. Charlie Yentsch addressed how the CZCS had improved, with substantially more detail than had ever been available before, the basic understanding of how the planktonic base of the oceanic carbon system worked: the interactions of light, nutrients, ocean circulation and plankton species that resulted in the variegated patterns of CZCS

---

[88] Paul C. Fiedler, "Seasonal and interannual variability of coastal zone color scanner phytoplankton pigments and winds in the eastern tropical Pacific", *Journal of Geophysical Research*, 99, 18,371-18,384, (1994).

[89] Carrie L. Leonard and Charles R. McClain, "Assessment of interannual variation (1979--1986) in pigment concentrations in the tropical Pacific using the CZCS", *International Journal of Remote Sensing*, 17, 721-732, (1996).

[90] Norman Kuring, Marlon R. Lewis, Trevor M. Platt, and John E. O'Reilly, "Satellite-derived estimates of primary production on the northwest Atlantic continental shelf", *Continental Shelf Research* 10, 461-484. 1990; Trevor Platt, Carla Caverhill, and Shubha Sathyendranath, "Basin-scale estimates of oceanic primary production by remote sensing: The North Atlantic", *Journal of Geophysical Research*, 96, 15,147-15,159, (1991); Alan Longhurst, Shubha Sathyendranath, Trevor Platt, and Carla Caverhill, "An estimate of global primary production in the ocean from satellite radiometer data", *Journal of Plankton Research*, 17, 1245-1271, 1995; Trevor Platt, Shubha Sathyendranath, and Alan Longhurst, "Remote sensing of primary production in the ocean: Promise and fulfillment", *Philosophical Transactions of the Royal Society of London B*, 348, 191-202, (1995).

[91] Janet W. Campbell, and Thorkild Aarup, "New production in the North Atlantic derived from seasonal patterns of surface chlorophyll", *Deep-Sea Research*, 39, 1669-1694, 1992; David Antoine, Jean-Michel André, and André Morel, "Oceanic primary production 2. Estimation at global scale from satellite (coastal zone color scanner) chlorophyll", *Global Biogeochemical Cycles*, 10, 57-69, (1996).

[92] James A. Yoder, Charles R. McClain, Gene C. Feldman, and Wayne Esaias, "Annual cycles of phytoplankton chlorophyll concentrations in the global ocean: a satellite view", *Global Biogeochemical Cycles*, 7, 181-193, (1993).

pigment that were becoming increasingly familiar to ocean scientists.[93] Yentsch focused his attention on the familiar North Atlantic, which had already been known to host a large-scale annual bloom of phytoplankton each spring. Though the general pattern observed by the CZCS was in accord with the theoretical understanding, the CZCS showed how rapidly phytoplankton populations increased across the basin, and also demonstrated that the development of high concentrations was not a uniform wave spreading from south to north, but rather developed as large scale patterns swirled around eddies and current patterns.

Two different phytoplankton groups: nitrogen-fixers and coccolithophorids – also intrigued researchers. Nitrogen-fixing plankton, typified by *Trichodesmium*, provide an alternate source of nitrogen than what is dissolved in seawater, and thus can provide sufficient nitrogen to trigger blooms – sometimes of the noxious variety. In 1993, Subramaniam and Carpenter (1993) suggested a method to detect *Trichodesmium* in CZCS data, an effort that would become increasingly important for applications of ocean color data in the 1990s.[94] Coccolithophorids were already known to mislead the CZCS pigment algorithm; graduate student Christopher Brown of the University of Rhode Island, working with Jim Yoder, determined a way to identify coccolithophorid blooms in CZCS data and evaluated their distribution around the world.[95] Brown's identification algorithm would also be an important element in improved data processing schemes that were being considered for the next generation sensor.

During his Ph.D. research, Brown had a particularly rewarding morning after nights of processing AVHRR data searching for coccolithophorid blooms. He came into the office, opened a new file, and found himself looking at an amazing, cloud-free image of the Gulf of St. Lawrence, with the center of the Gulf sporting a beautiful and large *Emiliania huxleyi* coccolithophorid bloom, larger than anything he had seen before in the Gulf of Maine.[96]

Pollution and the Enviroment

It is a not widely-known or welcome fact that industrial liquid acid wastes have been regularly dumped since the 1970s offshore of the East Coast of the United States, notably just a bit east of New Jersey's heavily-used beaches. Some of the earliest images from the CZCS clearly showed the acid waste sites, and also showed, alarmingly, that they were considerably closer to the coast than anyone had realized. (The images demonstrated that,

---

[93] Charles S. Yentsch, "CZCS: Its role in the study of the growth of oceanic phytoplankton", in Vittorio Barale and Peter M. Schlittenhardt, editors, *Ocean Colour: Theory and Applications in a Decade of CZCS Experience*, edited Kluwer Academic Publishers, Dordrecht, The Netherlands, pp. 17-32, (1993).
[94] Ajit Subramaniam and Edward J. Carpenter, "An empirically derived protocol for detection of blooms of the marine cyanobacterium Trichodesmium using CZCS imagery", *International Journal of Remote Sensing*, 15, 1559-1569, (1993).
[95] Christopher W. Brown and James A. Yoder, "Coccolithophorid blooms in the global ocean", *Journal of Geophysical Research*, 99, 7467-7482, (1994); Christopher W. Brown, "Global distribution of coccolithophore blooms", *Oceanography*, 8(2), 59-60, (1995), http://www.noc.soton.ac.uk/soes/staff/tt/eh/cbrown.html, (accessed 29 April 2009).
[96] Christopher W. Brown, ocean color history survey response, received 26 June 2008.

despite supposed requirements to dump at least 12 miles offshore, they were actually dumping within the 12-mile limit.[97]) Images of the acid waste dump from the CZCS became very familiar in the early stages of the mission. Jane Elrod of GSFC described in ocean-colorful detail some CZCS observations of the East Coast acid waste dump sites.[98] Conversely, while he was Program Manager at NASA HQ in the early 1980s, Ken Carder said that the pollution images from CZCS "turned off the phytoplankton people … it was hard to get [them] interested in primary production on a quantitative scale." [99]

The potential for remote sensing to shed light on pollution was discussed both during the CZCS mission, in the post-mission phase, and in the planning for the next generation sensors. While specific pollution events like oil spills may not be readily detectable – the East Coast acid waste dumps were a notable exception – the pervasiveness of nutrient pollution, leading to higher nutrient concentrations, eutrophication, oxygen stress, and harmful algal blooms – has been studied with significant interest. Frank Müller-Karger summarized both the promise and the requirements to use oceanographic and aircraft remote sensing for the monitoring of marine pollution.[100]

Climate Change

Having a global data set available created ideas and opportunities, and many of these ideas led to publications. The issue of anthropogenic climate change became both more newsworthy and an increasingly important topic in Earth science in the 1980s and 1990s. Ever-expanding computational capabilities allowed researchers to create more sophisticated climate models, indicating increasing concerns about global warming. In 1992 (the same year that Müller-Karger's paper on monitoring of marine pollution was published) then-Senator Al Gore published his book *Earth in the Balance: Ecology and the Human Spirit*, describing his personal view of the various environmental concerns facing Earth's human population.[101]

Thus, climate change and its effects on the oceans were also on the minds of oceanographers, and it seems that 1992 was a particularly noteworthy year for this research thrust. James Aiken, Gerald Moore, and Patrick Holligan addressed this topic in 1992; while lauding the data that was collected by the CZCS, the authors discussed what other data

---

[97] Gene Feldman, comments recorded at the Ocean Color Collaborative Historical Workshop, January 13-14, 2009, St. Petersburg, Florida.
[98] Jane A. Elrod, "CZCS view of an oceanic acid waste dump," *Remote Sensing of the Environment,* 25, 245-254, 1988; NASA GES DISC, "Classic CZCS Scenes: Chapter 8: Ocean Pollution", http://daac.gsfc.nasa.gov/oceancolor/scifocus/classic_scenes/08_classics_pollution.shtml, (accessed 29 April 2009).
[99] Kendall Carder, comments recorded at the Ocean Color Collaborative Historical Workshop, January 13-14, 2009, St. Petersburg, Florida.
[100] Frank Müller-Karger, "Remote sensing of marine pollution: A challenge for the 1990s", *Marine Pollution Bulletin,* 25, 54-60, (1992).
[101] Albert Gore Jr., *Earth in the Balance: Ecology and the Human Spirit,* Houghton Mifflin, Boston, MA, 416 pages, (1992).

needed to be collected for climate change research.[102] The relationship between the marine biota and the atmosphere was underscored by the authors' recognition that some species of plankton (notably coccolithophorids) produce a chemical compound called dimethyl sulfide (DMS). DMS molecules can act as cloud nuclei, and thus changing patterns of phytoplankton blooms could affect oceanic cloud cover, which would then affect primary production. The following year, Harris, Feldman and Griffiths examined potential relationships between climate change and primary production.[103] They noted the importance of ocean circulation to global primary production, and suggested that changes in circulation would feed back to affect plankton populations. They noted in particular that the current observational capabilities were insufficient to address climate change processes.

SeaWiFS

The routine availability of SeaWiFS data beginning in 1997, and also the availability of SeaDAS to allow "at-home" data processing, meant that researchers who had waited and waited to begin their remote-sensing analyses in earnest now had a vast field of data to work with, and powerful tools to use on it. At oceanographic science meetings starting in 1998 and into the next century (the American Geophysical Union convenes an Ocean Sciences meeting every evenly-numbered year), it was a common sight to see SeaWiFS imagery in many posters during the meeting, even on posters that had virtually nothing to do with remote sensing, but for which a SeaWiFS image provided excellent visual context.

Publications in journals began to proliferate. It is impossible to provide more than a sampling of these publications; the International Ocean Colour Coordinating Group, which was formed in 1996 (see Chapter 7), provides on online bibliography that is fairly comprehensive.[104]

The JGOFS program was able to conduct one process study, of the Southern Ocean, with SeaWiFS in orbit. Jefferson Keith Moore and colleagues found elevated chlorophyll corresponding to the succession of oceanic fronts in the Southern Ocean, and SeaWiFS data was found to be in fairly good agreement with shipboard measurements.[105] Another study of the Southern Ocean environment in the vicinity of South Georgia Island was led by

---

[102] James Aiken, Gerald F. Moore, and Patrick M. Holligan, "Remote sensing of oceanic biology in relation to global climate change", *Journal of Phycology*, 28, 579-590, (1992).

[103] Graham P. Harris, Gene C. Feldman, and F. B. Griffiths, "Global Oceanic Production and Climate Change," in Vittorio Barale and Peter M. Schlittenhardt, editors, *Ocean Colour: Theory and Applications in a Decade of CZCS Experience*, Kluwer Academic Publishers, Dordrecht, The Netherlands, pp. 237-270, (1993).

[104] IOCCG, "Ocean Colour Bibliography", http://www.ioccg.org/biblio.html. Due to size, this bibliography was not downloaded for reference.

[105] Jefferson K. Moore, Mark R. Abbott, J.G. Richman, W.O. Smith, T.J. Cowles, Kenneth H. Coale, W.D. Gardner, and Richard T. Barber, "SeaWiFS satellite ocean color data from the southern ocean", *Journal of Geophysical Research*, 26 (10), 1465-1468, (1999).

Rebecca Korb of the British Antarctic Survey, and they observed wide differences in productivity in different years, and productivity patterns influenced by the islands extending much further than indicated by ship sampling.[106]

Mati Kahru and Greg Mitchell of Scripps intensely scrutinized the California Current, observing a massive harmful algal bloom early in the mission, and documenting the influences of El Niño and La Niña on the current and coast.[107] El Niño and La Niña global variability were also the theme of the Behrenfeld et al. 2001 paper that had been highlighted by the Earth Science Update.[108]

One of the more widely-read papers stemming from the SeaWiFS data set was published in *Nature* in 2006. Led by Behrenfeld, a coterie of ocean color researchers examined the sobering topic of global warming and its effects on ocean productivity.[109] The results were not reassuring; the warming of the oceans occurring over the SeaWiFS mission caused a decrease in global ocean primary productivity averaging 190 teragrams of carbon per year. It may be difficult to grasp the magnitude of this effect because a teragram isn't a unit that is very familiar, in either the metric system or the English system, because of its size: a teragram may be more readily comprehended as 1 *billion* kilograms. This teragram-scale message from nature generated some attention in the media: quoted in the *San Francisco Chronicle*, co-author David Siegel said, “"What's amazing is this is the first time we see it on a global scale… We have an inkling now what will happen to the ocean's biology in future climates.” [110] Regarding this paper, Behrenfeld noted, “It clearly demonstrated the tight coupling between ocean chlorophyll/primary production and climate driven changes in upper ocean physical properties (SST, ENSO, mixed layer depth). The success of that study was entirely a reflection of the years of meticulous work conducted by NASA's ocean color processing group to ensure the highest quality products possible…” [111]

With SeaWiFS having provided a lengthening record of phytoplankton chlorophyll

---

[106] Rebecca E. Korb, Mick J. Whitehouse, and Pete Ward, “SeaWiFS in the southern ocean: spatial and temporal variability in phytoplankton biomass around South Georgia”, *Deep Sea Research. II. Topical Studies in Oceanography*, 51, 99-116, (2004); NASA GES DISC, “South Georgia: A View Through the Clouds”, http://daac.gsfc.nasa.gov/oceancolor/scifocus/oceanColor/south_georgia.shtml, (accessed 21 May 2009).

[107] Mati Kahru and B. Greg Mitchell, “Spectral reflectance and absorption of a massive red tide off Southern California”, *Journal of Geophysical Research*, 103(C10), 21,601-21,609, (1998); Mati Kahru and B. Greg Mitchell, “Influence of the 1997-98 El Niño on the surface chlorophyll in the California Current”, *Geophysical Research Letters*, 27 (18), 2937-2940, (2000); Mati Kahru, and B. Greg Mitchell, “Influence of the El Niño - La Niña cycle on satellite-derived primary production in the California Current”, *Geophysical Research Letters*, 29 (9), 1846-1849, (2002).

[108] Michael J. Behrenfeld, J. T. Randerson, Charles R. McClain, Gene C. Feldman, Sietse O. Los, Compton J. Tucker, Paul G. Falkowski, C.R. Field, C.B. Frouin, Wayne E. Esaias, Dorota D. Kolber, and Nathan H. Pollack, “Biospheric Primary Production During an ENSO Transition”.

[109] Michael J. Behrenfeld, Robert T. O'Malley, David A. Siegel, Charles R. McClain, Jorge L. Sarmiento, Gene C. Feldman, Allen J. Milligan, Paul G. Falkowski, Ricardo M. Letelier and Emmanuel S. Boss, “Climate driven trends in contemporary ocean productivity, *Nature*, 444, 752-755, (7 December 2006).

[110] Jane Kay, “Ocean warming's effect on phytoplankton”, *San Francisco Chronicle*, (7 December 2006).

[111] Michael Behrenfeld, email message received 3 April 2009.

concentrations, several groups attempted to take a longer-term perspective by attempting to connect the CZCS data with SeaWiFS data. The motivation for these efforts was to observe, if possible, any changes in the state of the ocean's biological state from the CZCS era to the SeaWiFS era. Watson Gregg of GSFC and his collaborators performed the first attempts to do this, comparing the CZCS 1979-1986 data to the first three years of SeaWiFS data (1997-2000), and he initially concluded that global phytoplankton chlorophyll concentrations had declined around 6%. This trend included such notable observations as a 30% decline in summer North Pacific chlorophyll concentrations.[112] Interestingly, when Gregg examined just the SeaWiFS record from 1998 to 2003, he observed an overall increase, driven by coastal chlorophyll increases.[113]

The other attempt to connect the CZCS era to the SeaWiFS era was led by David Antoine in France, working with André Morel and RSMAS researchers. This group attempted to make a CZCS-like dataset out of the SeaWiFS data by using the closest four SeaWiFS bands to the CZCS band set, and compared the data sets from 1979-1986 to SeaWiFS from 1998-2002. This particular period was strongly influenced by both the El Niño and La Niña events; in contrast to Gregg's early results, they reported a 22% increase in phytoplankton concentrations globally.[114] The differences in the results from the two groups show both the importance of choosing comparable time-periods, and more importantly, demonstrate how difficult it is to compare supposedly similar data from two very different remote-sensing instruments.

Jim Yoder of the University of Rhode Island, who had dunked himself at the SeaWiFS launch party, carried on extensive ocean color research with the new data from SeaWiFS. One of his primary collaborators was Jay O'Reilly of NOAA, who worked in the laboratory adjacent to URI's oceanographic laboratory. O'Reilly, a painstaking researcher who assisted with the Goddard DAAC's final data recovery effort for CZCS data by listing missing files on the optical platters that were rescued by one final turn of the last existing optical platter drive, also led the collaboration of 22 researchers that had produced the OC4v4 algorithm for SeaWiFS data, which became the global algorithm standard.[115] Yoder, O'Reilly and

---

[112] Watson W. Gregg, and Margarita E. Conkright, "Decadal changes in global ocean chlorophyll", *Geophysical Research Letters*, 29(15), 0.1029/2002GL014689 , (2002); Science Daily, "Satellites see changes in key element of ocean's food chain", http://www.sciencedaily.com/releases/2002/08/020812070050.htm, (accessed 24 December 2009); Watson W. Gregg, Margarita E. Conkright, Paul Ginoux, Jay E. O'Reilly, and Nancy W. Casey, "Ocean primary production and climate: Global decadal changes," *Geophysical Research Letters* 30(15): 1809, 10.1029/2003GL016889, (9 August 2003).

[113] Watson W. Gregg, Nancy W. Casey, and Charles R. McClain, "Recent trends in global ocean chlorophyll", *Geophysical Research Letters*, 32:L03606, doi:10.1029/2004GL021808; Science Daily, "NASA satellite sees ocean plants increasing, coasts greening", http://www.sciencedaily.com/releases/2005/03/050309142356.htm, (accessed 24 December 2009).

[114] David Antoine, André Morel, Howard R. Gordon, Viva F. Banzon, and Rober H. Evans, "Bridging ocean color observations of the 1980s and 2000s in search of long-term trends," *Journal of Geophysical Research*, 110 , C06009, doi:10.1029/2004JC002620, (2005).

[115] Jay E. O'Reilly, Stephane Maritorena, David Siegel, Margaret C. O'Brien, Diedre Toole, B. Greg Mitchell, Mati Kahru, Francisco P. Chavez, Peter Strutton, Glenn Cota, Stanford B. Hooker, Charles R. McClain, Kendall L. Carder, Frank Muller-Karger, Larry Harding, Andrea Magnuson, David

graduate student Stephanie Schollaert specialized in the western North Atlantic.[116] In the latter paper, they demonstrated the pervasive effects of human-produced haziness from East Coast cities on the ocean color data acquired over the adjacent Atlantic. Proving that SeaWiFS was not just an ocean color sensor, Petra Stegmann of the University of Rhode Island (now at the Pacific Fisheries Environmental Laboratory) studied African dust aerosols with SeaWiFS data.[117]

Mete Uz, collaborating with Yoder and Vladimir Osychny, investigated the effects of both current-induced eddies and the wide-ranging planetary Rossby waves on ocean productivity. In a widely-cited paper, they discovered that eddies and waves act as nutrient pumps to the surface, an important process influencing global chlorophyll concentrations and the ocean carbon cycle.[118]

Howard Gordon turned his attention briefly from atmospheric correction to the subject of calcite – specifically the calcite produced by coccolithophores.[119] Co-author William "Barney" Balch was soon to conduct one of the most ambitious at-sea experiments ever attempted to improve the quantification of this phenomenon (described in Chapter 6).

The area of the ocean with the least color – the southeastern gyre of the Pacific Ocean – even received scrutiny. Yves Dandonneau and his co-authors discovered that the extreme blue of these ocean waters is occasionally disturbed by greener Rossby waves passing through this nutrient-depleted region. The Rossby waves concentrate tiny bits of organic detritus in their zones of convergence, which was interpreted by SeaWiFS as higher chlorophyll concentrations, seemingly indicating phytoplankton growth where it is severely challenged.[120] The implications of this study and others, regarding factors which interfere with the accurate estimation of chlorophyll concentration in ocean waters, were discussed by

---

Phinney, Gerald F. Moore, James Aiken, Kevin R. Arrigo, Ricardo Letelier, and Mary Culver, "Ocean color chlorophyll *a* algorithms for SeaWiFS, OC2 and OC4: Version 4," in S.B. Hooker and E.R. Firestone, editors, *SeaWiFS Postlaunch Calibration and Validation Analyses (Part 3), NASA Technical Memorandum 2000-206892*, Volume 10, pp. 9-23, (2000). [Note: One of the author's first activities at the Goddard DAAC was to recover the CZCS missing files from the optical platters, using the last remaining optical platter drive.]

[116] James A. Yoder, Stephanie E. Schollaert, and Jay E. O'Reilly, "Climatological phytoplankton chlorophyll and sea-surface temperature patterns in continental shelf and slope waters off the Northeast U.S. coast," *Limnology and Oceanography*, 47, 672-682, (2002); Stephanie E. Schollaert, James A. Yoder, D.L. Westphal, and Jay E. O'Reilly, "Influence of dust and sulfate aerosols on ocean color spectra and chlorophyll-a concentrations derived from SeaWiFS off the U.S. East Coast," *Journal of Geophysical Research*, 108 (C6): 3191, doi:10.1029/2000JC000555, (2003).

[117] Petra M. Stegmann, "Characterization of aerosols over the North Atlantic Ocean from SeaWiFS", *Deep Sea Research Part II: Topical Studies in Oceanography*, "*Views of Ocean Processes from the Sea-viewing Wide Field-of-view Sensor (SeaWiFS) Mission: Volume 2*", 51(10-11), pp. 913-925, DOI: 10.1016/j.dsr2.2003.10.006, (May-June 2004).

[118] Mete Uz, James A. Yoder, and Vladimir Osychny, "Global remotely sensed data supports nutrient enhancement by eddies and planetary waves," *Nature*, 409, 597-600, (2001).

[119] Howard R. Gordon, G. Chris Boynton, William M. Balch, Stephen B. Groom, Derek S. Harbour, and Tim J. Smyth, "Retrieval of coccolithophore calcite concentration from SeaWiFS imagery", *Geophysical Research Letters* 28, 1587-1590, (2001).

[120] Yves Dandonneau, Andres Vega, Hubert Loisel, Yves du Penhoat, and Christophe Menkes, "Oceanic Rossby waves acting as a "hay rake for ecosystem floating by-products,", *Science*, 302 (5650) , 1548 – 1551, DOI: 10.1126/science.1090729, (28 November 2003).

Hervé Claustre and Stéphane Maritorena in the same issue of *Science*.[121]

Other areas of the ocean received due attention; Bosc, Bricaud and Antoine characterized the Mediterranean Sea with four years of SeaWiFS data.[122] Viva Banzon and colleagues at RSMAS provided a comprehensive look from space at the Arabian Sea monsoon in 2000.[123]

The daily observations of the variations of the color of the ocean provided truly unexpected surprises. Along the coast of the Gulf of California in Mexico, one particular agricultural region, the Yaqui Valley, has a very regular timing of irrigational watering for wheat production. In 2005, Michael Beman, Kevin Arrigo, and Pamela Matson reported observations of regular algal blooms in the Gulf of California corresponding to the times of these irrigations, indicating the excess nutrients washed from the fields provided nutrients for algal growth in the adjacent ocean waters.[124] This paper was one of the clearest indications that murky coastal regions adjacent to large rivers, such as the Mississippi in the United States and the Pearl River in China, are partly due to the supply of nutrients from agriculture.

Looking at the global ocean, Zhongping Lee of the Naval Research Laboratory and Chuanmin Hu of the University of South Florida characterized the global distribution of the clear Case-1 waters first described by André Morel.[125]

In 2004, the journal Deep Sea Research published two special issues that were fully devoted to scientific results from SeaWiFS. These two special issues, consisting of 33 research papers, were edited by Dave Siegel, John Marra, and Andrew Thomas.[126]

---

[121] Hervé Claustre and Stéphane Maritorena, "The many shades of ocean blue", *Science*, 302 (5650), 1514-1515, DOI: 10.1126/science.1092704, (28 November 2003).
[122] Emmanuel Bosc, Annick Bricaud, and David Antoine, "Seasonal and interannual variability in algal biomass and primary production in the Mediterranean Sea, as derived from four years of SeaWiFS observations", *Global Biogeochemical Cycles*, 18 (1), GB1005, doi:10.1029/2003GB002034, (January 10, 2004).
[123] Viva F. Banzon, Robert E. Evans, Howard R. Gordon, and Roman M. Chomko, "SeaWiFS observations of the Arabian Sea southwest monsoon bloom for the year 2000", *Deep Sea Research II*, **51,** 189-208, (2004).
[124] J. Michael Beman, Kevin R. Arrigo, and Pamela A. Matson, "Agricultural runoff fuels large phytoplankton blooms in vulnerable areas of the ocean", *Nature*, 434, 211-214, (10 March 2005).
[125] ZhongPing Lee and Chuanmin Hu, "Global distribution of Case-1 waters: An analysis from SeaWiFS measurements", *Remote Sensing of Environment*, 101, 270-276, (2006).
[126] David Siegel, John Marra, and Andrew Thomas, editors, "Views of Ocean Processes from the Sea-viewing Wide Field-of-view Sensor (SeaWiFS) Mission: Volume 1", *Deep Sea Research Part II: Topical Studies in Oceanography*, 51 (1-3), (January-February 2004); David Siegel, John Marra, and Andrew Thomas, editors, "Views of Ocean Processes from the Sea-viewing Wide Field-of-view Sensor (SeaWiFS) Mission: Volume 2", *Deep Sea Research Part II: Topical Studies in Oceanography*, 51 (10-11), (May-June 2004).

ABOUT THE AUTHOR

James Acker grew up in Des Plaines, Illinois, a suburb of Chicago. He gained a love of both science and the water from his parents; his father Robert was a chemical engineer, sailor, and swimmer, and his mother Julie was a singer and socialite who grew up in the coastal town of Winthrop, Massachusetts. He initially became fascinated by volcanoes and the ocean by reading his father's back issues of *National Geographic.* He attended Maine West High School and then Lawrence University in Wisconsin, where he majored in chemistry. Though he had wished for a career in oceanography, he had decided on a career path in analytical chemistry until he met with Lawrence alumnus Peter Betzer, an oceanographer at the University of South Florida. Betzer convinced him that he could be a chemical oceanographer, which led him to grad school at the USF Department of Marine Science. Unter the tutelage of Betzer and Robert Byrne, he did research on the dissolution of biogenic carbonates in the deep ocean (which provided opportunities to visit three far-flung "K" islands on research cruises: Kwajalein, Kodiak, and Kerguelen). Following grad school, he did postdoctoral research under the guidance of Owen Bricker at the U.S. Geological Survey.

Answering an advertisement for the position of Earth Observing System Oceanographic Liaison brought him to NASA Goddard Space Flight Center in Greenbelt, Maryland. In this position, he met many of the oceanographic scientists who would be part of the SeaWiFS and MODIS mission teams. He worked directly with the SeaWiFS Project on their Technical Memorandum series, authoring a retrospective on the CZCS NET. Then he joined the Goddard Distributed Active Archive Center (DAAC) to conduct user services for SeaWiFS data. In this position he became acquainted with numerous ocean color scientists around the world, and maintained close contact with the SeaWiFS and MODIS Projects. Despite the changes in the archive, he continued working at the DAAC (now the Goddard Earth Sciences Data and Information Services Center, GES DISC) with the Giovanni system, led by Greg Leptoukh. Leptoukh convinced him to propose a ocean color history book in response to the call from the NASA History and Science Divisions, which led to this manuscript.

Acker is an avid swimmer, sports fan, science reader, moviegoer, and advocate of the Giovanni data system. He is married with three children and lives in the Baltimore suburbs of Maryland. He has published 42 research publications as an author or co-author. His proudest research achievements using ocean color data were using SeaWiFS data in Giovanni to provide support for the accuracy of a circulation model of the northern Red Sea, and observing the transport of carbonate sediments from Bermuda, the Bahamas, Cuba, and an atoll in the South China Sea. The latter was a process that he had proposed to observe with SeaWiFS data, but which had not been observationally confirmed until SeaWiFS observed sediment plumes caused by Hurricanes Floyd and Gert in 1999.

INDEX

Made in the USA
San Bernardino, CA
19 February 2019